Studies in History and Philosophy of Science

Volume 40

More information about this series at http://www.springer.com/series/5671

Edmundo Balsemão Pires • Joaquim Braga
Editors

Bernard de Mandeville's Tropology of Paradoxes

Morals, Politics, Economics, and Therapy

Springer

Editors
Edmundo Balsemão Pires
Faculdade de Letras-Universidade de Coimbra
Largo da Porta Férrea
Coimbra, Portugal

Joaquim Braga
Faculdade de Letras-Universidade de Coimbra
Largo da Porta Férrea
Coimbra, Portugal

ISSN 0929-6425 ISSN 2215-1958 (electronic)
Studies in History and Philosophy of Science
ISBN 978-3-319-19380-9 ISBN 978-3-319-19381-6 (eBook)
DOI 10.1007/978-3-319-19381-6

Library of Congress Control Number: 2015949926

Springer Cham Heidelberg New York Dordrecht London

Printed on acid-free paper

Springer International Publishing AG Switzerland is part of Springer Science+Business Media (www.springer.com)

Preface

In the epoch that goes from 1650 to 1800, from Lisbon to St Petersburg, evolved conceptual lines and themes that allow us to identify the semantic trends of 1800. Here, one finds the most significant self-descriptions of modern science and the conceptual foundations of the descriptions of the social and political systems of the Old Continent. This period is usually identified as the Age of Enlightenment and characterized in the motto adopted by Kant – *sapere aude!* An ethics of knowledge and the will was fixed and recognized around this expression, meaning autonomy of the will and freedom of judgment and suggesting an understanding of human nature and its progress. But the motto included also a perspective about the historical evolution continuing themes present in the "quarrel of the ancients and the moderns."

If Kant's appropriation of the Latin motto has assisted several historians of science, philosophy and of cultural creations, in a broad sense, in the signaling of the bifurcation of this epoch, between tradition and innovation, it is no less true that the author of the *Critique of Pure Reason* can hardly serve as an illustration of the great variety of directions of the period.

From the point of view of moral and political thinking, the diversity of the orientations present in the Age of Enlightenment is not just a result of the positions of the authors regarding the demanding issues of a period in transition from the political forms of the *Ancien Régime*. In this particular field we can observe the diversity of the reception lines of the ethical-political concepts of the ancient world, Platonism, Aristotelianism, Stoicism, and Epicurism and different perspectives on happiness and virtue that arrived from antiquity to modern times and helped here as terminological sources of inspiration for the new ethical and political challenges. Of course, in the seventeenth and eighteenth centuries, the difficulties of the ethical theories are not reducible to the interpretative problems of the classics. They were especially acute in so far as they also result from the crisis of Christianity as a central component of the social and political transformations. The religion as a source of social conflict gains special significance, in a period in which the secularist trends or the teachings about tolerance disturb old convictions, everywhere.

In Spinoza's *Ethics* but more particularly in the *Theological-Political Treatise* was at stake a reconceptualization of the History of the Revelation of the monotheism, in which the metaphysical dualism of thought and extension, of spirit and matter, was addressed at the source of the philosophical concept of God, the human soul with its drives, emotions, and desires. In addition, the *Theological-Political Treatise* gave Spinoza the opportunity to deal in depth with the meaning of the miracles in the History of God's Revelation, which also attracted Pierre Bayle. We know the importance of P. Bayle in the European reception of Spinoza. As a result of P. Bayle's criticism, a controversy about Spinozism, atheism, and pantheism was kindled throughout Europe. The History of Revelation discussed by Spinoza is much more than one episode in the Hermeneutics of the authorship of the sacred texts. His ideas implied larger consequences in the Metaphysics of Nature and Moral Theology. P. Bayle recognized these wider consequences in the article "Spinoza" of his *Dictionaire Historique et Critique*.

The importance of P. Bayle in the distinction between religious belief and public morality is essential, because P. Bayle, who inspired B. de Mandeville, thought that atheists could act morally even not having religion. Although this particular idea was common to Shaftesbury, there are aspects of P. Bayle's insight that lead to other, more radical, directions.

In his *Various Thoughts on the Occasion of a Comet*, P. Bayle posed the question of the source of popular superstition as well as the problem of the justification of the popular piety. This enquiring is similar to Spinoza. His approach shows that men can destroy the underpinnings of the superstitions on the influences in human life of the passage of comets without minimally affecting the rudiments of morals and customs. This conclusion was a step in the differentiation between nature and morality based on the difference between metaphysical judgments about supernatural consequences and practical judgments.

The doctor, writer, and philosopher Bernard de Mandeville (1670–1733), famous due to the publication of the poem *The Grumbling Hive, or Knaves Turn'd Honest* (1705), later included in the *Fable of the Bees: or, Private Vices, Public Benefits* (1714), cultivated the literary genre of the satire in line with a contemporary European mode. Employing his literary competence, he articulated in a popular and very comprehensive way serious theoretical problems on moral theory, on the social effects of vice and virtue, the origin and nature of human society, the conditions for the wealth and ruin of nations, or on self-preservation of the individuals and the interest of the species.

B. de Mandeville deepened P. Bayle's distinction. His acquaintance of skepticism in the line of Montaigne, the anti-rationalist French ideas, and the "erudite libertinism" is easily traceable. But it was the singular form of the combination of these influences that made the originality of B. de Mandeville.

The seventeenth and eighteenth centuries were traversed by the theoretical problems regarding the epistemological status of history, in relation to natural history and the religious and institutional history. In this aspect, a so-called conjectural history has developed in addition to the demands of textual criticism of the Biblical Hermeneutics (La Peyrere, Spinoza, and Richard Simon) and the increasing attention to human and people diversity on the planet (J.-F. Lafitau), before moving to the idealizations of cycles and repetition (J. B. Vico) or progress and perfectibility (Turgot, Condorcet).

In 1687, Fontenelle wrote his *Histoire des Oracles*, work in which he discusses the meaning of the writing of history and the combination in the narrative style of the facts and the imagination. The essay to differentiate the factual from the imaginary and to explain their fusion in the historical narrations continues the themes of La Peyrere, Spinoza, P. Bayle, and the biblical modern criticism.

In his theory of society, B. de Mandeville will reveal to us the whole range of combination of fiction and appearance and the concealment of self-interest on the basis of human actions. In addition, continuing theses from the "erudite libertinism" he showed how the historical narrative mirrors the interests of those who narrate the episodes.

In his work, the themes running in the undercurrents of the European Enlightenment (on morality, society, religion, and history) are reevaluated in a way very instructive for us, because it brings a more precise focus on the diversity of directions present in the history of ideas and concepts of the period. We can understand better why what is called "Enlightenment" should not be identified with a belief in human perfectibility or in a linear progress of the human species thanks to reason and to education.

In particular, the criticism of B. de Mandeville to the system of moral philosophy of Shaftesbury has the merit of exhibiting the crisis of the harmonist views of the "Age of Enlightenment" regarding nature and human nature. Among other examples, his ideas about the relationship between religion, church, and state in his *Free Thoughts on Religion, the Church and National Happiness* or the criticism of the philosophical theories on the impact of virtue and vice in society points to a social and political world in turmoil, far from the rational harmony of Shaftesbury. But, despite the facts of the social life which point to the opposite of a harmony, it was not easy to demonstrate with philosophical arguments the absence of any agreement between an intimate self-liking, society, and the virtue "according to nature." The merit of this proof is entirely due to B. de Mandeville. He teaches us how to figure a dynamic social order born of the combination of contradictory forces instead of a harmonic unity.

The author of the *Fable of the Bees* disassembled the dichotomies of the traditional moral thinking to show that the outcomes of the social action emerge as new, non-intentional effects from the combination of moral opposites, vice, and virtue, in such a form that these lose their moral significance.

Modern society is a tropism made of many combinations of opposite meanings.

Assuming this last perception we were led to the title of this book *Bernard Mandeville's Tropology of Paradoxes: Morals, Politics, Economics, and Therapy*. The work of this great writer, philosopher, and physician is woven by the awareness of the paradoxical nature of modern society and the challenges that this recognition brings to an adequate perspective on the historical world of modernity.

This book integrates studies of some of the best specialists in the thought of B. de Mandeville and of other philosophers and historians of Modern Thought, who accepted the challenge of rethinking his legacy on the occasion of the passage of 300 years since the publication of the *Fable of the Bees*.

Coimbra, Portugal
May 25, 2014

Edmundo Balsemão Pires

Contents

Chapter 1
Men Become Sociable by Living Together in Society: Re-assessing Mandeville's Social Theory

Malcolm Jack

Abstract In the first part of *the Fable of the Bees*, Mandeville's emphasis is on developing a passionate theory of man in society, greatly influenced by his reading of French sceptical literature, mediated through his mentor, Pierre Bayle. That analytical tradition provided a psychological theory of man as an egotistical creature pursuing his need for recognition, respect and pre-eminence as much as one who must fight for survival in the Hobbesian state of nature. Mandeville distils and uses the central, sceptical notions of *amour propre* and *amour de soi même* to explain why men act in certain ways, including ways that may be labelled altruistic. This kind of analysis is made from first principles, although its principles are claimed to be derived from observation of the human condition.

In Part 2 of *the Fable* Mandeville becomes interested in the origin and progress of society, much in the mode that came to dominate Scottish Enlightenment thought in writers such as Adam Ferguson and Lord Monboddo. Mandeville's treatment of the origin of language is particularly interesting in this context. While this "conjectural history" is alleged to be based on empirical observation, it can also be seen as a logical construct, a model which is used to explain the historical evolution of society from primitive barbarism to the polite, commercial refinement of eighteenth-century Europe.

The two strands of his thought may be seen as a link between the sceptical literature of seventeenth century France and the eighteenth-century conjectural history of the Scottish Enlightenment which ultimately underlay the emergence of modern, social science.

Keywords Psychological theory • Passions • Pride • Social evolution • Conjectural history • Social science

---

M. Jack (✉)
Independent Scholar
e-mail: malcolm.jack@btinternet.com

E. Balsemão Pires, J. Braga (eds.), *Bernard de Mandeville's Tropology of Paradoxes*, Studies in History and Philosophy of Science 40, DOI 10.1007/978-3-319-19381-6_1

1.1 Introduction: Background

Mandeville would, no doubt, have been amused to be remembered in the company of religious luminaries and saints but both Calvin and St. Augustine are two names that immediately come to mind when we consider the satirical essayist's dark portrayal of human nature. Sharing something of Calvin's belief that those *not* of the Elect were doomed to a life of drudgery, Mandeville certainly considered that in mock-Augustinian spirit, he was describing the frailties and defects of human nature and searching out the base motives that underlay all action, however virtuous it appeared to be.

In his spirited *Preface* to his *magnum opus* he tells us in a bantering tone: " 'Twas said of Montaigne, that he was pretty well versed in the Defects of mankind, but unacquainted with the Excellencies of human nature: if I fare no worse, I shall think myself well used" (Mandeville 1924: 2: 189).

Mentioning Montaigne, after alluding to Calvin and St Augustine is only to point to the tip of the iceberg of Mandeville's cosmopolitanism for in addition to referring to a panoply of classical figures known to all educated Europeans of his time, he himself makes reference to Erasmus, La Rochefoucauld, Spinoza, Gassendi and Bayle. F.B. Kaye has shown that parallels in his writing with the work of Machiavelli, Charron, La Placette, Leibniz and Abbadie indicate a familiarity with all those authors. Mandeville's first English works included translations of La Fontaine's *Fables* and a burlesque poem in imitation of Scarron. Paying due deference to those who have traced close links between his psychological egoism and the mechanical system of Hobbes, (one of the English or Latin predecessors he also refers to along with Bacon, Locke, Temple and Steele and others) we need to remind ourselves that we are dealing with a quintessentially Continental mind, one schooled in the cosmopolitan and eclectic environments of Rotterdam and Leiden where he was educated. At Rotterdam he attended the Erasmian School where his great mentor, Pierre Bayle, taught; at Leiden he studied philosophy, as well as medicine and, according to the university records, registered for his fair share of the students' tax free beer and wine. His linguistic skills were already considerable – as well as the staple Latin and Greek of the schools, he also was fluent in French. Before long, English was to become the virtual mother tongue of this native Dutch speaker and he shows off his linguistic dexterity by translating French works into his adopted tongue.

This background in the Netherlands or United Provinces to use it proper name, was a seething cauldron of intellectual dispute between traditional Aristotelians and modern Cartesians (Descartes himself had taken up residence in the Netherlands in the early seventeenth century) which affected the teaching of both the subjects – philosophy and medicine – for which Mandeville inscribed in Leiden in 1685. In his treatise of 1689 on the question of whether animals have souls, Mandeville defends the Cartesian contention that animals are mere automata based on the principle that they have no rational faculty but purely react to the circumstances of their environment. It is a position that he moved away from in later writings when he adopted the more recognisably Aristotelian position restated by Bayle in his *Dictionary* articles which described a closeness of behaviour between men and animals. Of course in the hands of a libertine thinker, bringing men and animals closer may be the beginning of denying men any special attributes especially if analysis shows that the rational faculty plays little part in human motivation. Man is a natural being, an inhabitant, even if a somewhat superior one, of the animal world.

Nor was the newly qualified medical man from Leiden inexperienced in political polemics. According to Dutch scholars, notably Rudolph Dekker both Mandeville and his father, Michael were involved in what became known as the "Costerman riots" in Rotterdam in 1690, which began after the summary execution of a citizen, Cornelius Costerman by the authorities when he confessed to murdering a tax collector (Dekker 1992: 481–98). The pros and cons of what the authorities had done raged in the city; young Mandeville was drawn into the argument perhaps penning the pamphlet *The Sanctimonious Atheist* revealing a penchant for radical and polemical views. Dekker speculates that as a result of being branded with the reputation of being a trouble maker, Mandeville may have decided to leave his native country. For whatever reasons, he took up permanent residence in England, where he married an English lady and settled to practice medicine.

1.2 Psychological Theory of the Passions

Let me turn to my theme of Mandeville's social theory – suffice to say at this stage that however much Mandeville may have become English by living in London, practising his medical art and engaging in Grub Street polemics, his mind had already been forged in the very different intellectual climate of the Dutch Republic by the time he crossed the channel to his adopted country.

How then do we begin with our examination of his social theory? The answer must be to start with his psychological assessment of man's passionate nature or what he called "anatomising the invisible part of man" (Mandeville 1924: 1: 145) and of which F. A. Hayek said he was most proud (Hayek 1966).

A good starting point is given in his later work, *the Origin of Honour* (1732) where he says "All Creatures are swayed and wholly governed by their Passions... even those who act most suitable to their knowledge, and strictly follow the Dictates of their Reason, are not less compell'd to do so by some Passion or other, that sets them to work, than others who bid Defiance and act contrary to Both, and whom we call the Slaves to their Passions" (Mandeville 1732: 31).

Thus Mandevillean man is compulsively motivated by his appetites. Even when he consciously follows a reasoning process, it is a passion that ultimately determines his behaviour. Mandeville is of course following in the tradition of the French *libertinage érudit* whose advocates rejected rationalism as a way of explaining human behaviour. Jacques Abbadie talks of analysing human nature "en entrant dans le detail des ses passions" (Abbadie 1692: 259): only by an extensive examination of the passions underlying human behaviour will the philosopher come to an understanding of what makes men behave in the way that they do.

1.3 The Predominant Passion: Pride

A central feature of this literary psychology was that one passion was primary or predominant. Mandeville followed this aspect of the tradition by making pride or *amour propre* his ruling passion. He defines it in this way "Pride is that natural faculty by which every mortal that has any understanding over-values, and imagines better things of himself than any impartial Judge, thoroughly acquainted with all his Qualities and Circumstances, would allow him" (Mandeville 1924: 1: 124).

The passion of pride therefore consist in the pronounced self-regard that each individual has for himself which is reflected in his "extraordinary concern" (Mandeville 1924: 1: 124) with what others think of him. This concern will prompt him to act in various ways including acting in ways that seem to benefit others rather than himself. If the inward manifestations of pride include narcissistic concerns with the appearance of the self (vanity); its outward expression in striving for the good opinion of others may have beneficial social effects.

Both the narcissistic and more outward looking manifestations of pride have important economic effects. Fashionable clothes, ever changing, provide an important and continuous stimulant to trade and to manufacture. This effect operates at all levels of society since each class will seek to emulate the one above it. "The druggist, Mercer, Draper, and other considerable shopkeepers find no difference between themselves and the Merchants, and therefore dress and live like them" (Mandeville 1924: 1: 129).

Feeling this pressure from below, the merchants' wives dress in the style of 'women of quality;' these superior beings of quality are then propelled to seek even greater refinement so as to be distinguished from their social inferiors.

The constant striving after status is, according to Mandeville much more significant in social advancement than the activities of any small group of virtuous men acting for the public good. Leaders and politicians need to understand this: "For to say, that if all Men were truly Virtuous, they might, without regard to themselves, consume as much out of Zeal to serve their Neighbours and promote the Publick Good, as they do now out of Self-love and Emulation, is a miserable Shift and an unreasonable Supposition" (Mandeville 1924: 1: 133).

Furthermore in their pursuit of 'comfort and ease', man will adhere to patterns of behaviour which appear entirely disinterested. Such behaviour will include adherence to a prevalent code of honour, a subject Mandeville returned to in his last major work, *The Origin of Honour* (1732).

Mandeville was driven to refine his definition of pride as self-love, according to F.B. Kaye, because of the incisive criticism that Bishop Butler made in his *Sermons* of 1726. Butler's argument amounted to saying that the reduction of all human behaviour to an explanation in terms of a selfish principle is a mere tautology since, by definition, all actions of the self must be selfish. Furthermore not only is such a definition merely tautologous but it fails to differentiate between principles of action "proceeding from cool consideration of self-advantage" from those (benevolent) actions which run counter to the advantage of the agent (Butler 1855: 241). Benevolent actions, according to Butler, cannot therefore be reduced to self-love.

Mandeville's redefinition, whether in reaction to Butler or not, amounts to his adoption of a distinction between pride and self-love, known in the French *moraliste* tradition of the seventeenth century with which he was familiar but, as Mikko Tolonen has pointed out, in a new and morally-neutral tone (Tolonen 2013: 28). In the Dialogues of *Part II of the Fable* (1729) Horatio challenges Cleomenes to be clearer about the difference between self-love and self-liking. Cleomenes defines self-liking (*amour propre*) in this way:

> I will endeavour to explain myself better. I fancy that, to increase the care in Creatures to preserve themselves, Nature has given them an instinct, by which every Individual values itself above its real Worth; this in us, I mean in Man, seems to be accompanied by a Diffidence, arising from a Consciousness, or at least Apprehension, that we do overvalue ourselves: it is this that makes us so fond of approbation, Liking and Assent of Others: because they strengthen us in the good Opinion we have of ourselves. (Mandeville 1924: 2 129/130)

Self-love (*amour de soi même*), on the other hand, is given to all animals for the purpose of self-preservation; in later terminology, it is the instinct for survival. Moreover self-love, as Abbadie had said in respect of *amour de soi même,* is a legitimate form of self-interest whereas self-liking, often excessive, is corrupt (Abbadie 1692: 263). We shall see that in Mandeville's hands, the corrupt passion nevertheless becomes of supreme importance in making man a social creature.

1.4 Other Passions

If pride is the predominant passion, it by no means the only one and, at times, its predominance seems to be challenged. Lust or sexual desire, linked to shame by "artful moralists" (Mandeville 1924: 1: 145) is an important passion however much it is left unspoken of in civil society. Without lust, conventionally defined as a vice, there would be no continuance of human existence let alone of society. On the other hand, progress from a dreary state of nature where lust is prevalent partly relies on another passion, that of Avarice, the acquisitive instinct which, when leading to the successful accumulation of wealth, is not only an important economic driver but also a determinant of an individual's status in society. Fear, so succinctly described by Hobbes as "Aversion, with the opinion of hurt from the object" (Hobbes 1909: 43) is given a more subtle psychological twist by Mandeville. "The Passion that is rais'd in us when we apprehend that Mischief is approaching us ... The Disturbance it makes within us is always more or less violent in proportion, not of the Danger, but of our Apprehension of the Mischief dreaded, whether real or imaginary" (Mandeville 1924: 1: 254).

Fear, as we shall see, is an important driving force is converting savage man to a belief in the value of living in a community where common dangers can be more easily resisted than in a solitary state.

On the other hand, Mandeville's definition of pity "a fellow feeling and condolence for the Misfortunes and Calamities of Others" (Mandeville 1924: 1: 254) is less cynical than Hobbes's pity which is little more than *schadenfreude*, or the escape from a fear that the same calamity may befall the person pitying the fate of another.

1.5 Motivation: Psychological Derangement

I am not suggesting that Mandeville's earlier theory of human motivation bore no resemblance to Hobbes's in a general sense for it did share a belief that man was a creature who acted out of sense of self- interest, an egotist guided towards pleasure and away from pain by instinct. In Hobbes's account the "causes" of that behaviour result from the direct effect of the stimulation of the senses by external phenomena. The process he describes is a mechanical and materialistic one in which sense impression are implanted upon the brain by specific things happening that "cause" the human to react in one way or another, appearing as effects as in the workings of natural phenomena. There is not much room here for dispositional or habitual behaviour since that would suggest action uninitiated by external stimulation.

While Mandeville accepts the egoistic basis of Hobbes's system he adds to it a refinement that derives from the French, sceptical background with which we have already seen him to be familiar. For although for Mandeville the passions "within ... unknown to them [men], govern their Will, and direct their Behaviour" (Mandeville 1924: 2: 139) there is superimposed on that pattern a theory of psychological derangement or escape into illusion which makes Mandeville more ready to tackle dispositional or habitual behaviour than Hobbes. Playing off the passions against each other and against themselves is an important part of the political management of large, complex societies where needs far exceed those of life in a savage state.

This Mandevillean derangement or self-deception takes three main forms in his theory. The first is the blinding of each individual to his own defects, a feature that leads him to overvalue his own worth so that even "the meanest wretch puts an inestimable value upon himself" (Mandeville 1924: 1: 54). It is self-liking that leads men in this direction and makes them such bad judges in their own cases. The second feature of this derangement is the ability of man to keep hidden from himself the real motives for his action. Cleomenes states this aspect clearly in *Part 2 of the Fable of the Bees* "I believe moreover, that a Gentleman so accomplished, all his Knowledge and great Parts notwithstanding, may himself be ignorant, or at least not well assured of the Motive he acts from" (Mandeville 1924: 1: 54).

This second feature of Mandeville's derangement, suggesting that the causes which operate on man are either unrecognised or repressed (because he says, it is too mortifying for an individual to undertake self-examination into these "Hidden springs") (Mandeville 1924: 2: 79) brings him closest to the Freudian notion of the subconscious.

The third feature of Mandeville's cognitive derangement is the extension of the Baylian idea of an inconsistency in human behaviour which makes rational behaviour, according to principles, difficult if not impossible. This idea is made clear in *The Origin of Honour* (1732) where Horatio remarks that fear of death holds back men from being able to act in a rational way. Cleomenes, speaking for Mandeville, replies that this is not so. Man certainly does fear death and the beyond but throughout his life behaves as if he has no thought of the future. Immediate concerns over survival and status among his fellows push aside such thoughts. His mental make-up renders man incapable of acting on the logic of his impulse (i.e. the fear of death); he has not the consistency to act rationally according to a principle (Mandeville 1732: 18).

Mandeville's elaboration of this theory of self-deception, shorn of any moral connotation, can nevertheless be traced to the French *sceptiques* as well as to Erasmus as Irwin Primer has argued (Primer 1993). Abbadie talks of the way in which *amour propre* distorts the individual's value of himself (Abbadie 1692: 259). La Placette goes further adding that it suppresses any ideas of fault or weakness in an individual and fools the individual into thinking that behaviour which is in no way meritorious is indeed meritorious (La Placette 1697: 28). Both these writers were building on the legacy of Montaigne (a favourite of Mandeville as we have seen), particularly in his Pyrrhonian work, *Apologie de Raimond Sebond* (1967) where he exposes the utter fallibility of the human mind and its inability to know anything with certainty. However, if this French sceptical tradition first made Mandeville wary of speculative systems, his training as a medical doctor would have confirmed his antipathy to the Cartesian system of his youth. He tells us over and over again that it is essential to proceed by observation – and incidentally pours scorn on doctors who fail to attend to their patients (see Hilton 2010) – and, in the English tradition of Newton, Locke and Sydenham, to base reasoning about the human condition on an empirical footing.

Considering Mandeville's psychological theory of self-deception enables me to turn to his theory of the origins of society for it remains key to understanding what we can discern as his two approaches to man's success as a social animal.

1.6 Origin of Society: Phase 1 Dextrous Management

The first strand in Mandeville's social theory is that "dextrous management" by which the lawgivers and politicians must convert "private vices, into publick benefits" (Mandeville 1924: 2: 319). This, he warns, is no easy task; it will involve the exercise of great skill based upon a thorough understanding of the passions which, as we have seen, drive human behaviour. Cleomenes emphasises the difficulty facing the prospective leaders:

> To preserve Peace and Tranquility among Multitudes of different Views, and make them all labour for one Interest, is a great Task; and nothing in human affairs requires greater Knowledge, than the Art of Governing. (Mandeville 1924: 2: 318)

What is more the process of forming a coherent society will be a slow one for grim though the conditions of the state of nature are (and Mandeville agrees in this respect with Hobbes) fear of one another and a natural indolence will prevent men being led out of it into civil society. What the politicians have to do, and E.J. Hundert reminds us that they are not a group of public spirited individuals but rather a "cunning minority", (Hundert 1994: 19) is to exploit the potential that the ruling passion of pride for turning men into social beings, content to accept restrictions on their individual freedom of action.

Although like most political theorists of his time, Mandeville uses the state of nature as both a conjectural tool and a distant but vague historical reality, in whichever sense it is a condition in which only the potential for social life exists. In describing natural man, Mandeville resorts to the analogy of childhood. Just as a child is amoral and undisciplined by any social restraint, natural man is a creature for whom morality has no meaning and he too, is unaffected by social duties or responsibilities. It is only by a long process of education and indoctrination that children gain a sense of vice and virtue; natural man too is subject to a long process of adaptation to social life. That process is supervised by politicians who have to coax men into accepting the benefits of living communally, in particular by the use of flattery. And as A.J. Lovejoy pointed out, Mandeville recognised that this would be a long, historical process, using the myth of the lawgivers mainly for satirical purposes (Lovejoy 1968: 176).

What Mandeville does say, in his early account, is that the principal method by which the leaders gain political control involves the fabrication of a sophisticated political myth in which the whole human species is divided into two groups – on the one hand, the angelic men who pursue the public interest rather than their own private interest and, on the other, the brutish creatures who merely pursue their own interest, without regard to any other. The actions of the first group, the angelic men, were declared virtuous by the politicians since their actions benefitted society as a whole; the actions of the second group were vicious since they only benefitted the agents themselves. The first group had exercised self-restraint in the public interest; the second group had merely indulged their appetites. Virtue and Vice are thus harnessed toward social ends by making them terms of approbation and disapprobation respectively. So Mandeville says were "the first Rudiments of Morality, broach'd by skilful Politicians, to render Men useful to each other as well as tractable" (Mandeville 1924: 1: 70).

Men are thus tutored to social behaviour by persuasion and the most powerful weapon in the armoury of the politicians is the resort to flattery. From the earliest days of childhood, throughout adult life, man's pride (self-liking) is played upon by parents, teachers and ultimately in society, by skilful politicians. To augment his already over-valued sense of self-worth man will be influenced by the opinion of others: rulers, as Machiavelli had shown, employ stratagems to enhance the status of individuals by singling them out for honours.

Mandeville's myth, standing alone, would have placed him firmly in the Machiavellian republican tradition which was transmitted to England through James Harrington's *The Commonwealth of Oceana* (1700). But, in fact, having explained the origin of the moral distinction between disinterested action and selfish action in the public sphere, he then turns the whole theory of virtue on its head by insisting that it is only as a result of vice, for example in the form of an avaricious pursuit of wealth, that the great hive of bees (Britain) has any chance of maintaining, let alone enhancing, its greatness.

This heterodox approach had already been set out in the *Female Tatler* dialogues of which Mandeville's contribution ran from November 1709 to March 1710, that is, between the appearance of the *Grumblin' Hive* in 1705 and the first part of *the Fable of the Bees* in 1714. The Oxford Gentleman, Mandeville's spokesman in the *Female Tatler* dialogues sets out the politician's task in this way:

> The jarring discord of contraries makes the harmony of the whole, it is the business of the skilful politician, to make everything serve in its proper place, and extract good from the very worst as well as from the best. (Mandeville 1709: no. 64)

This juggling of the passions is pithily expressed in Pope's lines in Epistle II of the *Essay on Man* where he says:

> Passions, like Elements, tho' born to fight
> Yet, mix'd and soften'd, in his work unite. (Pope 1967: 253)

The difficult task of creating social bonds is made easier by certain economic consequences of men's behaviour. While honest citizen are praised for their virtue, in respect of progress: "they are of little use, and the country might soon be overstock'd with them, for where they abound, poverty must ensue, therefore happy is the land, whose constitution is so well fenced with wholesome laws, that fear and prudence may supply the place of honesty" (Mandeville 1709: no. 64).

In fact the politician must rely on the greed of the avaricious in their relentless pursuit of material gain and the proud who seek status and pre-eminence in society. They both stimulate trade and employment and with it, wealth.

The wise lawgiver will also understand the role played by the "idle rich" in cementing the social contract. The Oxford Gentleman gleefully proclaims:

> Madam, said he, it is unquestionable, that the greatest and most immediate benefactors to human society, are the idle favourites of blind fortune, that having more money left them than they know what to do with, take no other care than to please themselves, and studying as well to create new appetites as to gratifying those they feel already, are given over to all sensuality, and value neither health nor estate in the purchase of delight. (Mandeville 1709: no. 62)

1.7 Social Theory Phase 2: Evolution

These are ingredients of the great hive described in *the Fable* but how has society advanced to this stage? The answer takes us to the second main strand in Mandeville's social theory, namely that gradual evolution of society which is principally described in Part II of *the Fable* and which links Mandeville to later Scottish writers such as Adam Ferguson and Lord Monboddo.

The key factor in this evolutionary account is the same pride or self-liking which the skilful politicians exploit. The struggle for survival, in which man scrapes together what he needs to subsist, is undertaken in the family context which provides primitive man with his first taste of social life while also furnishing him with the means to indulge his appetite for praise. Mandeville says:

> An untaught man would desire that everybody that came near him, to agree with him in the Opinion of his superior Worth, and be angry as far as his Fear would let him, with all that should refuse it: He would be highly delighted with him, and love every body, whom he thought to have a good opinion of him, especially those, That by Words or Gestures should own it to his Face. (Mandeville 1924: 2: 133/4)

This early social background thus provides for the satisfaction of self-liking as well as self-love since men would be flattered by members of their families. Nevertheless, family relations could be difficult: a parent's natural affection for his child could be overturned by any act of defiance on the part of the child. Only a combination of love and fear, which Mandeville called "reverence" (Mandeville 1924: 2: 202), would ensure that the child behaved well according to the parent's dictates. The sentiment of reverence would later be transferred to rulers acting in the same role as the head of the family units that preceded the state.

Mandeville is at pains to emphasise the gradualness of the evolutionary process from family living to social living. He takes the view that even the crudest forms of society, such as those of the North American Indians or the Pacific islanders, would have taken a long time to develop. Man's natural environment would not necessarily induce him toward social living, based as it would have been on nomadic subsistence. It was only when that condition became dangerous and uncomfortable in the extreme that man would be driven by fear to seek a wider communal way of life. Uneasy alliances of family groups would continue for many epochs before society could be welded from them. Savage man's capricious way of life, taking what he needed for the day and his lack of concern for the future made him a reluctant social being. Before any real progress toward society could be made, external pressures would have to be considerable and one invention, at the very least, would have had to be made.

The first of the external pressures threatening man would have been the danger of wild animals which would have led him to a closer defence of the family. That would require co-operation and a degree of organisation. For the first time primitive man would start to make plans with others, realizing that strength of numbers was an important part of meeting the threat of attack. Larger family groups or alliances would be formed to defend all their members. This first step towards communal living would be a tentative one and might not last longer than the danger itself lasted. Among the families and groups forced together, a spirit of competiveness would grow. Some would claim superiority because of their greater prowess in dealing with the external threat; rivalry about their relative importance could lead to a breakup of the loose bonds keeping them together. When the danger from without was less, these tensions could lead to conflict. A period of internecine family war could continue for a long time, in which men lived miserably. The evidence from travel accounts of primitive people outside Europe suggested this very pattern.

Nevertheless man would slowly learn from these experiences. He would come to understand that the threat from his fellow men might be as great as the threat from wild animals unless some system of order was formulated. Codes of conduct would need to be devised but they could not be until man had acquired the faculty of speech.

1.8 Origin of Speech

Mandeville's treatment of the development of speech is an original insight. Rather than accepting the orthodox view that speech was a God-given faculty Mandeville developed the notion that learning to speak was a gradual and empirical process, arising at first from sign language. In a savage state, where man's needs were simple, "The Want of Speech is easily supply'd by dumb Signs; and it is more natural to untaught Man to express themselves by Gestures, than by Sounds" (Mandeville 1924: 286).

By way of crude gestures and facial expressions, together with tokens such as weeping, laughing, smiling or sighing, primitive man communicated with his fellows. Only after some time of living together with others on a daily basis would he "find out sounds" (Mandeville 1924: 288) to match ideas that he wished to share. Gradually men discovered the flexibility of the voice and the possibility of articulating words. The art of language developed from these crude origins. But a further step in social progress was the invention of writing so that codes and laws could be recorded since oral agreements could be misunderstood or forgotten. Although Mandeville rates this acquisition as the third important step to bringing about social cohesion (the first two being fear of wild beast and fear of other men) he does not go into much detail about it at this point but often refers to the importance of law in the social context in other places.

Mandeville supplements his account of these necessary conditions for establishing civil society with details of many other advances, such as the discovery and use of implements in agriculture and in manufacture, the invention of money as an exchange mechanism, what has become known as the division of labour whereby different specialized functions are carried out by different categories of workers and professionals. Once again the political management of this complex pattern of evolution is in the hands of leaders who must master what Rousseau was to call the art of darkness.

Throughout his account of social evolution, Mandeville stresses the fragility of the historical process: each stage in development is long and drawn out and there can be lapses into a state of near anarchy as groups joined together fall out with one another. The development of civilization is, on the one hand, a series of adaptations on the part of man to the random course of history; on the other, his attempt to build upon what he has learned. Leaders need to understand the intricacies of these processes and what might or might not be possible at any stage of development. Mandevillean man is therefore guided yet self-taught. External pressures operate to make him co-operate with his fellows and it is within society that he can most fully express his craving to gain applause and act for the common good. The passions, so long condemned by moralist as vices, are the very springs of social advancement.

1.9 Conclusion

Mandeville's two strands of social theory, the evolutionary account I have just been considering and the earlier account of the myth making of the lawgivers, describe the psychological adaptation of man to his environment. They are ways of accounting for his sociability, a quality haunting the imagination of Enlightenment philosophers, intent on explaining why man is a social animal, best suited to living in communities. In order to help them in that task contemporary philosophers ransacked history, ancient and modern, to support their conjectures about the origins of social institutions. While I agree with Mikko Tolonen that the strands of Mandeville's social theory are distinct stages in his thinking, I am less convinced that we have to read them (and therefore the two parts of *the Fable of the Bees*) as completely separate works (Tolonen 2013).

The reason for my view is that there seems to me to be a pervasive psychological theory running through all Mandeville's thinking on man in society although the emphasis certainly changes. Mandeville begins on the sceptical, *moraliste* side in interpreting man as a social being, an approach which plays down the role of reason as instrumental in bringing about progress until men are actually in civil society and begin to see the benefits of following common rules of justice, as David Hume was later to describe. It is true that applications of these psychological theories to historical study tended to be made within a moral framework. Those who supported the theory of continuous progress saw progress in terms of moral improvement, enlightenment and later happiness. In their eyes however slow advances were, they steadily brought modern man towards the good life in the good society.

Mandeville's adaptation of the theory of continuous progress was radical because he insisted on measuring that progress in strictly non-moral terms. His modernist credentials were already plainly set out in the *Female Tatler* where he says:

> When I... compare the meanness as well as the ignorance of the infant World, and yet unpolished Nations of Africa and America, to the Knowledge and Comforts of human life which the more civiliz'd Countries, and more especially the polite parts of Christendom enjoy, I can hardly forbear thinking how we are indebted to all that ever invented anything for the public good: It is they who have actually meliorated their kind, and from that grovelling state and despicable condition in which we now see the Negroes and other Savages, raised their posterity to the enjoyment of those blessings we have among us. (Mandeville 1709: no. 62)

There can be no clearer statement on the side of "politeness" than that. Primitive societies are characterised by squalor and discomfort: the idea that such an existence could be an object of admiration or a model for living is brushed aside with contempt. There is no trace in Mandeville of any nostalgia for the state of nature which, in the Hobbesian tradition, is bleak and harsh. And at the root of his progressive ideology is a theory of the evolution of society which is based on a psychological analysis of a lapsed Augustinian, convinced that the depraved creature of his study could be a successful citizen of commercial, polite society.

References

Primary Sources

Abbadie, J. 1692. *L'art de se connoitre soi même*. Rotterdam.
Butler, J. 1855. *Ethical discourses*. Syllabus by Dr. Whewell, ed. J.C. Passmore. London.
Harrington, J. 1700. *The Oceana of James Harrington and his other works*. London.
Hobbes, T. 1909. *Leviathan*, ed. By W.G. Pogson Smith. Oxford: Oxford University Press.
de La Placette, J. 1697. *Nouveux Esscus de Morale*. Amsterdam.
Mandeville, B. 1709–1710. *The female tatler*. London.
Mandeville, B. 1732. *An enquiry into the origin of honour, and the usefulness of Christianity in war*. London: Brotherton.
Mandeville, B. 1924. *The fable of the bees*, ed. F.B. Kaye, 2 vols. Oxford: Oxford University Press.
Montaigne, M. 1967. *Oeuvres complètes*. Paris: eds. Du Seuil.
Pope, A. 1967. *Poetical works*. Oxford: Oxford University Press.

Secondary Sources

Dekker, R. 1992. Private vices and public virtues revisited: The Dutch background of Bernard Mandeville. *History of European Ideas* 14: 481–498.
Hayek, F.A. 1966. Dr Bernard Mandeville Lecture on Mastermind. *Proceedings of the British Academy*, London: vol. LII.
Hilton, P. 2010. *Bitter honey: Recuperating the medical and scientific context of Bernard Mandeville*. Bern: Peter Lang.
Hundert, P. 1994. *The enlightenment's fable: Bernard Mandeville and the discovery of society*. Cambridge: Cambridge University Press.
Lovejoy, A.O. 1968. *Reflections on human nature*. Baltimore: John Hopkins University Press.
Primer, I. 1993. Erasmus and Bernard Mandeville: A reconsideration. *Philological Quarterly* 72(3): 313–335.
Tolonen, M. 2013. *Mandeville and Hume: Anatomists of civil society*. Oxford: Voltaire Foundation.

Chapter 2
Bernard de Mandeville and the Shaping of Conjectural History

Frank Palmeri

Abstract This paper argues that Bernard Mandeville's *Fable of the Bees, Part II* (1729), is the first work to bring together all the elements of conjectural history, a form that rose to prominence in the Enlightenment, between 1750 and 1800. Conjectural history builds on the natural law tradition of Hobbes and Locke, but it is non-contractual and considers a longer span of history. In both these respects, Pufendorf's account of early society opens a way that is taken by later conjectural historians. Of these, Mandeville's *Fable, Part II*, is the earliest to present a naturalistic, non-contractual narrative of the early stages of civil society. Although Vico's *New Science* exhibits some of these features, it remains tied to a providential and cyclical view of history.

Keywords Conjectural history • Natural law • Pufendorf • Vico • Philosophy of history

In 1723, Bernard Mandeville added a conjectural "Enquiry into the Origin of Moral Virtue" to the collection of essays he had published in 1715 as the *Fable of the Bees*. An inquiry, in the spirit of this short speculative essay, into the origins of conjectural history itself might find in the dialogues of Mandeville's *Fable of the Bees, Part II* (1729) the first account of early society to bring together the major elements of the emerging form of conjectural history in a single work. The *Fable, Part II* would thus inaugurate a genre that assumes a distinctive prominence and influence in European thought between 1750 and 1800; in the *Fable*, it makes an appearance more than 20 years before Turgot's inaugural speech at the Sorbonne, which has been considered by some as the first statement of a stadial conjectural history. To support this view, it will be important to distinguish the conjectural historical narrative in Mandeville's *Fable, Part II* from the speculative accounts of the foundation of society in the natural law theories of Hobbes, Locke, and Pufendorf; doing so can bring to the fore several features in the *Fable, Part II* whose presence characterizes and informs almost all later instances of conjectural

F. Palmeri (✉)
College of Arts and Sciences, University of Miami, Miami, USA
e-mail: fpalmeri@miami.edu

E. Balsemão Pires, J. Braga (eds.), *Bernard de Mandeville's Tropology of Paradoxes*, Studies in History and Philosophy of Science 40, DOI 10.1007/978-3-319-19381-6_2

history. Vico's *New Science* – whose first version was published 3 years before the *Fable, Part II* – bears a close relation to, but does not embody, the conjectural form.[1] Pufendorf's work more than Vico's constitutes a bridge or passage from the natural law theories to conjectural histories.

In the seventeenth century, natural law theorists of the origins of human society such as Hobbes and Locke posited that at some unspecified point in the past, humans emerged from a primitive condition and formed themselves into a society by means of a contract. According to Hobbes, the solitary, asocial humans living in a natural condition entered into a compact among themselves, each man motivated by fear of his neighbors. Since no covenant could be enforced without a power over the contracting parties, the contract bound the people to surrender their natural right of self-defense into the hands of a sovereign, whose responsibility would be to guarantee the life and property of every member of society, but whose authority could not be questioned. Hobbes acknowledges that the condition of nature, the war of all against all, may never have existed, at least not over the whole world at the same time, but he contends that the only alternative to such a sovereign is the breakdown of society into civil war (Hobbes 1981: 187).

In Locke's view, although free, propertied individuals in a natural condition did not live in fear, yet they decided to come together to form a community for the more "secure Enjoyment of their Properties, and a greater Security against any that are not of it" (Locke 1988: 331). To do so, they entered into a compact among themselves to accept a power that would protect their property and security. Significantly, these "Freemen" (Locke 1988: 333) compact among themselves but also with the political authority that they set up to protect themselves and their property; therefore, unlike in Hobbes, in the event that the government fails in accomplishing its purposes, they retain the right to overturn it and replace it with another. Locke, even more explicitly than Hobbes, calls this thought experiment a "Conjecture" and a "Hypothesis" (Locke 1988: 334–5), and nonetheless defends it strongly. To the objection that history gives "very little account of Men, that lived together in the State of Nature" (Locke 1988: 334), he responds that we may as well suppose that the armies of Xerxes were never children because we do not hear of them until they were already men and soldiers.

Both of these theories contain several elements that would contribute to the form of conjectural history. They depend on speculation concerning the prehistory of human social life in times for which no documents, records, or remains survive. In addition, they both avoid appealing to providence for an explanation of the earliest developments of society. However, they also adopt features that will prove to be incompatible with the genre. Most significantly, both depend on a foundational contract as a crucial explanatory part of their conjecture. Significantly, also, neither Hobbes's nor Locke's account represents a history of gradual changes taking place

[1] I am not aware of evidence that either Vico or Mandeville influenced the other's thought or use of form.

over hundreds of generations and proceeding through several stages to the modern era. Rather, both provide only a conjectured historical moment – that is, not a history at all.

Samuel Pufendorf provides a detailed and extensive discussion of the passage from a state of nature to society in the *Law of Nature and Nations* (1672), explicitly revising Hobbes's conjectural account concerning the natural state, the reason for government, and the content and function of the multiple founding contracts.[2] For Pufendorf, the original state of human life, and the natural state, does not consist of a war of each against all, a life that is "solitary, poor, nasty, brutish, and short," but rather of a confluence of individuals and families who came together to protect themselves from wild beasts (Pufendorf 1717: 104–9; bk. 2, ch. 2, sec. 2–4).[3] In this state, government has not been formed, but people are sociable and peaceful. If these early humans had really been at war with each other, they would not have survived; but they did survive, and therefore must have lived on terms of friendship, bound together as reasonable creatures, and not by an agreement or covenant among themselves (Pufendorf 1717: 110–15; 2.2.6–10).[4]

What moved the heads of families to form civil societies, then, was not the need to escape from a state of war in order to survive. Rather, the growth of agriculture, trade, and urban life were producing luxury, competition, and inequality among men, with the potential to lead to conflict, and the fathers established government to provide against future evils that could be apprehended. Moreover, the founding actions did not consist of a single, all-encompassing compact, as in Hobbes, but of a series of agreements: first, the decision to set up a state and a provision for how to proceed, then a decision on a particular form of government, and finally a covenant with the one or the group who became sovereign, and who must have agreed to care for the well-being and safety of the new citizens in return for their obedience (Pufendorf 1717: 458–70; 7.2.6–10). In the absence of monuments and documents recording these foundations, this conjectural account, Pufendorf contends, may take us as close as possible to the succession of events. In any

[2]Istvan Hont traced the way that Pufendorf's revisions of Hobbes's state of nature – incorporating commerce in the earliest stage, and distinguishing between the founding of society and of the state – prepared the way for the stadial narratives of the Scottish Enlightenment (Hont 2005). I do not conceive of conjectural history as being limited to accounts that specify exactly four stages based on means of subsistence; this conception of the genre as "four-stage theory" owes much to Meek 1976. On the theological underpinnings of Pufendorf's natural jurisprudence and its role in histories of early society in the Scottish Enlightenment, see Moore and Silverthorne 1983.

[3]In support of this picture of earliest man, Pufendorf cites Lucretius, *On the Nature of Things*, Bk. 5, and refers as well to the similar account of Diodorus Siculus, Bk. 1, ch. 5. Wiktor Stoczkowski provocatively argues that later conjectural, evolutionary, and anthropological accounts of the earliest human societies and their stages of development almost all conform to the paradigms established by traditional or common-sense anthropology, often traceable to ancient writers (Stoczkowski 2002).

[4]Pufendorf thus maintains a thesis that has only recently regained currency, as a result of work in contemporary evolutionary biology – that humans more often benefit from being honest and loyal than from cheating and lying (Pufendorf 1717: 145; 4.4.2). See de Waal 1996.

case, he concludes, such agreements *must have* taken place at least tacitly in the institution of commonwealths. Pufendorf thus retains the language of contract, but he takes a significant step away from earlier forms of the concept by multiplying the founding agreements, and acknowledging them possibly to have been only implicit, not formal. In addition, his delineation of a sequence of covenants takes a step in the direction of elaborating a series of stages of social development.

In *The Whole Duty of Man* (1675), his condensation of the *Law of Nations*, Pufendorf loosens the contractual framework in his account of the origins of human society by ascribing the formation of societies larger than the family to an inherent sociability that prompts men to be on good terms with their neighbors instead of regarding them as enemies (Pufendorf 2003: 168–74). Rather than base his argument ahistorically on a single founding moment, Pufendorf conceives of the development of institutions of exchange and of government over an extended period, ever since the emergence of human society. However, he does not speculate about or specify historical stages that would or must have been traversed over that time: he presumes that goods were exchanged by the earliest men in society, and that trade persisted through all the intervening years.

Having given an account of the major approaches of the natural law theorists to the founding of society, we can at this point enumerate the primary distinguishing features of the genre of conjectural history, based on analysis of more than two dozen examples of the genre in the later eighteenth century. Most importantly, of course, this kind of historical narrative traces the origins of society back to a time before the existence of documents and other remains, indeed in some cases, like Mandeville's, to a time before the emergence of articulate language. It can thus only be speculative or inferential. The form often makes use of a rhetoric of the necessary conditional tense, as in: "The origin of speech *must have been* to satisfy wants" (Mandeville 1988: II, 289). The authors of conjectural or hypothetical histories engage in such speculative returns to earlier times as an alternative to accepting providential accounts based on scripture and religious doctrine. The form is thus strongly non-providential or naturalistic; it is also non-contractual. As I have shown, unlike natural law theory, conjectural narratives aim to provide plausible narratives of slow historical developments, not thought experiments focused on a single founding moment of contract. Dugald Stewart, who named the genre in his *Life of Adam Smith*, maintained that the form could provide an account that was true even if it differed from what actually happened (Stewart 1980: 296).

Because the form adopts such a long view of early society, conjectural narratives almost all trace stages or periods in the development of social life. In each stage, institutions belonging to various spheres of life exhibit a common structure. In fact, a variety of conjectural history is devoted to analyzing transitions between stages for which no clear evidence exists, as in the account that Adam Smith gives of the transition from feudal to merchant society in the third book of the *Wealth of Nations*. Conjectural narratives presume that human actions often have unintended and unplanned consequences, that humans make their history but without knowing in advance what course that history will take. They thus often give evidence of the heterogeneity or incommensurability of actions and results. Finally, as they trace the stages of society or of an institution such as religion from the earliest days to historically documented and even modern times, conjectural histories exhibit ambivalent attitudes to modernity, secularization, and the thesis of historical progress. In sum, conjectural narrative adopts a speculative, naturalistic, non-contractual explanation of early social forms that usually falls into stages and shows the unintended consequences of human actions.

Mandeville, in his conjectural "Enquiry into the Origin of Moral Virtue," argues for the human, naturalistic origins of morality, claiming that cunning politicians persuaded people it was for their own good to overcome rather than satisfy their desires. "The Moral Virtues," he writes famously, "are the Political Offspring which Flattery begot upon Pride" (Mandeville 1988: I, 51). It was by being manipulated and flattered into believing it is moral to delay one's gratifications that savage man was "broke[n]" in, taken from the wild state and domesticated.

Mandeville revises and greatly expands his account of the origins of social institutions in the dialogues between Cleomenes and Horatio that constitute the *Fable, Part II* (1729), where he abandons the thesis that cunning politicians duped other humans by appealing to their pride. In Dialogues Five and Six, he provides not just a brief provocation, but a plausible reconstruction of the beginnings and earliest development of social institutions, including language, religion and morals, kinship, poetry, and music.[5] In *Part II*, he makes a claim, like Hobbes, for the importance of fear in determining the early humans to establish a society and government. However, it is fear of large wild beasts that Mandeville believes brought men together. He has Cleomenes call this his "Conjecture, concerning the first Motive, that would make Savages associate: It is not possible to know any thing, with Certainty, of Beginnings, where Men were destitute of Letters; but I think, that the Nature of the thing makes it highly probable, that it *must have been* their common Danger from Beasts of Prey" (Mandeville 1988: II, 231; emphasis added). The fear of other men, on which Hobbes laid so much weight, he considers to have been only secondary in importance to the fear of wild animals and the need to defend against them (Mandeville 1988: II, 266–7).

Rather than ahistorically imagining a founding contractual agreement, Mandeville expands the temporal dimension to take into account that hundreds of generations must have elapsed from the tentative beginnings to the gradual consolidation of any social institution.[6] Moreover, when he imagines one of these other social forms, it is again fear – this time of thunder, lightning, and unseen powers –

[5] I am thus in agreement with Mikko Tolonen, who considers Mandeville's *Part II* as a conjectural history (Tolonen 2013: 77–99), and discusses Mandeville's account of the unintended, non-contractual development of society through stages over a lengthy period of time. My emphasis here, however, is on Mandeville as the first thinker to put together these elements of the form we now recognize as conjectural history.

[6] Malcolm Jack draws attention to Mandeville's emphasis on the extreme length of time required for the gradual processes of socialization of early mankind, as well as the relation of Mandeville's account of early men to the descriptions of tribal people in European travel narratives of the preceding two centuries (Jack 1989: 53–62). Jack's analysis of Mandeville is particularly valuable for placing the *Fable, Part II* in relation to a pair of later conjectural histories that see social

that must have provided the foundation for the first ideas of religion among savage men. Like Vico, who believed early men expressed their awe of thunder in their earliest language and beliefs, Mandeville speculates that early humans must have been motivated by fear in forming the beliefs with which they made sense of a threatening world.

It is only after having laid down the original foundations of society in the emotion of fear and not in the rational prevision of benefits attending a social state that Mandeville has Cleomenes acknowledge that men do form contracts and that they agreed in one of their contracts to institute a form of government. However, before there could be a compact to create a government, people must have had a concept and system of law; for there to have been laws, written language must have existed; and before there could be writing, spoken language had to have been practiced for generations; and even spoken language itself must have developed only slowly from gestures and inarticulate sounds again over generations (Mandeville 1988: II, 269, 287–90). In other words, Mandeville traces back from the end point of government the various stages through which human societies must have proceeded in order to reach recognizable political institutions. He thus practices the most distinctive strategy of conjectural history concerning a period for which no documents can be found to serve as evidence. Once these developments have taken place, yet another step is needed for a historically recognizable society to emerge: as Cleomenes, says, "No number of Men, when once they enjoy Quiet, and no Man needs to fear his Neighbour, will be long without learning to divide and subdivide their Labour" (Mandeville 1988: II, 284). Mandeville here formulates an early recognition of the importance of the division of labor, which generates divisions between ranks and eventually classes. Finally, metallurgy, enabling the production of tools and weapons, would contribute to the construction of a complete society; again, though, for perhaps scores of generations, the use of flints, shells and wood hardened in fire would have preceded use of metal products.

Although the *Fable of the Bees, Part II*, does not delineate a distinct set of stages through which all social institutions passed as a unit so that they were always congruent with each other, it consistently shows the series of steps and the time that must have elapsed in the passage to modern commercial society.[7] Without naming Locke and other natural law theorists, Cleomenes criticizes their "Absurdities" in "alledging as the Causes of Man's Fitness for Society, such Qualifications as no Man ever was endued with, that was not educated in a Society, a civil Establishment, of several hundred Years standing" (Mandeville 1988: II, 301). The *Fable, Part II* makes use of all the other characteristic strategies of the conjectural histories, elaborating a speculative, inferential vision based on little or no evidence, of a naturalistic, non-contractual, and extended historical process, which is caused

---

development as a deeply ambiguous process of both progress and decline – Rousseau's *Discourse on Inequality* and Ferguson's *Essay on the History of Civil Society*.

[7]Martin Otero Knott emphasizes Mandeville's shift away from "theoretical stories" of social contracts to a "conjecturally speculative . . . history of society" in the *Fable, Part II* (Knott 2014: 40).

by men who do not foresee the results of their actions, but who through their unremarkable responses to their environment take unrecorded, laborious, and time-consuming steps toward the institutions of society.[8]

The one other work that might be considered with the *Fable of the Bees, Part II* as the first full-length conjectural history would be Vico's *New Science*. Vico actually employs the necessary conditional – to designate what *must have* been the case in the earliest times – more often than Cleomenes in the *Fable*; he thus indicates the speculative nature of his history and the paucity of evidence on which it is based, consisting mostly of imaginative though often strained etymologies of Latin words and phrases.[9] Beginning with the giants who roamed the earth after the flood, in what he refers to as the "so-called state of nature" (Vico 1984: 352), Vico gives a naturalistic account of the original steps in the constitution of society: fear of thunder and meteorological phenomena first gave savage men a conception of the gods, and with that idea came religion. This originary set of beliefs revolved around divination – the attempt to read the will of the divinities (Vico draws an etymological connection between the two) – as well as reverence for and desire to placate the gods. From a feeling of shame before the gods, which had been instilled by providence, men and women adopted more regular unions, and marriage was born. Once marriages were formally contracted, families moved out of caves, and settled on land near sacred springs and plots that became burial sites (Vico 1984: 181–86). Thereafter, settlements became permanent, trees were cleared by burning, and agriculture became established. However, even though an elaborate social life had developed, properly political institutions only appeared when the families were divided into the strong and the weak, with the latter becoming serfs and the former nobles.[10] The struggle of the serfs to obtain just and equal treatment in the commonwealth determined the shape of the history of the prototypical republic of Rome. This history has for its subject the heroic age concerning which some documents and monuments survive – the earliest written laws, as well as the narrative of their interpretation, supplements, and expansion in the histories of early Rome, especially Livy's. Here, Vico leaves behind speculative or conjectural history proper.

At no point do the savages who first become cyclopean families, and then eventually republican patricians, rely on contracts or agreements for social development.[11]

---

[8]J. A. W. Gunn argues that Mandeville does not deny the existence of providence as the ground for the "mysterious realm of origins" or the ultimate answer to the problem of good and evil in the world (Gunn 1975: 117). Even if granted, this is a very attenuated concept of providence, especially as compared with the prominence of the concept in thinkers such as Vico, or, among later conjectural historians, Kames and Herder. E. J. Hundert argues for the naturalistic character of Mandeville's analysis (Hundert 1994: 84, 113).

[9]In this, his practice resembles that of Nietzsche, in the *Genealogy of Morals*, Essay 1 (1888), a later conjectural history.

[10]The early appearance in Vico's account of the division between the strong and the weak parallels the originary status of this opposition in Nietzsche's *Genealogy of Morals*, Essay 1.

[11]The concept of the cyclopean family is based on the isolated, primitive, cave-dwelling single families of Cyclops, as depicted, for example in the *Odyssey*. The cyclopean family also figures as

Perhaps most strikingly, in Vico's account, each stage of human history – whether the age of gods, heroes, or men – possesses an organic unity, a holism that binds together various cultural institutions and forms. Thus, for example, in the earliest age, the youth of the world, to which Vico devotes the greatest part of his attention, the gods rule human affairs; theology is poetic; language is ritual, gestural, or onomatopoetic; and the mode of knowledge is concrete and sensory. This homology, along with the other features just observed, brings Vico's work into accord with the form of conjectural history.

On the other hand, Vico consistently appeals to providence to explain the course of the gentile nations. These appeals have a force and frequency greater than would be required of a pretense to disguise a naturalistic method. Consider, for example, the title of Book 2, Chap. 5: "It is Divine Providence that Institutes Commonwealths and at the Same Time the Natural Law of the Gentes" (Vico 1984: 234–36).[12]

Moreover, in Vico's conception, history takes the form of a cycle through the course of the three ages and then a recourse to a new age of reflective barbarism. But it is difficult to see this cyclical structure as the expression of a naturalistic account of social development: even if there were regressions in history as a result of excessive reason and philosophy, it is not clear why these declines would need to return the society all the way to the beginning and a new age of barbarism. Vico's *New Science* is still committed to a view of history as the work of divine providence, which takes not an unpredictably, disorderly progressive course, but reveals a neater, indefinitely and repetitively cyclical shape. Since Vico departs in significant ways from the essential features of conjectural history, I would suggest that Mandeville's *Fable* more closely realizes the elements of conjectural history for the first time.[13]

Twenty-five years after the final edition of Vico's *New Science* (1744), German conjectural histories such as Herder's *Ideas for a Philosophy of History of Mankind* (1784–1791) and Kant's "Idea for a Universal History with Cosmopolitan

---

the first stage of human society in Henry Maine's influential history of Roman legal institutions, *Ancient Law* (1861).

[12]Mark Lilla argues that from his earliest works through the final *New Science*, Vico is committed to a view of history in which providence exercises a directing hand, shaping human nature for beneficial results. On this account, the *New Science* falls short of providing a naturalistic account of the genesis and course of human societies (Lilla 1993). Leon Pompa characterizes one of Vico's many cryptic statements about the world of gentile history as a "strange mixture of claims" which asserts that this historical world is created by man but also that it is the work of a superhuman mind that acts for human ends (Pompa 1990: 161). Karl Löwith argues that Vico "never intended to discard revelation" (Löwith 1949: 135), but for Löwith this retention of providence somehow balances perfectly with a history of merely human actors, so that Vico combines theology and philosophy of history in a way that is not matched by any modern historical thinker.

[13]James Thomson's *Liberty* (1735–1736) provides another instance, nearly contemporaneous with Vico's, of the idea that societies develop from primitive beginnings through stages to a possible state of luxury and decline, after which the cycle begins again in another country. See *Liberty*, Part II: *Greece*, ll. 3–85 and 391–420 (Thomson 1986). Mandeville's narrative is notable for its avoidance of the Polybian cyclical or helical return from luxury or corruption to a newly primitive state.

Intent" (1784) picked up his thought in works that saw historical development in organic and holistic terms. A few decades earlier and a quarter century after Mandeville's *Fable, Part II*, conjectural history became a prominent and distinctive Enlightenment genre in the French and Scottish traditions: in Rousseau's *Discourse Concerning the Origins of Inequality* (1755), Hume's *Natural History of Religion* (1757), Adam Smith's *Lectures on Jurisprudence* (1762) and *The Wealth of Nations*, Book 3 (1776), Adam Ferguson's *Essay on the History of Civil Society* (1765), William Robertson's "A View of the Progress of Society in Europe," from his *History of Charles V* (1769), John Millar's *Origin of the Distinction of Ranks* (1771), and Condorcet's *Sketch for a Picture of the Historical Progress of the Human Mind* (1795). Following this Enlightenment tradition, several nineteenth-century works re-appropriated and revised the conventions of this naturalistic, non-contractual form. The first two books of the second edition of Malthus's *Principle of Population* (1803), take shape as a conjectural history of the checks to population from starvation and disease in tribal societies around the globe, pursuing an argument whose elements Mandeville had already outlined in the *Fable, Part II* (Mandeville 1988: II, 245, 254). From there, and throughout the nineteenth century, this form shaped the emergence of thought in political economy, anthropology, and sociology. Mandeville's *Fable, Part II*, thus stands at or near the beginning of this genealogy of the disciplines of social knowledge in European thought.

References

de Waal, Frans. 1996. *Good natured: The origins of right and wrong in humans and other animals*. Cambridge, MA: Harvard University Press.

Gunn, J.A.W. 1975. Mandeville and Wither: Individualism and the workings of providence. In *Mandeville studies: New explorations in the art and thought of Dr. Bernard Mandeville*, ed. Irwin Primer, 98–118. The Hague: Martinus Nijhoff.

Hobbes, Thomas. 1981 [1651]. *Leviathan*. New York: Penguin.

Hont, Istvan. 2005. The language of sociability and commerce: Samuel Pufendorf and the theoretical foundations of "four-stages theory". In *Jealousy of trade: International competition and the nation-state in historical perspective*, ed. Istvan Hont, 159–184. Cambridge, MA: Harvard University Press.

Hundert, E.J. 1994. *The enlightenment's* Fable: *Bernard Mandeville and the discovery of society*. Cambridge: Cambridge University Press.

Jack, Malcolm. 1989. *Corruption and progress: The eighteenth-century debate*. New York: AMS.

Knott, Martin Otero. 2014. Mandeville on governability. *The Journal of Scottish Philosophy* 12: 19–40.

Lilla, Mark. 1993. *G. B. Vico: The making of an anti-modern*. Cambridge, MA: Harvard University Press.

Locke, John. 1988 [1689]. *Two treatises of government*, ed. Peter Laslett. Cambridge: Cambridge University Press.

Löwith, Karl. 1949. *Meaning in history*. Chicago: University of Chicago Press.

Mandeville, Bernard. 1988 [1732]. *The fable of the bees: Or, private vices, publick benefits*, ed. F.B. Kaye, 2 vols. Indianapolis: Liberty Fund.

Meek, Ronald. 1976. *Social science and the ignoble savage*. Cambridge: Cambridge University Press.

Moore, James, and Michael Silverthorne. 1983. Gershom Carmichael and the natural jurisprudence tradition in eighteenth-century Scotland. In *Wealth and virtue: The shaping of political economy in the Scottish enlightenment*, ed. Istvan Hont and Michael Ignatieff, 73–87. Cambridge: Cambridge University Press.

Pompa, Leon. 1990. *Human nature and historical knowledge: Hume, Hegel, and Vico*. Cambridge: Cambridge University Press.

Pufendorf, Samuel. 1717 [1672]. *The law of nature and of nations: Eight books*, ed. Jean Barbeyrac, trans. Basil Kennett. London.

Pufendorf, Samuel. 2003 [1691]. *The whole duty of man, according to the law of nature*. Indianapolis: Liberty Fund.

Stewart, Dugald. 1980 [1794]. Account of the life and writings of Adam Smith, LL.D. In *Essays on philosophical subjects*, ed. W. P. D. Wightman and J. C. Byce, 269–351. Indianapolis: Liberty Fund.

Stoczkowski, Wiktor. 2002. *Explaining human origins: Myth, imagination, and conjecture*. Cambridge: Cambridge University Press.

Thomson, James. 1986 [1735–1736]. *Liberty*. In *The castle of indolence and other poems*, ed. James Sambrook. Oxford: Clarendon Press.

Tolonen, Mikko. 2013. *Mandeville and Hume: Anatomists of civil society*. Oxford: Voltaire Foundation.

Vico, Giambattista. 1984 [1744]. *The new science*, trans. Thomas, Goddard Bergin and Max Harold Frisch. Ithaca: Cornell University Press.

Chapter 3
Mandeville and the Eighteenth-Century Discussions About Luxury

Edmundo Balsemão Pires

Abstract Luxury entails a public differentiating use of objects and commodities, which is grounded on the overlapping of the spending with commodities and the ostentation of perceptible signs stimulating social imitation. In the eighteenth century, the debates on luxury emphasized the importance of the scrutiny of the power of imagination as intimately related to the contagious and mimetic character of the use of luxury objects. Thus, "luxury" represents a conceptual and, more generally, a semantic momentum in the evolution of the description of the society grounded on the influence of imitation. The several textual testimonies of the luxury debates in the eighteenth century, including Bernard de Mandeville's contribution, attest the epistemological perplexity regarding the status of the psychic side of the definition of commodities, or economic goods, as utilities responding to specific human needs: the escalade of the emulation in the acquisition of luxury objects seems to deny any relevant connection of luxury to basic needs. This justifies the reference to imagination as the psychic source of the needs connected to luxury consumption and the absence of a direct correspondence between luxury goods and needs. B. Mandeville's views on the theme of luxury and overconsumption in his *Fable of the Bees* and other writings are a privileged starting point for the explanation of these aspects of the evolution of the modern commercial society.

Keywords Luxury • Commercial society • Imitation • Consumption • Utility

---

This text is a contribution to the discussion of the theme of luxury as an aspect of the evolution of the modern society and modern semantics. It develops perspectives on the semantic of luxury from B. Mandeville's theoretical positions regarding the subject. The paper addresses again some problems already scrutinized by the authors that have studied the theme of luxury from T. Veblen (*The Theory of the Leisure Class*, 1899), Guglielmo Ferrero (*The Evolution of Luxury*, 1901) and W. Sombart (*Luxus und Kapitalismus*, 1913) to J. Sekora (*Luxury: the concept in Western Thought*, 1977) and C. Berry (*The Idea of Luxury*, 1994).

E. Balsemão Pires (✉)
Faculdade de Letras-Universidade de Coimbra, Largo da Porta Férrea,
3004-530 Coimbra, Portugal
e-mail: edbalsemao@icloud.com

E. Balsemão Pires, J. Braga (eds.), *Bernard de Mandeville's Tropology of Paradoxes*,
Studies in History and Philosophy of Science 40, DOI 10.1007/978-3-319-19381-6_3

3.1 Luxury's Semantic Layout

In the Eighteenth Century the descriptions of society, actions, commodities or human needs that included luxury as a predicate were overdetermined. The concept itself is a product of discursive overdetermination. Recent work on the eighteenth century use of the notion, Maxine Berg's, Elizabeth Eger's or Woodruff Smith's books,[1] for example, illustrate the intersecting of semantic drifts from the discussions on economic conditions of the nations; on the value of sumptuary consumption; on taste and the differences between rude and civilized manners; on the social impact of private voluptuous experiences; on the transformation of the form of commodities to accommodate taste or more generally the implications of the thoughts on the "refinement in arts".

But, in all these aspects luxury exemplifies a semantic reaction to the riddle of social causation and to the enigma of the emergence of society from individual actions. From this point of view, the analysis of luxury is relevant for the study of the evolution of the descriptions of modern society and for the investigation of the meaning of a double encapsulation of psychic representations of society and of communicative replicas of the psychic images.

The evolution of modern luxury shows semantic trends evolving towards patterns differing from the luxury of the ancient societies even when common traits seem preserved. The essential divergence is to be sought in the importance acquired by the meaning of luxury as fashionable merchandise in modern times. The mercantile character of the products of luxury is absent in the archaic or ancient societies. Guglielmo Ferrero's difference between the ancient, dirty luxury essentially associated to the function of signalling the belonging to a superior social rank and a modern clean, comfortable and democratic luxury (Ferrero 1901) is only explainable as a consequence of vast transformations in the social differentiation of society, in the influence of the mimetic drifts of the social interaction in the formation of a society of masses and in its semantic effects.

I'll argue that the general theoretical design of Bernard de Mandeville's work is the best frame to integrate the main semantic characteristics and changes of the theme of luxury in the eighteenth century, in the epoch of the crisis of the ideologies responsible for the sumptuary restrictions and the emergence of mass consumption.

If we look to the layout structure of the descriptions of luxury in its semantic components we conclude that the semantic of modern luxury is made of the articulation of two main indexical designators: an index measuring *quantity* changes in commodities consumption and the index of the variations in the spread of a social *epidemic*.

Both indexes were persistently articulated in the judgments about luxury in the eighteenth century. The combination of both designators is notorious in public discussions and polemics and in many theoretical descriptions. This may be a

[1] Berg and Eger (2003); Smith (2002); Berry (1994).

characteristic of luxury in the moral discourse but the indexes operate under a specific descriptive layout that organizes the other semantic components of the descriptions and adjust them to what is felt as alarm situations. It is this semantic layout of the modern luxury that explains the difference of luxury and excess or luxury and waste.

3.1.1 The Index of Excess

Quantitative luxury indexes define the value of the distinction between normal or standard consumption and excess or waste. The discussions try to clarify what is normal or excessive regarding some general rules such as the "appropriate to life". The platonic list of the essential goods was the frame of the definitions of a standard in eating, dressing and dwelling. The distinction normal/excessive obeys to two codes: the "code of nature" and a code related to the social rank. The sumptuary laws and their later apologists follow a combination of the natural *per se* and the natural according to the rank in order to establish the equivalent to an objective *scale of luxury* suitable to anticipate deviations. In the *Remarks* of the *Fable of the Bees* B. Mandeville addresses many times the theme of the proportion of the social waste. But, modern luxury is not a theme exclusively related to the amount of an appropriate quantity in consumption.

3.1.2 The Epidemic Index

Frequently, luxury is described as a social disease and the imitation is identified as its means of propagation. What is the object of imitation? The enjoyment of the other or any gratification through the senses allegedly caused by the possession of certain things is such object. A possession hypothetically followed by enjoyment followed by envy is the basic causal scheme associated to imitation in luxury. It defines its driving potential. The index faces the problem of the control of the extent of social imitation. Luxury was regarded as epidemic and addictive. The justification for this general impression is in the fact that in societies defined by strict rules for social mobility the problem "who imitates who?" or better "who can imitate who?" was very pertinent. Imitation was delimited by social expectations related to the positions of the individuals in social ranks. The code of this index is articulated with the distinction confined/unconstrained. For the processing of this distinction in a *scale of mimetic luxury* the code can appropriate the definition of a standard of living according to the social ranks, in order to establish deviation lines. In *An Enquiry into the Origin of Honour*, in the fourth dialogue, Cleomenes asserts the existence of two "sins" that persist in Humankind in all civilized nations: luxury and pride. Symptomatically, he conceives both as epidemic diseases (Mandeville 1732: 205).

The formal layout of the manifold descriptions and communication on luxury in the eighteenth century results from the combination of these two indexes and scales.[2]

Especially in the case of the discussions of the eighteenth century we need to be also attentive to the fact that with "luxury" one communicates about a social order that has lost its guarantees in nature or in the nature's psychic counterparts with the associated causation model. I'll suggest a provisional formula to understand such unsteadiness of the new conditions, partially resulting from the globalized trade: *Luxury is the transmittable perception that the consumption of a commodity is not exclusively based on a necessity of life.*

This turns luxury into something puzzling according to the utilitarian definition of a commodity across the psycho-physiological corridor of the human needs and its linear causation model of human action. The discussion of utilitarianism is inevitable in the context of an adequate account of the social effects of luxury.

3.2 Trade and Luxury in "Commercial Society"

From the artisanal production intended for a limited market made of known consumers and anticipated habits of consumption the modern industry evolved to a production projected for anonymous consumers and unknown habits; from a local to an inter-local market. The increasing importance of long distance commerce puts more abstraction in the production of commodities than ever before. The anonymous market is anonymous not only because one doesn't know personally who is selling and buying but because the capacity to anticipate the concrete needs, desires and tastes of the consumers, qualitatively and quantitatively, is missing. The proportion between needs and commodities becomes structurally absent. From this imbalance results that industry and its relation to commerce are increasingly defined as artificial and abstract. However, it is this mutual abstraction that is an incitement for the habits of consumption beyond the basic needs of nutrition, shelter and clothing. Long distance commerce creates needs for new commodities and commodities for absent needs. This equation explains the growing importance of fantasy in modern consumption and its embedment in the new industrial-commercial configuration. Modern production gives man commodities whose consumption is not fixed in the recognition of a subjective lack but depends on the socialisation of needs, which is ruled by schemes of the imagination. Imaginary processes and fictional constructions about needs constitute the psychic resonance of the autonomy of the commodity in the economic communication regarding the biopsychic configuration of the needs. When this general imaginary circumstance conspicuously emerges it agrees with the formation of the modern idea of taste with the correspondent public and a market for luxury products.

[2]Read a description of the popular literature on luxury published in magazines and newspapers along eighteenth Century, in Britain in D. Zylberberg work (Zylberberg 2008).

Imagination and "goods of imagination" represent forms that exemplify the disentanglement between the immediate psychic contents of the needs and the social response to them in the commercial society. This explains the conservative critique of the proliferation of the refined products, but also the importance of the identification of luxury with the artificial and refined forms. Refined forms or "goods of imagination" define the consumption of commodities in a society that can't avoid the untying of the psychic representation of wants from the varied communicative solutions to these wants.[3]

William Temple's views on the prosperity of the Dutch in his *Observations upon the United Provinces* (1687) predisposed the negative mood regarding luxury and prodigality. W. Temple estimated as a condition for wealth the abstinence of excessive consumption, when he wrote about the Dutch: *they furnish infinite Luxury, which they never practise; and traffique in Pleasures, which they never taste.*[4]

On the other hand, Nicholas Barbon's ideas in *A Discourse on Trade* (1690) were probably influential in B. Mandeville's defence of the beneficial effects of international trade. Indeed, the economic doctrine of the *Discourse* is an inspirational source of the modern free trade and a predecessor of B. Mandeville in the approval of the import of foreign luxury products, in the analysis of the honorific dimension of the use of adorns and in the distinction between habits that are vices for the single man but useful for trade and those that are absolute vices or absolute virtues.

In the late seventeenth century with the exception of Saint-Evremont, Pierre Bayle and N. Barbon who approved the use a certain degree of luxury, mercantilist theories were dubious regarding the theme. What was considered positive about the international trade for the improvement of industry and internal consumption was negative if demanding a significant import of luxury goods from abroad. In 1691, Dudley North published a *Discourse upon Trade*. Here, the author discussed the relation of the interest rate and the increasing or decreasing in luxury consumption, criticised the isolation of nations from the international trade and the application of sumptuary laws.

But, in the same year the writer and moralist Jean Fran du Tramblay in *Les Nouveaux Essais de Morale* (1691) condemned the use of luxury beyond the limits of the natural use of the objects regarding their utility or considering "l' utile à la condition". In his writings the problem of the conciliation of utility and luxury is obvious. Fénelon in *Les Aventures de Télemaque* (1699) explicitly refers the danger of imitation across the different ranks as a cause of social disorder, laziness and

[3]Hundert (1994: 28–40). In many cases, the imaginary character of the needs associated to luxury gives the subject of the moral discussions on the utility or social convenience of the consumption of some goods. A positive evaluation of the economic use of imagination is many times accompanied by a positive appreciation of luxury. The degree of moralization of the discussions on luxury changes also with the value given to the imagination in the formation of the human needs. See also the work of M. Hilton for an historical account of the demoralisation/moralisation of the discussions on luxury: (Hilton 2004).

[4]A discussion on William Temple's evaluation of the Dutch model of consumption and B. Mandeville's reaction is in the work of A. Bick (2008).

effeminate manners. The author is attentive to the social impact of the mimetic spreading of luxury. Tramblay and Fénelon combine the two indexes of the formal layout of the description of luxury in a prototypical manner.

The 1730s are the turning epoch in the flourishing of the writings on luxury. In England the episodes of the "South Sea Bubble" partially explain the incidence. In France the motives were created with the content of chapter IX of J. François Melon's *Essai Politique sur le Commerce* (1734), his ideas on the socio-historical relativity, the general beneficial character of luxury and the critique of the sumptuary laws (Jennings 2007) aligned with Voltaire's 1736 vindication in *Le Mondain*. The reactions to D. Hume's views on luxury and commerce explain some textual production of the 1750–1760s. But many arguments against luxury and its apologists (B. Mandeville) were disseminated through John Dennis's pamphlets about *Vice and Luxury. Publick Mischiefs* (different dates, 1724) focused on B. Mandeville's *Fable of the Bees*. If we consider J. Dennis's attacks the author of the *Fable* seems to be like a catalyst figure in the evolution of the discussions on luxury in the 1730s. After J. Dennis, later in the 1750s and the 1760s, we can follow the moralistic critique of the *Fable* and the importance assumed by his author in the discussions about luxury in John Brown's *Estimate of the Manners and Principles of the Times* (1757–1758) and in Thomas Cole's *Discourse on Luxury, Infidelity and Enthusiasm* (1761). One year after J. Dennis's pamphlet, in 1725, a book title ascribed to George Bluet was published with the purpose of a serious examination of B. Mandeville's doctrines. In the section III of the book the author describes the notions of wealth and luxury of the *Fable of the Bees* and rejects its assumptions. Due to a combination of the thesis of the agrarian origin of wealth and a mercantilist position regarding the balance of imports and exports, the author supports the implementation of sumptuary laws for domestic consumption and for the regulation of foreign trade. He condemned B. Mandeville's apparent confusion of wealth and luxury, his interpretation of W. Temple's report on the progress of the Dutch nation and quoting Saavedra Fajardo's emblems suggests the example of Portugal and Spain as nations destroyed by the abundance of money and the laziness caused by luxury (Bluet 1997: 227–382). Examples from the popular literature and pamphlets, such as Erasmus Jones's *Luxury, Pride and Vanity, the Bane of the British Nation* (1736), represent the importance of the theme of luxury in this period.

In the second half of the eighteenth century[5] the defence of the "natural order" by Mirabeau in *L'Ami des Hommes* (1756) and by Quesnay in the article "Grains" (1757) justifies the attack on the "decorative luxury" and led the Physiocrats to a censure of luxury as the responsible for the excessive spending of the aristocracy with the court life despite urgent needs of renovation in the productive capacity of the land. In Mirabeau's *Philosophie Rurale* (1763) inspired in Malebranche the physiocratic doctrine condemned the artificial multiplication of the needs through the imagination. The index of excess of the formal layout of the description of

[5]The discussions on luxury in France changed substantially with the Revolution. See the work of J. Shovlin (2000) about this evolution.

luxury is here predominant, but it is also evident the preoccupation with the social impact of superfluity. A Physiocratic line of argumentation is noticeable in the late eighteenth century in Schlettwein's work *Grundfeste der Staaten oder die Politische Oekonomie* (1779). Here, the author describes luxury as a social disease created by the multiplication of goods of the imagination for the satisfaction of a disturbed sensibility.[6]

The several textual testimonies of the luxury debates in the eighteenth century, including B. Mandeville's contribution, attest the epistemological perplexity regarding the status of the psychic side of the definition of commodities, or economic goods, as utilities responding to specific human needs: the escalade of the emulation in the acquisition of luxury objects seems to deny any relevant connection of luxury to basic needs. This justifies the reference to imagination as the psychic source of the needs connected to luxury consumption and the absence of a direct correspondence between luxury goods and needs. The Physiocrats and J. J. Rousseau have deplored the ambitions of an artificial society made of artificial needs far from the natural rhythms of the agrarian production, but they followed a line of critique already present at the end of the seventeenth century, in the popular, satirical, use of the terminology related to "luxury" or in the writings of J. F. du Tramblay or Fénelon, for example.

It is an historical assumption of this paper the notion that in the eighteenth century the debates on luxury departed from the crisis of the justifications of the "old luxury", connected to a sharp separation of the ranks and to the signs of social distinction typical of the *Ancien Régime*. A "new luxury" is already in formation in the habits of consumption in the seventeenth century in many European countries, in France, England and in the Dutch Republic, related to the modes, the comfort and the growth of the urban centres (de Vries 1999; Peck 2002). This novelty explains a change in the public attention to the communicative conditions of luxury as a social epidemic spreading across all the ranks.[7]

Along this semantic evolution the underlying theoretical difficulty is how to place luxury in the correlation of the individual action and the social causation in its autonomous cycle of causes and effects.

[6]The hesitation concerning the real contribution of luxury to the progress of Mankind is remarkable also in Turgot's "Sketches of a Universal History". *Le luxe outré, où la vanité fait accumuler les ornements, parce qu'elle les considère moins comme ornements que comme signes d'opulence. Étouffe le goût. On ne cherche plus le plaisir que font les choses aux sens et à l'esprit, on ne rentre plus en soi-même: on n'écoute plus que la mode. (...) Arts, vertus, tout est infecté de cette erreur; de là, les fausses vertus de beaucoup de philosophes* (Turgot 1844: 657).

[7]In Erasmus Jones's *Luxury, Pride and Vanity, the Bane of the British Nation* (1736, with a second edition in 1750), a libel against the "pleasures and novelties of the Times", the contagion of luxury across the ranks, due to imitation, and its complement in the theatrical character of the social habits associated to waste, is a repeated censure.

3.3 B. Mandeville's Perspective: Luxury and Honorific Behaviour

3.3.1 Honour and Virtue

In *An Enquiry into the Origin of Honour*, B. Mandeville conceived morality under a meaning similar to virtue. Virtue applied to actions means a predicate whose qualities contribute to the felicity of the individual and the happiness and peace in the society. In order to describe the use of the words *moral* and *ethics* the author traces their etymology from the Greek and Latin. The Greek root for virtue comes from *ares* that points to the god of war. Also in Latin it is possible to recognise the military reference in *vir*, which is the ground of the word *virtus. Fortitude* is a common foundation of the moral vocabulary (Mandeville 1732: iii). Accordingly, *virtue* signals the behaviour that is suitable to ensure the "conquest of ourselves". With the meaning of "power over something" or "over someone" the name *virtue* is applied to substances (animals, vegetables or minerals) provided with qualities capable to produce specific results that reveal the same plastic power.

From these general premises B. Mandeville came to a conclusion regarding the use of the term moral. "Moral" is a semantic extension to the manners of the "virtues" recognised in some types of actions. This means that "virtue" is not a moral intrinsic quality of an action. "Virtue" is a general, natural and human quality in substances or in actions, which represents a capacity to rule organic, psychic or social conditions. But, if this is the real origin of virtue and moral it is not possible to abstract our ideas of moral or virtue from the conditions of the application of the force or power consequential to the application of the virtue itself. This justifies B. Mandeville's rejection of the identity between moral ideas and eternal ideas.

Defined according to this "force" what expresses a virtue is always a capacity to modify human passions: *it curbs, regulates or subdues some passion that is peculiar to Human Nature*. Virtue is not directly grounded in the nature of God, simply because God doesn't have passions or forces shaping passions.

If virtue entails some sort of subordination of human passions this also means that *in virtue is self-denial*. What is needed is a proper definition of self-denial without incurring in a general, coercive, and preconceived renunciation of the spontaneous force that is present in virtue.

The themes discussed along the *dialogues* in *An Enquiry into the Origin of Honour* are honour and self-liking. The first dialogue led directly to the notion of self-liking as a *value that all individuals set upon their own persons*. The equilibrium of self-liking is not always the same and from its lack results suicide. The excess is a vice called excessive pride. The moderation in the use of self-liking is a social virtue Horatio calls "love of praise". The good instruction of this one reveals self-liking in a condition of perfect equilibrium between its internal drive and the external, social, context.

The word honour is of a very ancient origin. It evolved from Antiquity but saved the same meaning. It means to please and to gratify someone on account of the self-liking (Mandeville 1732: 92). Societies that recognized the importance of nobility reveal the link between social praise, honorific ceremonies and the stimulation of self-liking. The desire of applause is not isolated from self-liking. Instead of being abstracted from it, it is circularly connected to it.

The moralists ignored the basic role of the dissimulation in the acquisition of a right to social applause, but we will see that the dissimulation of the "love of praise" is vital to rule the depth and opportunity of social applause in the generation of self-liking.

In the *Remarks* of the *Fable of the Bees* the author has defined love as a composite passion, made of contradictory passions "blended in one". Self-love includes the complexity of love and employs the blended resulting structure of the compound passion to the self and to self-affection.

The inner connection of self-love and honour in the self-liking stability is a recurrent theme in B. Mandeville's writings and characterises a pillar of his interpretation of the true reasons of the actions morally justified, the conservation of the *appearance of virtue* in conversation and society and the interpersonal structure of action. Later, the same analytical frame was reassessed in T. Veblen's concept of the "leisure class" (Veblen 1918).

The recognition of the circular relation of self-love, honour and self-liking and its social conditions is intimately connected to the critique of the ancient Ethical Systems in the *Remarks* of the *Fable of the Bees*, regarding the identification of the "true" motivation of the human actions. Here, the author has depicted critically the Epicureans and the Stoics. But particularly his criticism of stoicism is very tough. It is stoicism that epitomizes an undifferentiated self-denial and the identification of virtue with a vague control over passions as the true moral predicates of the motives of human actions (Brooke 2012: 159 and ff.). However, a disinterested self-denial with no impact in the ruling of self-liking is everyday contradicted by the facts of life. The stoic ideal of the Wise or the model of the Mendicants Religious Orders denote more the *appearance* of self-denial than an authentic rejection of the world. The desire of glory and sovereignty is always present, but disguised. This explains why men don't remember the ascetic for his poverty but for his glory.

Memory is the faculty that reinstates the truth of the moral disguise of the ascetic. The ascetic may live according to the promise of the complete denial but surely he also figures the glory after death. Penitence and martyrdom are extreme sources of glory and honour, instead of being representatives of humility or religious virtue. This justifies in the *Remarks* the reference to Montaigne's formula "they believe what yet they do not believe". This condition is normally described in the moral concept of *hypocrisy*.[8]

[8]Motivated by B. Mandeville's distinction between a "malicious" and a "fashionable hypocrisy" in the *An Enquiry into the Origin of Honour*, the social reflective structure in the self-concealment of hypocrisy was recently reassessed by David Runcinam in line with his own distinction of a first- and second-order hypocrisy (Runcinam 2008: 45–73). Both distinctions seem able to identify levels of dissimulation and/or dissimulation awareness of the agents. However, the difference of first- and second-order hypocrisy depends on the type of social interaction, on the intervention of

Honour represents a compliment to someone who deserves it according to the views of someone else. If the compliment is accorded the person who gives it knows that the revered person fortifies the self-liking. Shame is the contrary of honour. But shame also reveals the operation of self-liking in Human Nature. This is the reason that explains that pride and shame are not separate passions but two expressions of a single passion, which is self-liking in positive and in negative tones.[9]

From these premises B. Mandeville concluded: *Honour (...) is a technic word in the Art of Civility, and signifies a means which Men by conversing together have found out to please and gratify one another on account on a palpable passion in our nature, that has no name, and which therefore I call self-liking* (Mandeville 1732: 14).

The reflective character of self-liking is clear but it is also evident the circumstances adequate for the development of this passion and its effects in society. *Conversation* is the context responsible for the reflexivity of the passion and for its causation across social relations. Due to his appropriation of the theme of "conversation" B. Mandeville gives his account of the honorific behaviour, explains the relation to oneself as a relation dependent on the social interaction and also denotes the unintentional structure of social causation.

Honour is a pillar of the social bond even if religion seems incapable to maintain normal relations between men in civil society. But the real source of honour however is a particular fear. The fear of shame is the justification for the pursuit of honour (Mandeville 1732: 40). The social mechanism responsible for the establishment of a strong connection between shame, the affection of self-liking and the desire of honour is the education. Through education men are trained to be sensitive to shameful situations. The efforts of the moralists must be focused on the development of educational measures to strengthen the capacity to feel dislike with oneself in particular situations and actions. The strengthening of this faculty has remarkable anticipatory results and creates the frame for conducts of avoidance.

The formation of honour comes after such educational conditions. There is a connection between the invention of honour and the consolidation of the society. To the end of the consolidation of the social bond the contribution of honour is much more efficient than the invention of virtue. In the passage where the author compares honour and virtue is obvious he is referring to a reflective device with no moral specific content. What interests him is the reflective functioning of the social bond resulting from the anticipation of shame or

---

social imitation to coordinate the actions of the agents and on the structure of the society. A serious discussion of the nature of social causation can contribute to the understanding of the formation of the "person" as a social agent and the relative extent of a second-order hypocrisy. From the point of view of the formation of society as an autonomous cycle of communications the notion of hypocrisy is only provisional.

[9]About the meaning of shame read the "Remark C" in volume I of *The Fable of the Bees* (Mandeville 1924 I: 87): *First, to define the Passion of Shame, I think it may be call'd a sorrowful Reflexion on our own Unworthiness, proceeding from an Apprehension that others either do, or might, if they knew all, deservedly despise us.*

from the flattering behaviour conducing to honour. The role of the virtues in the establishment of the social bond according to this structure is ancillary.

The virtuous man acts according to his virtue and needs no social approval. The man who acts motivated by honour may disguise himself in a virtuous man and claim absolute respect from the others as if he was a virtuous person. Honour is more directly connected to social expectations and deals more directly with the social circumstances than virtue. The distinction between both men is mainly due to the absence of proclamation of the virtue by the virtuous man and the need for public proclamation in the case of the honour. The first is not a social condition; the other demands undeniably the social relation to others.

Many essays were made in the past in order to accommodate the principle of honour with the principle of virtue. One of these efforts came from Christianity. How to conciliate the principle of honour with the Christian humility? The efforts leading to the conciliation changed the Christian Church itself. The Church combined sacred rites with signs of vainglory, mixed pomp and superstition (Mandeville 1732: 47). This can explain why the actual Church as a social institution is not based exclusively on the virtue but also on honour and pride.[10]

The practice of duelling was a theme for the author of the *Fable*. He described the duel as an institution defined by the rules of self-liking and honour and the historical abuse of the duels was for him a symptom of the social value acquired by the culture of honour and the "instinct of sovereignty" (Mandeville 1732: 68–69). The legal perspective on injuries takes the singular case of the injury as a singular instance of the universal law and universal prescription. But the culture of honour takes the injury in the sense of a personal aggression, an immediate individual injury. Thus, the injury as an aggression directed to the self-liking of someone stimulates a singular response and a person-to-person ritual for the regulation of the conflict.

Originally, Christianity was the opposite of the principle of honour. The concept of honour is signified in the formula: "we are idols to our selves" (Mandeville 1732: 89). According to the Christians duelling must be definitely removed from the society through the elimination of the passions that cause it such as vainglory, honour and courage. B. Mandeville on the other hand argues that the elimination of these passions would be very negative for the society. He claims that passions of this sort are essential for the reproduction of society and no positive results are expectable from a destruction of honour.

[10]Christianity can be useful in war because it combines the incitement to virtue with the practice of honour. The religious enthusiasm is an additional ingredient that contributes to strengthen the belief on the courage and fortitude of the Christian soldiers in war (Mandeville 1732: 160–161; 164).

3.3.2 Honour and Luxury

It is not easy to identify a uniform or coherent critique of the aristocracy and the aristocratic sense of virtue in the mandevillean critique of Shaftesbury's "politeness".[11] Indeed, B. Mandeville's use of the notions of honour, pride and shame of the aristocratic culture has a descriptive content and applies to the depiction of human society as he observes it without moral categories. Certainly, he rejects the aristocratic hypocrisy, but he employs to the society the categories of the aristocratic *decorum* deprived of their normative moral function and attributes, as if honour, pride and shame were behavioural or social mechanisms for the coordination of actions and expectations and not expressions of the morality of the aristocratic society. Here, B. Mandeville's method consists partially in a demoralisation of the moral categories.

Aristocratic virtues are an historical manifestation of the etiquette of honour leading to the distinction of a particular social rank. But self-liking, honour, pride and shame are perpetual in human societies and operate in Mandeville as social mechanisms for the coordination of actions. The commercial society displays new features of the coordination of actions based on these passions. The reasonable imperative is to organise the moral discourse and politics in order to acknowledge the universal value of such passions, their importance in the regulation of social life and the use of them in favour of the improvement and prosperity of the nations.

Honour and Luxury are aspects of the same semiotic structure. Luxury is a palpable face, a visible condition of the honorific practices.

The examination of luxury leads the author of the *Fable of the Bees* to the inclusion of pride at the same side of luxury in the list of the "promoters of trade". Both are intimately connected to the effects on the self of the self-representation of the "opinion of the others". Thus, B. Mandeville gives an explanation for the epidemic index in the semantic of luxury.

The reason for the use of the common definition of "promoters of trade" is not accidental, but quite essential. The semiotics of luxury is directly connected to the honorific behaviour. However, it is not very common the association of trade with pride and honour.

Luxury communication across the large public spaces is another aspect of the contagion of luxury, different from conversation and direct emulation. The *Remarks* point to the effects of the public use of signs of honour on other men or to those aspects of the public facade that are not necessarily in accordance with the social status of the person, such as the use of handsome apparel to gain special attention

[11] I'll not depict here the complete constellation of B. Mandeville's political background and his own position regarding the social and moral discussions and the projects of moral reformation of manners in England, in the epoch. On this particular subject see the work of M. M. Goldsmith (1976). I am only concerned with the framework of Mandeville's critique of the correspondence or continuity of the private and the public aspects of agency in the tradition of Shaftesbury's "civic morality".

from the others. Luxury in apparel stimulates the honorific behaviour in the public, the applause and also the emulation. This inspires pride and improves self-liking. The resulting cycle of luxury, honour, pride and emulation will be fuelled by its own elements in orbital causation. The cycle itself represents the society on the stage. It is important to remember that B. Mandeville had in mind the "large and populous cities" as the ideal anonymous public for the performance of the trendy persons who desire to be "esteemed by a vast majority". The concept of fashion is explicitly used in the context of the description of the emulation and the "various shiftings and changings of modes" characteristic of the urban centres open to the foreign trade.[12] Fashion develops as a semiotic outcome of the coupling of trade, pride and luxury. Its psychic vehicle is a "desire of being observed", which is not focused on the inner self but on the publicly displayed *persona* (Hundert 1994: 185). Such "desire of being observed" doesn't elect specific observers but depends on the anonymity and erratic attendance at the public spaces.

The concept of taste is implied in the *semiosis* of luxury. Important semantic and philosophical aspects of Shaftesbury's civic "politeness"[13] are combined with the aristocratic trend of this cultivation of the public appearance, convenience, "property" and taste. But B. Mandeville observes taste and property from a second-order perspective.

In the second part of the *Fable* Cleomenes explains a theory about a direct correlation of self-liking and the use of ornaments. Cleomenes's theory applies to animals and humans. But in the case of humans the self-liking is connected to the use of exterior symbols, ornaments and the culture of polite manners. Consequently, contesting again Shaftesbury, politeness is not resultant from religion or morality. Fashion, good manners, ornaments, taste and the exhibition of luxury are *various methods of making ourselves acceptable to others, with as little prejudice to ourselves as is possible* (Mandeville 1924 II: 107). Cleomenes adds to this an "historical proof" of the association between luxury and politeness in the ancient societies, Greece, Rome and the Eastern Nations.

The depiction B. Mandeville made of the religious ceremonies, especially canonizations, is a combination of Moral Psychology and Semiotic of Luxury. He argues that the ceremonial splendour of the religious practice of canonizations is engaged to cause admiration in common people. The inducement of piety is the final objective. But this sentiment only arises under certain semiotic stimulus. These are coupled with signs of excess and prodigality, which also have their relation to miracle. Solemnity is surrounded by pomp. Faith is only inferred to give legitimacy to the whole of the ceremony.

---

[12]The combination of the growth of global trade, the constitution of a global web of tastes and products of manufactures and the fascination with the exotic characterises a culture of luxury in the Eighteenth-Century, which transformed the pre-modern categorisation of luxury and consumption typical of the "sumptuary laws" (Berg 2004; Eaton 2006; McCabe 2008).

[13]On the semantic relations of taste, beauty, civility and commercial society in Eighteenth-Century Britain, see the work of Robert W. Jones (1995).

He described the evolution of the Church from primitive Christianity to his own time as a progression from austerity of manners to the ritualization of honorific practices.

The Christian doctrine of self-denial, humility and austerity of manners is used nowadays as rhetorical means to obtain the admiration of the multitude.

Cleomenes compares the actual Church with a fabric, a manufacture distributed in many branches in order to develop the dualistic structure of the Church: the mystical and excessive and the austere and altruistic.

What is said in the articles of the doctrine and in the humility of the good examples is inverted in the ceremonial practice of the Church and the Pope. The proclamation of austerity of manners is the method to ensure that the multitude believes in the wonders of the Saints and in the infallibility of the Pope, but also the right technique to proclaim the need for self-denial of the multitude and the virtue of obedience. The wonders and the Pope's authority are both grounded in semiotic proofs, which are given to the multitude as signs of glory, praise and abundance, precisely as luxury.

The duality of humility and splendour of the Church reproduces in manners the distinction between self-denial and self-liking. It exemplifies the duality of the code of honour.

B. Mandeville didn't reject the idea of luxury adjusted to the social rank as generally accepted but in some cases vaguely recommended in his epoch. His originality regarding the theme of luxury is not to be found in this particular aspect. His comments about the social expansion of luxury as a result of the complex scheme of social causality with the respective emergent properties are much more significant.

He noticed that after the first edition of the *Fable* the ideas of this book concerning luxury were condemned. He defends himself from the critics saying that he never sustained luxury as an end in itself. He conceives the propensities to avarice or to prodigality as the expressions of general conditions in mankind and in the structure of the national societies and he claims that such circumstances as indolence, self-gratification, good-nature, a jovial temper, youth, folly, arbitrary power, easy money, plenty of provisions and uncertainty of possessions are the right conditions for prodigality. On the other hand, if he considers adequate the introduction of sumptuary laws to control the levels of employment of poor people in a country in trouble, he thinks derisory its application to an opulent kingdom.

The author of the *Fable* contests the traditional definition of luxury as any consumption of goods beyond the basic needs of the living beings according to the platonic division of the basic needs. Against the view that continues the ideal vision of a bare, naked life, B. Mandeville sustains that the products of fancy and wit are since the beginnings of human societies present in the satisfaction of the needs. This general aspect of his doctrine on luxury will be addressed again in

J. Dennis's critique of the Remark L of the *Fable*.[14] In the claim that luxury is a vice because it is an excess beyond the degree of nature the traditional moral theory on luxury uses two standards to estimate a practice or a commodity as "luxuries". The first standard evaluates the possible excess considering the basic needs of a living man generally envisaged. The second standard takes into account the social custom defining what is acceptable as excess according to the social rank. This means a double standard and a double use of the concept of the degree of excess according to nature. What is beyond these limits is inappropriate luxury that contributes to infect public life with a fever transmitted through emulation and imitation across all the ranks of society, effeminising and enervating people and extinguishing the military virtues, which are the source of the soldier's discipline in war.

B. Mandeville didn't appreciate both standards in the same way or under the same evaluative yardstick. He excluded the existence of the first absolute standard but he established limitations to the imitation across ranks. His response to the double standard of the luxury definition may be used as a proof of B. Mandeville's own hesitation regarding this particular issue.

The imitation across the social ranks is dangerous for the conservation of social differences, which are essential for the reproduction of the asymmetric structure of the honorific behaviour. If B. Mandeville describes honour as the very foundation of society its suppression as a consequence of the luxury's fever should be noticed as ruinous.

In *Remarks* L, M, P and Q, B. Mandeville continues his views with a critical evaluation of a common doctrine ("a received notion") that sustains that "Luxury is as destructive to the wealth of the whole Body Politic, as it is to that of every individual person who is guilty of it" (Mandeville 1924 I: 108). The "received notion" about the effects of luxury in personal life and in society emphasizes the negative aspects of the corruption, the effeminate of manners, avarice and rapine. The common impression identifies luxury with a dreadful symptom of excess in consumption. But B. Mandeville also takes into account the doctrine that stated the undesirable consequence of luxury in the balance of trade especially in those nations that have no gold or silver "of their own growth". Against this reproach he asserts that nations should be ruled by prudent policies. He adds that under the right equilibrium of imports and exports "no nation can ever be impoverished by

[14]In many points of his remarks and letters J. Dennis takes luxury as equivalent to a vice. At the end of "Public Mischiefs" he explicitly declares that private vices are all included in luxury (Dennis 1997: 172) reducing the whole to a part. Ignoring the singularity of the social causality and establishing a continuous line from the individual actions and purposes to its social effects in interaction, J. Dennis says that the subtitle of the *Fable* is a contradiction (Idem: 165) and contrary to the revealed as well against the natural religion. The critique mentions various examples from the Ancient Testament and Ancient History to prove that luxury, private and public, is the opposite of virtue and that it can never be for the public benefit (Idem: 168–169). Notwithstanding the ingenuity of the arguments, J. Dennis's text *An Essay upon publick Spirit*, London, 1711, epitomizes the general design of the moral and religious vision of society and "manners" that identifies in the growth of luxury a symptom of decadence and a consequence of the adoption of a foreign way of living.

foreign luxury". It is easy to identify here a preference for the mercantilist approach of the international trade and it is this preference that justifies a moderate apology of luxury.[15]

B. Mandeville's analytic efforts were focused on the distinction between what develops from the intentionality of the individual actors or their direct benefits with the actions and what results for the society from the convergence of many indeterminate, unanticipated actions. From this distinction comes a different view on how luxury affects the individual and how it affects society.

The "received notion" about luxury sustains that it is a vice for the individual *and* for the society. The justification for this observation is given by the causal connection linking individual agency and its social outcomes that is allegedly proved. Thus, the current moralist description of the effects of luxury in the individual life illustrates how luxury works in social life: if dreadful consequences for the individual were observed, the same effects will be displayed at the social stage. The moral concept of luxury stresses the thesis of the linear causality connecting individuals and society.

But in order to follow this connection the moral vision of society should postulate what the luxury has added to the natural life of men representing simultaneously the vice as an excess. The hypothetical difference between a life according to nature or to the "social condition" and a voluptuous life is crucial in the depictions of the moral view. Thus, a semiotic of the voluptuous life begins here, attentive to the signs of excess in buildings, furniture, equipages, clothes and food. However, Luxury is historical, relative and a gradual product of the improvement of human experience with nature and society. It is a product of education and civility. The comments in the *Remarks* of the *Fable* concerning the luxury of the poor are very instructive about the role of the social evolution in the transformation of the judgements on what defines luxury in the objects of daily use, social equipment, as hospitals, food or dressing.

3.3.3 Luxury and Social Causation

B. Mandeville's work can be described as an exploration of the moral meaning of the *double reflexive encapsulation* of communication and consciousness in the modern conditions of the commercial society. In the language of morals, of vice and virtue, he tried to explain why social causality in the historical setting of the commercial society could no longer be understood as an expansion of the intentionality of the individual agency with its moralised psychic predicates.[16]

[15]About this aspect, see also the thesis of Thomas A. Horne (1978: 51–75).

[16]It is probable that B. Mandeville's medical formation contributed to his new approach to social causality and to the use of the terminology of the "emergent properties". One can find analogies with the model of chemical emergence in biological processes in many passages of his work. This means that he was attentive to the difference of chemical and mechanical causality in the

Luxury is a subject of considerable importance in his work because it exemplifies at the same time the autonomy of the effects of communication in modern society due to its autonomous series of nexuses, which the international trade catalyses, and its imaginary resonance or psychic sedimentation of communication and the failures of a moral control over the social effects of luxury. I'll explain my thesis regarding B. Mandeville's notion of luxury according to this general view.

B. Mandeville's criticism of the moral abstract censure of luxury rejects the model of the linear causality in two ways.

Firstly, he says that the vices are related to the appetites and "as long as men have the same appetites, the same vices will remain".[17] This means that even men with scarce resources can develop a vice of luxury proportionally to their social condition. The brute and poor and the elegant, expensive and artificial men can destroy their lives with abuses and excesses in a similar way. The suggestion corroborates the hypothesis that there are not conclusive signs of luxury, if luxury is defined as excess in consumption.

Secondly, according to the general theory of modern society and social causality of his author, the *Fable of the Bees* claims that vices (and luxury) can contribute to the general welfare and the prodigality of life of some people and is a real incentive for an industrious and energetic society.[18] On the other hand, B. Mandeville anticipates some J. M. Keynes's arguments about the relative macroeconomic impact of individual and public savings in the growth of the general wealth of a nation. Indeed, both contest the equivalence of saving applied to families and saving applied to nations. The repercussion of frugality and the effects of a modest avarice in private affairs differ from the economic rules to be applied to the increasing of wealth in a country. In the case of nations the level of wealth proportional to the inhabitants shall decrease dramatically if the aim is to render the country frugal.

These two lines of critique support each other. Both presume that it is not possible to infer consequences for the stability of the society from descriptions of the relative instability of the human appetites. From the appetite nothing stable or unstable follows directly for the society. This explains why vices of the appetite

---

physiological domain. The differences in the two models of causality entail the rejection of an additive representation of the cause -> effect sequence. In the physiological causal sequences, given the conditions the effects produced are never a simple prolongation of the same causes. The discussion of the appropriate model for the description of digestion engaged the attention of the young B. Mandeville. Apparently, he conceived literary imitation as an analogous of digestion, continuing a *locus communis* of the classics. But the analogy may have unexpected consequences if one shifts the concept of imitation from the literary field to the social. For a detailed discussion of these topics, read A. P. McKee's Ph. D. Dissertation (McKee 1991).

[17]See *The Fable of the Bees,* I "Remark L", p. 111.

[18]At the end of "Remark L", B. Mandeville summarized his own ideas regarding the beneficial contribution of luxury to a Nation's progress. Idem, *op. cit.*, "Remark L", p. 113: (...) *with a wise Administration all People may swim in as much foreign Luxury as their product can purchase, without being empoverish'd by it (...) a wealthy Nation may live in all the ease and plenty imaginable and in many parts of it, shew as much pomp and delicacy, as human wit can invent, and at the same time be formidable to their neighbours.*

or eccentricities of taste produce under appropriate conditions good consequences for the wealth of the nations. Of course, this new approach puts B. Mandeville at the opposite side of the ancient tradition that conceives the virtuous actions and the social virtues as those which are in accordance with nature, from Plato's *Philebus* to Cicero's book III of *De Oficiis* and the stoic tradition renewed in Shaftesbury's systematisation of the "whole and parts" moral model in his *Enquiry concerning Virtue, or Merit* (Shaftesbury 2001).

In the *Remark B* of the *Fable of the Bees,* I, B. Mandeville tells a story about two individuals Decio and Alcander, both sugar merchants (Mandeville 1924 I: 85 and ff.). The story describes Alcander's efforts to gain advantage in the sugar business from a privileged information he had in time t1 concerning the forthcoming arrival of an important amount of sugar coming from "West India", much more substantial than formerly expected. Alcander tries to sell the sugar to Decio at the current prices. He expects a price decrease as a consequence of the increasing offer. If he sells at the current market price he will take advantage of the dropping. But at time t2 Decio received the news of a tempest at Sea and of shipwrecks of the navies carrying the sugar. Alcander doesn't take into account the news and Decio accepts to buy from him the sugar at the current market prices. Decio wins this game. The lesson B. Mandeville infers from the story concludes *that neither of them would have desired to be done by, as they did to each other* (Mandeville 1924 I: 86).

The moral of this story also tells that B. Mandeville's concept of "benefits" to the society shouldn't be understood as intentional good services of the individual intended actions. In other words, if it is a duty for the members of the society to be good in the stoic sense this ideal society couldn't subsist "for any considerable time without the vices of man", according to the Remark T (Mandeville 1924 I: 163).

The *Fable* employed the language of morals and of causation to describe self-organized processes in modern society, which are independent from a psychic, conscious, regulation or control.

Mobilizing the terminology of causality the following is said in "Remark G" and in the "Vindication" of the book.

> The short-sighted vulgar, in the chain of causes, seldom can see farther than one link; but those who can enlarge their view, and will give themselves leisure of gazing on the prospect of concatenated events, may in a hundred places see good spring up and pullulate from evil, as naturally as chickens do from eggs. (Mandeville 1924 I: 100; 241)

This general recognition of the complexity of social causation was compromised in some parts of the *Fable* due to the use of an appeasing discourse. The image of the "skilful politician" guiding the private vices to the general benefit of the society is one of these occurrences. But the essential is the perception of a different structure of causation in the case of social action.

On the other hand, the mutual dependence of needs and services in the society explains why men want a social life. However, this outcome is not intentional or planned. It is originated in a way similar to the contagion of luxury in society, due to the expansion of trade. It is a product of an evolutionary adaptation of the subjective construction of the aims of the individual actions to the structure of society.

In *A Search into the Nature of Society*, the author is clear about the motives of his disagreement with the moral doctrine and the identity of "pulchrum et honestum" in Shaftesbury's *Characteristics*. He disapproves the thesis of the natural propensity of men to society and the definition of virtue as a penchant to act in pursuance of the public good in accordance to an inner affection, which also would agree with the natural beauty. In Shaftesbury's views the contrary of virtue is vice as a violent movement in conflict with the general welfare. B. Mandeville first remark points to the distinction between social value and the intrinsic value of commodities, such as paintings. Social value is always exposed to variations, which can't be easily put under a moral control. He discusses scarcity as a source of value. On the other hand, endorsing a sceptical reasoning possibly inspired by P. Bayle he mentions the diversity of constitutions of men and nations as the explanation for the variety in taste and moral ideas. This proves that there isn't a uniform natural tendency to a natural beauty and virtue in human nature or a universal "living according to nature" defended by Shaftesbury in his moral system.

Following the stoic common doctrine and establishing his own concept of "moral sense", Shaftesbury in the *Enquiry concerning Virtue, or Merit*, along the division on "whole and parts" (book I, part II, section I) and after examining the main metaphysical doctrines devoted to the problem of the organizing principle of the universe advanced his views on the ethical consequences of the metaphysical ideas about the harmonious universe or "system of nature" made of congruent parts and wholes. The "whole and parts" model was a relevant contribution for the understanding of the nexus of individual actions and social causality. In the ethical dimension, the actions that contribute to our own good and self-conservation are virtuous if contributing also to the system of the whole or to the good of the others. According to the *Enquiry*..., this harmony entails the recognition of an underlying agreement between the self-interest and the virtue. The individual interest is not the antagonist of the common good. The pursuit of the first doesn't entail the sacrifice of the second. The proof for such agreement is the judgement of the reflective "moral sense" in every individual member of the species acting according to its own notion of self-interest. The direct purpose of the *Enquiry*... was the refutation of T. Hobbes's views on the violent human nature relentlessly driving the individuals to the end of self-preservation excluding a natural human community. Shaftesbury contested the principle of the irrational structure of the living nature of humankind and included human community in the core of his system of nature.

The discussions on luxury are a good exemplification of a more general difficulty in the assumption of the autonomy of the society regarding nature and morality. Despite his criticisms of the stoic norm of "living according to nature" when B. Mandeville evokes the passions and the innate principle of sovereignty he restates the theoretical context that endorsed the use of nature as a discernment tool in the analysis of modern society. He is in an ambiguous position regarding the need for a *natural history of society*. The ambiguity of his position leads us to imagine that his use of the concept of nature has a restrict function in the whole of his system of ideas. The case with the narrative of the origin of the society from the families is illustrative. He uses nature conditionally or negatively as a tool in his proof of the thesis that there is not a stable definition of nature. Thus, nature is used against itself. A concept of nature is needed in order to show that from nature nothing with a prescriptive content can follow in morals, aesthetics or society. B. Mandeville shows that Shaftesbury's concept of nature is rhetorical. But the confirmation of this depends on a conditional use of the same concept.

B. Mandeville's central objection to the doctrine of the *Characteristics* is in the statement of the impossibility of virtue in the terms of Shaftesbury's own notion without the self-denial of the individual. Virtue according to Shaftesbury's *Enquiry*... realistically means a condemnation of the concrete men with their vices or self-interest. Instead of the analysis of the concrete men, their needs and tendencies it vaguely promises a moral reform of the human nature. As such, abstinence is not positively observed because from self-restraint nothing can clearly be inferred from the real men to the actual condition of a society. This means that the inference of society or common good from virtue defined as a form of self-denial is mistaken and the wrong way to deal with the mechanisms of social causation. According to the implicit assumption of the criticism one shall conclude that if evil disappears from human nature society ceases also to exist, because in the actual condition of society this one is composed of vices as well as of virtues and the causality of social actions is made of both. A more radical inference from B. Mandeville's critical appreciation points to the thesis of a non-moral foundation of society. Despite other criticisms, in this thesis B. Mandeville follows T. Hobbes, La Rochefoucauld and a Hobbesian line of social thought differing from the neo-stoic Shaftesbury and the principle of "living according to nature".

He improves his concept of the veiled self, opposed to the social persona, which is the true subject of honour in the place of the subject of virtue and reveals how the dissimulation escorts the gentleman's behaviour: *in the midst of pomp and distinction he lives in, he never appears to be entertained with his greatness, but rather unacquainted with the things he excels in.* Later in the dialogue he adds the essential: *the passion of pride is at its heights when more concealed.*

He also asserts that the description of the real egoistic motivation of the actions doesn't entail the recommendation of vice. It is this particular idea that outwardly aims the reconciliation with Shaftesbury's civil Ethics that is the most ingenious. He declares that the building of society is not an aftereffect of moral behaviour but of the dissimulation of the real motives and interests of the self. When Cleomenes declares that his aim is analogous to Shaftesbury's advancement of civility, he is really saying that self-interest is suitable for the improvement of the whole of the society and even more effective than a devout proclamation of virtue. The decisive is the combination of a veiled self-interest with polite manners. Politeness is a practice of flattering the pride of others and concealing our own. But it is only a social virtue and a good rule for self-liking if this concealment is effective.[19]

[19]This means that the psychic and the social dimensions of action are not connected in a mechanical way or according to an additive principle. (Cf. Jones 2011).

In part II of the *Fable* the description of the semiotic mechanism of dissimulation of the passions and social causation in the case of pride and luxury leads Cleomenes to the comparison with the *castrati* in the Opera: *luxury will rend a nation flourishing and private vices result in public benefits in a similar manner that a castration results in the strengthening of the male voices* (Mandeville 1924 II: 79). This analogy proves that the semiotic game of the self and the *persona* through dissimulation and manifestation is a basic condition of the scheme of social causation from individual actions. In luxury, the object of imitation is nothing more than the simulant of the Thing or the enjoyment repeatedly emulated. Luxury is the heir of penury and penury the outcome of an ungoverned luxury, which the critics acknowledged but without knowing the underlying mechanism of the social causation and the dissimulation.

B. Mandeville's connection of luxury, the Nation's wealth and social causation through emulation is a step to the understanding of utility beyond the physiological model of the economic needs of the classic tradition and the Physiocrats and to the recognition of the emergence of properties from complex causation in the case of actions and society.

This can be seen as an adjacent consequence of his medical model of the social interaction, which is responsible for a *cause – symptom* model of the *alien behaviour – self-liking* connection. The same model is used again in the description of the effects of mixtures and combinations of moral contraries, such as avarice and prodigality. In the part II of the *Fable of the Bees* the author conceives the mixtures of moral components as new substances provided with a power to induce in society a completely new effect that wouldn't result from the separated elements. He compares the autonomous elements with poisons and the mixture with a "good Medicine" due to its beneficial impact in the society as a whole. The comparison with chemical causality helps him in the depiction of the emergent qualities that develop from basic or elemental structures taking a different configuration and producing different effects. The emergent qualities represent the social effects and the basic elements the individual agency with the respective moral evaluation of the action's ends. In the part II of the *Fable* the emergent qualities are compared to the effects produced by the combination of many grapes in the production of wine through fermentation. There is only wine if fermentation occurs and fermentation follows only if many grapes are fused. "Vinosity", he says, is a consequence of fermentation. However, fermentation is something that can't be defined exclusively by the addition of units of grapes to previous units. Horatio declares that sociability of men is like fermentation causing vinosity (Mandeville 1924 II: 137). Men are sociable because they live together in multitudes learning particular habits. The habits we associate to social life are indeed effects of the social life itself and circularly produced by conversation. One of the roles of conversation is the reproduction of schemes for the coordination of action and communication.

The brief reflections on the divine Providence, chance and the origin of Evil made at the end of the part II of the *Fable* point to a concept of nature that is not previously tuned in accordance to the human, finite, representation of the ends of the creation. The vision of an incomprehensible nature of God supports the corollary that society is not previously adjusted to a comprehensible plan. Society is not attuned to nature or nature to society in agreement with a universal visible plan. Men don't have intuitively access to the "plan of God" or to a plan of the human society. In society they simply react to each other according to the rules of self-liking. From the complex sequences of these reactions an order emerges as from the fermentation of the grapes "vinosity".

His later defence of Luxury against G. Berkeley's attacks in *A Letter to Dion* (1732) points to the thesis that politeness and education of manners as well as commercial advancement and generally consumption depend upon "qualities, we pretend to be ashamed of". From this descriptive statement he comes to a "terrible paradox".[20] The paradox rests on the double allegation of a positive outcome of the vices for the progress of society and on the other hand the need for a moral suppression of the vices. Vice (and luxury) is necessary and desirable according to a "certain proportion".

A "certain proportion" is assessed through the evaluation of the general state of society. *This evaluation doesn't entail a moral judgement about society, but instead a social judgment about morals.*

Thus, the physician and philosopher proposed an unusual type of observation of society. Such model of observation adopts the principle of the autonomy of society regarding the moral calculations of the conduct of its individual members. Here, B. Mandeville is more revolutionary than commonly supposed and his account of luxury is a good illustration of his perspicuous mind.

References

Berg, Maxine. 2004. In pursuit of luxury: Global history and British consumer goods in the eighteenth century. *Past & Present* 182: 85–142.

Berg, Maxine, and Elizabeth Eger (eds.). 2003. *Luxury in the eighteenth century. Debates, desires and delectable goods*. New York: Palgrave Macmillan.

Berry, Christopher J. 1994. *The idea of luxury. A conceptual and historical investigation*. Cambridge: Cambridge University Press.

Bick, Alexander. 2008. Bernard Mandeville and the 'Economy' of the Dutch. *Erasmus Journal for Philosophy and Economics* 1(1): 87–106.

Bluet, George. 1997. An enquiry whether a general practice of virtue tends to the wealth or poverty, benefit or disadvantage of a people. In *Private vices, publick benefits? The contemporary reception of Bernard Mandeville*, ed. J. Martin Stafford, 227–382. Solihull: Ismeron.

Brooke, Christopher. 2012. *Philosophic pride: Stoicism and political thought from Lipsius to Rousseau*. Princeton: Princeton University Press.

[20]In *A Letter to Dion* B. Mandeville confirmed the positive contribution of luxury to the progress: *I own, Sir, it is my opinion, and I have endeavour to prove, that Luxury, tho' depending upon the vices of Man, is absolutely necessary to render a great Nation formidable, opulent and polite at the same time* (Mandeville 1997: 585).

de Vries, Jan. 1999. Luxury and calvinism/luxury and capitalism: Supply and demand for luxury goods in the seventeenth-century Dutch republic. *The Journal of the Walters Art Gallery* 57: 73–85.

Dennis, John. 1711. *An essay upon publick spirit*. London.

Dennis, John. 1997. Vice and luxury, publick mischiefs: Or, remarks on a book entituled the fable of the bees. In *Private vices, publick benefits? The contemporary reception of Bernard Mandeville*, ed. J. Martin Stafford, 137–208. Solihull: Ismeron.

Eaton, Natasha. 2006. Nostalgia for the exotic: Creating an imperial art in London 1750–1793. *Eighteenth-Century Studies* 39(2): 227–250.

Ferrero, Guglielmo. 1901. The evolution of luxury. *International Journal of Ethics* 11(3): 346–354.

Goldsmith, M.M. 1976. Public virtue and private vices. Bernard Mandeville and English political ideologies in the early eighteenth century. *Eighteenth-Century Studies* 9(4): 477–510.

Hilton, Matthew. 2004. The legacy of luxury. Moralities of consumption since the 18th century. *Journal of Consumer Culture* 4(1): 101–123.

Horne, Thomas A. 1978. *The social thought of Bernard de Mandeville – Virtue and commerce in early eighteenth century England*. London/Basingstoke: The Macmillan Press Ltd.

Hundert, Edward. 1994. *The enlightenment's fable. Bernard Mandeville and the discovery of society*. Cambridge: Cambridge University Press.

Jennings, Jeremy. 2007. The debate about luxury in eighteenth- and nineteenth-century French political thought. *Journal of the History of Ideas* 68(1): 79–105.

Jones, Erasmus. 1750^2. *Luxury, pride and vanity, the bane of the British nation*. London.

Jones, Robert W. 1995. *The empire of beauty: The competition for judgement in mid eighteenth-century England*. Ph. D., The University of York.

Jones, David W. 2011. Bubbles and bees: Historical exploration of psychosocial thinking. *Journal of Psycho-Social Studies* 4(2): 1–24.

Mandeville, Bernard. 1732. *An enquiry into the origin of honour and the usefulness of Christianity in war. London.*

Mandeville, Bernard. 1924. *The fable of the bees or private vices, publick benefits*, ed. F.B. Kaye. Oxford: Clarendon Press. The First Volume.

Mandeville, Bernard. 1924. *The fable of the bees or private vices, publick benefits*, ed. F.B. Kaye. Oxford: Clarendon Press. The Second Volume.

Mandeville, Bernard. 1997. A letter to Dion. In *Private vices, publick benefits? The contemporary reception of Bernard Mandeville*, ed. J. Martin Stafford, 577–611. Solihull: Ismeron.

McCabe, Ina B. 2008. *Orientalism in early modern France: Eurasian trade, exoticism, and the Ancien Régime*. New York: Berg.

McKee, Anthony F. Patrick. 1991. *An anatomy of power: The early works of Bernard Mandeville*. Ph. D. thesis, Glasgow.

Peck, Linda Levy. 2002. Luxury and war: Reconsidering luxury consumption in seventeenth-century England. *Albion* 34(1): 1–23.

Runcinam, David. 2008. *Political hypocrisy: The mask of power, from Hobbes to Orwell and beyond*. Princeton: Princeton University Press.

Shaftesbury, Anthony Ashley Cooper. 2001. An inquiry concerning virtue, or merit. In *Characteristicks of men, manners, opinions, times*, vol. II, ed. Idem, 3–102. Indianapolis: Liberty Fund Inc.

Shovlin, John. 2000. The cultural politics of luxury in eighteenth-century France. *French Historical Studies* 23(4): 577–606.

Smith, Woodruff D. 2002. *Consumption and the making of respectability 1600–1800*. New York/London: Routledge.

Turgot, Anne-Robert-Jacques. 1844. Plan de deux discours sur L' histoire universelle in Idem, *Oeuvres de Turgot*. Paris, vol II.

Veblen, Thorstein. 1918. *The theory of the leisure class: An economic study of institutions*. New York: B. W. Huebsch.

Zylberberg, David. 2008. *Ruined by pleasure? Luxury in eighteenth-century British discourse*. M. A. dissertation, Calgary, Alberta.

Chapter 4
Sex, Money, and Feelings: Mandeville's Dialogue with Sentimental Drama

Laura J. Rosenthal

Abstract "Sex, Money, and Feelings" suggest the possibility of a dialogue between Bernard Mandeville and the reformist playwrights. In response to attacks on the stage as immoral, Colley Cibber and Richard Steele reformed and sentimentalized the English stage. The vices they attack in their plays – duelling, luxury spending – emerge as key points in Mandeville, who defends them. In a passage meant to reveal the foolishness of the speaker, one of Cibber's foppish characters even anticipates Mandeville's argument by suggesting the benefits to society of his own spending. In the wake of the extreme popularity of Richard Steele's play *The Conscious Lovers*, the hero of which loves and supports a woman with no ulterior motive, Mandeville defends prostitution as a social benefit and continued to attach chastity and frugality as dangerous to the economy.

Keywords Bernard Mandeville • Sentimentalism • Theater • Colley Cibber • Richard Steele • Dueling • Luxury • Prostitution • Sexuality

In his *Modest Defence of Public Stews* (1724), Bernard Mandeville offers the least sentimental view of coupling imaginable, taking such a rational approach to this emotional issue that he can't resist the comic potential of the topic. (The treatise is authored, for example, by "Phil-porney.") The point, however, is serious: Mandeville advocates for the institutionalization of sex work, thus disentangling sexual labor from moral condemnation and removing blame from practitioners. Mandeville arrives at this solution, however, through a highly conventional sense of gender roles. He argues that a government-run system of prostitution would protect virtuous women from seduction, leaving the labor of sexual service to women who have already been polluted.[1] But there is a twist in his conventionality: as elaborated in *Fable of the Bees* (1714), in Mandeville's view virtuous women differ little from

[1]On Mandeville's sacrifice of polluted women for the sake of virtuous ones, see Mandell 1992.

L.J. Rosenthal (✉)
English Department, University of Maryland, College Park, MD, USA
e-mail: Lrosent1@umd.edu

E. Balsemão Pires, J. Braga (eds.), *Bernard de Mandeville's Tropology of Paradoxes*, Studies in History and Philosophy of Science 40, DOI 10.1007/978-3-319-19381-6_4

prostitutes. For Mandeville, women with the reputation for virtue only possess what he designates as "artificial virtue": that is, they understand that avoiding sexual activity and avoiding the appearance of sexual desire will benefit them in the form of both economic stability and approval from others. Women who figure this out work hard to maintain their artificial virtue, which bears little relation to the true virtue described by contemporaries, which Mandeville doesn't directly define but which seems to mean something like the avoidance of non-marital sex out of a belief that it would be wrong. So while true virtue and artificial virtue look identical in practice, Mandeville finds their distinction important enough to explore at length. Unlike either a moralist or smug libertine revealing the hypocrisy of virtuous women to embarrass or condemn them, however, Mandeville points out women's lack of virtue as a matter of fact, and even elaborates on the social benefits of their desires and vices. Thus, there is no room in either direction for sentimentalism in Mandeville's representation of coupling: there is no admiration for virtuous women and no pity for the fallen.

Mandeville, as E.J. Hundert has argued, documents the emerging contradiction between early capitalism and traditional Christian values and doctrine. The new economy, Hundert shows, depends on practices traditionally considered problematic for Christians, and Mandeville for the first time fully confronts this contradiction (Hundert 1994).[2] Mandeville's fully economic view of sexuality represents an extreme version of this, and is complicated by his assumption of heterosexual coupling as a physiological male need. For Mandeville, sex for men is not a luxury. It has a somewhat different status for women, however, and this imbalance leads Mandeville outside of the strictly economic. From an economic standpoint, artificial virtue and true virtue function identically and, as noted, appear as the same. What, then, we might ask, is Mandeville's stake in elaborating on this distinction? In turning to *Fable of the Bees*, this question becomes even more pressing, as this more prominent work lacks any kind of proposal for change: the author points out vice, but unlike his opponents he does not advocate for restraint or reform. While other philosophers of the eighteenth century advocate practices and commitments, Mandeville offers close analysis of personal behavior, but doesn't offer any recommendations, either ethical or strategic. Mandeville does not berate women for their sexual desires or suggest that they should rid themselves of them. He does not suggest that they should set down their artificial virtue, nor does he really suggest that they should refine this artificial virtue or that more women should adopt it. In this respect, Mandeville's discussions of female sexuality remain consistent with his analysis of virtue and vice in general.

Given that Mandeville's critique strays outside the strictly economic and into close analyses of coupling, I want to suggest that he not only presents sexuality in

[2]Hundert observes that Mandeville "introduced into the heart of European social understanding a series of arguments designed to sustain the radically unsettling conclusion that the moral identities of his contemporaries had been permanently altered by a previously unacknowledged historical transformation" (Hundert 1994: 14).

a remarkably, even comically unemotional way, but is actually engaged in a serious dialogue with a movement in which distinctions between true and artificial virtue would be crucial: not so much the activist Christian reformers like the Societies for the Reformation of Manners, but the more subtle representations offered by sentimental and reform drama.[3] Later in the century novelists would produce sentimental romance plots, but when Mandeville was writing the most prominent developments in this area were on the stage. Two years before the publication of the *Defence*, Richard Steele's groundbreaking sentimental play, *The Conscious Lovers* (1722), had taken London by storm and altered the way playwrights represented coupling on stage. While Steele's play may not have been the first sentimental comedy, it nevertheless marks a turning point given the author's own high-profile expressions of his attempt to reform the stage and the play's attempt to define a mode of heterosexual attraction insistently uncompromised by financial gain or sexual payoff.[4] In Mandeville's *Defence*, by contrast, nothing remains outside of the motives of financial gain or sexual payoff.

Mandeville was working through his most important ideas at a time when the theater had come under attack and was in the process of reforming. Mandeville published "The Grumbling Hive" (1705) in the midst of an extensive pamphlet war started by Jeremy Collier in 1698 with his *Short View of the Immorality and Profaneness of the English Stage*, a work that generated an extensive series of attacks and counterattacks, resulting in an increased scrutiny over how actors and actresses represented moral propriety on stage. While it may be true, as Robert D. Hume has argued, that sentimentalism did not simply drive libertine plays off the stage (Hume 1976: 9), theater managers nevertheless called deliberate attention to their reformed offerings, and reform was in the air. In this midst of this high-profile controversy, Mandeville expanded "The Grumbling Hive" into *The Fable of the Bees*, which extends the argument of the poem, publishing the first edition in 1714. Another edition of *The Fable* appeared in 1723 and the "Defence" appeared in 1724. Steele's *Conscious Lovers* began its influential 18-day run in 1722 and introduced new elements to the possibilities of sentimental theater. Theatrical reform, however, had begun even earlier: in 1696, Colley Cibber's *Love's Last Shift* had anticipated Collier's objections to the cynicism and licentiousness of the stage with a reformed view of coupling that insisted on the advantages of marriage over libertinism. Cibber's play announced its own difference from earlier comedy through the strategic repurposing of Restoration stock figures: his rake, for example, recommits to an early marriage he had left to pursue other sexual adventures. Steele was also developing new forms of sentimental comedy. While Mandeville published

[3]As Elaine M. McGirr points out, sentimental and reform drama have long been elided, and there are important distinctions between them. In her view reform comedy includes a "satiric lash" and they "do not elicit tears" (McGirr 2013: 386). For my current purposes, though, the distinction is less important, since both "reform" and "sentimental" comedies offer views of human possibility that Mandeville rejects.

[4]On *The Conscious Lovers* as self-conscious innovation, see Peter Hynes 2004, especially 142–145.

The Fable before *The Conscious Lovers* appeared on stage, Steele had already produced *The Funeral* (1701), *The Lying Lover* (1703), and *The Tender Husband* (1703–1704), in addition to publishing *The Tatler* (1709–1711) and *The Spectator* (1711–1714) as well as *The Christian Hero* (1701). When Mandeville was writing, then, Steele's reputation for a certain kind of refined sentimentalism and theatrical reform had already been established. Steele in particular moved comedy away from Cibber's emphasis on reform and into the realm of sentimental feelings.

Richard Steele, widely acknowledged as one of the innovators in sentimental drama, is one of the few other authors who Mandeville singles out for extended commentary in *The Fable.* In a passage worth quoting at length, Mandeville accuses Steele of manipulating his audience by falsely representing human beings as better than their true natures: "When the Incomparable Sir Richard," Mandeville writes, "in the usual Elegance of his easy Style, dwells on the praises of his sublime Species, and with all the Embellishments of Rhetorick, sets forth the Excellence of Human Nature, it is impossible not to be charm'd with his happy Turns of Thought and Politeness of Expression":

> But tho' I have been often moved by the Force of his Eloquence, and ready to swallow the ingenious Sophistry with Pleasure, yet I could never be so serious, but reflecting on his artful Encomiums I thought on the Tricks made use of by the Women that would teach Children to be mannerly. When an aukward Girl, before she can either Speak or Go, begins after many Intreaties to make the first rude Essays of Curt'sying, the Nurse falls in an ecstacy of Praise *There's a delicate Curt'sy! O fine Miss! There's a pretty Lady! Mama! Miss can make a better Curt'sy than her Sister* Molly! The same is echo'd over by the Maids, whilst Mama almost hugs the Child to pieces; only Miss *Molly*, who being four Years older knows how to make a very handsome Curt'sy, wonders at the Perverseness of their Judgment, and swelling with Indignation, is ready to cry at the Injustice that is done her, till, being whisper'd in the Ear that it is only to please the Baby, and that she is a Woman, she grows proud at being let into the Secret, and rejoicing at the Superiority of her Understanding, repeats what has been said with large Additions, and insults over the Weakness of her Sister, whom all this while she fancies to be the only Bubble among them. These extravagant Praises would by any one, above the Capacity of an Infant, be call'd fulsome Flatteries, and, if you will, abominable Lies, yet Experience teaches us, that by the help of such gross Encomiums, young Misses will be brought to make pretty Curt'sies, and behave themselves womanly much sooner, and with less trouble, than they would without them (Mandeville 1988: 1: 52–53).[5]

Learning to courtesy, which stands in this passage for a range of polite forms of behavior, is achieved in Mandeville's view by excessive flattery, either for the younger child's less accomplished curtseys or the older child's privilege of being let into the secret that the younger is being indulged – itself a form of indulgence. Steele resembles the adult women here when he represents humans as benevolent, which flatters readers and audiences into believing this about themselves. Anthony Pollock has taken this passage to suggest Mandeville's insightful critique of the ideological commitments of *The Spectator*, exposing the manipulations of that periodical. *The Spectator*, Pollock suggests, does not document a polite society, but instead bears

[5]Future references from this edition and cited in the text.

witness to a rather violent and chaotic one, attempting to manipulate its readers into a kind of passive propriety rather than political engagement (Pollock 2009: 55–74).

Certainly Mandeville was engaging Steele's journalism in this passage, but we might extend Pollock's insights to the possibility that Mandeville was in dialogue with Steele's theatrical productions as well. Steele already had a reputation as a dramatist as well as an author of periodical essays. Further, as Elizabeth Samet has pointed out, there is also something insistently theatrical about the way Mandeville presents his case throughout his work, and he turns to tragedy in order to explain sympathetic response (or lack thereof) (Samet 2003). The training scene itself, in fact, describes a highly theatrical moment in which a young girl performs for an audience as she refines her own self-presentation. Similarly, in Steele's play *The Tender Husband*, both of the plots follow the male manipulation of women into proper behavior. In the main plot, the older Clerimont manipulates his wife into virtuous, restrained behavior by having his own mistress pose as a foppish man who tries to seduce her. In the second plot, the younger Clerimont manipulates an heiress in marriage. Her violation consists of a passionate attachment to romances that distort her expectations for courtship (a plot device that Charlotte Lennox would later elaborate with great success in *The Female Quixote*).

The possibility that Mandeville was responding in the *Fable* and in the *Defence* to the high-profile controversies over the stage and its increasing embrace of sentimentalism might be further supported by the frequency with which Mandeville refers to drama in elaborating on his argument. In his extended attack on stoicism, Mandeville argues that Cato killed himself out of pride rather than virtue (1:214). Cato, of course, would be part of any discussion of stoicism in the eighteenth century; nevertheless, while Mandeville was finishing his *Fable*, Joseph Addison's tragedy *Cato* was breaking records for its popularity and profits. First appearing on stage in 1713, a year before Mandeville published the first edition of the *Fable*, *Cato* was the talk of the town. Addison's own emergent sentimentalism prompted a somewhat ambivalent version of Cato's story. While generally admiring Cato, the play nevertheless suggests that his suicide might have been misguided and also specifically notes that Caesar had planned to forgive him and his followers. Addison addresses this through Cato's own ambivalence about suicide, perhaps the most controversial part of the play. In Addison, Cato feels that he is doing the best thing for all concerned, although he also considered the possible impiety of the act. For Mandeville, however, the stoic merely acted about of pride. It would have been impossible for anyone in 1713 London to think about Cato without thinking about Addison's play.

There are also references to other popular plays scattered throughout the *Fable* that show a familiarity with and stake in the genre. When looking to describe sloth, Mandeville refers his readers to the stage, showing detailed familiarity with performance:

> Mr. Dryden has given us a very good Idea of superlative Slothfulness in the Person of a Luxurious King of *Egypt*. His Majesty having bestowed some considerable Gifts on several of his Favourites, is attended by some of his chief Ministers with a Parchment which he was to sign to confirm those Grants. First, he walks a few Turns to and fro with a heavy

> Uneasiness in his Looks, then sets himself down like a Man that's tired, and at last with abundance of Reluctancy to what he was going about, he takes up the Pen, and falls a complaining very seriously of the Length of the Word *Ptolemy*, and expresses a great deal of Concern, that he had not some short Monosyllable for his Name, which he thought wou'd save him a World of Trouble.

Mandeville does not mention the title of the play, *Cleomenes,* assuming that his readers will recognize it by the description. In the preface to part 2 of *The Fable*, Mandeville compares himself as a conqueror to Dryden's Almanzar (2: 10). He refers to Farquhar's popular play, *The Recruiting Office*, when describing life in the military (1:217). In explaining the power of custom, Mandeville observes that while in real life women who dress in men's clothes would incur serious disapproval, on stage even the most virtuous women do not object (1:137). Mandeville mocks those who argue that Gay's comedy *The Beggar's Opera* would lead to an increase in crime (2:6). The first dialogue in part two of *The Fable* considers theater and opera at length, suggesting an abiding interest in those forms. Given the rich engagement with the stage evidenced throughout *The Fable of the Bees*, then, it is not unreasonable to think that Mandeville refers to Steele as a sentimental dramatist and that his attack in general might be related not just to Christian reformers in general, but also to the specific reformation of the stage, spearheaded by Richard Steele himself.

Sentimental drama advocates for the reforms that Mandeville, at the heart of *The Fable*, held up for scrutiny. In particular, early sentimental drama seeks to stop men from dueling, to stop women from indulging in luxury spending, and to stop men from undermining their masculinity by spending money like women. Richard Steele's 1703 play *The Lying Lover,* takes up the problem of the duel. *The Lying Lover* had a respectable run, although it was not revived later and has been overshadowed by Steele's more famous work (Gollapudi 2011: 186–188). But it attracted attention at the time for its moral ambitions. At the heart of the plot is a nearly tragic duel between Young Bookwit and Lovemore that anticipates the averted duel in *The Conscious Lovers*, mistakenly aroused, as in Steele's later play, by jealousy. But unlike in *The Conscious Lovers,* in which the sentimental hero manages to avert violence, in the earlier play Young Bookwit, having had too much to drink, triumphs over his friend Lovemore and thinks he has killed him. Latine, another friend of Young Bookwit, offers to take responsibility for the murder so his friend can escape punishment. Happily, Lovemore had only been wounded and forgives Young Bookwit in light of his sincerely repentance. The most intensely sentimental moment of the play, however, takes place when Young Bookwit laments his actions while in prison. Latine suggests that he think about God's forgiveness, but Young Bookwit is consumed with regret and despair in a scene that is unusual for comedy at the time:

> How can I hope it! –
> No – I must descend from Man

Grovel on Earth, nor date look up again!
Oh Lovemore! Lovemore! Where is he now?[6]

The Lying Lover is not the first play of this period to feature reconciliation after a duel. In Aphra Behn's *The Rover*, Don Pedro becomes reconciled with Belvile after having crossed swords with him. What distinguishes *The Lying Lover* as sentimental and reformist, however, is the emphasis on the immorality of the duel, on Bookwit's profound regret, and on the affection between the men that allows them to forgive each other. In Behn's play, the characters must reconcile, but there is no sense that they violated morality in a regrettable way by their actions. This particular lesson in curtseying around dueling figures prominently in *The Fable*, a point to which I will return.

Sentimental and reformist comedies also take luxury as a key target. While, as Shirley Strum Kenny notes, Richard Steele's *The Tender Husband* returns to the Restoration comic style in certain ways (Kenny 1971: 194), it nevertheless offers some reformist morality of the kind that Mandeville would attack. In the main plot, Clerimont Senior has engaged his mistress Lucy to disguise herself as a gallant in order to seduce his new wife, Mrs. Clerimont. Clerimont senior proposes this as a solution to the difficult situation in which he has found himself. Unhappy with his wife, Clerimont can neither "Live at ease with her, or quite discard her" (1.1.21) until he catches her in a seriously compromising situation. Having traveled with him to France and Italy, Mrs. Clerimont has picked up cosmopolitan habits that disturb her husband for their worldliness and expense. She has become vain and flirtatious, indulging in what she sees as European-style "innocent freedoms." But while Restoration husbands mostly worried about becoming cuckolds, Clerimont Senior is concerned that his new wife spends too much money on expensive foreign indulgences. In discussing his scheme, in fact, he does not distinguish between his wife and his mistress in applying a cost/benefit analysis to each: "Dear Lucy, as you have been a very faithful, but very costly Wench to me, so my Spouse also has been constant to my Bed, but careless of my Fortune" (1.1.9-10). Mrs. Clerimont falls victim to the seduction plot, but Steele does not let the philandering husband go unreformed either. When Clerimont jumps out of a closet to attack his wife's seducer, she faints, then shortly awakens to see her husband in her seducer's arms. Both vow to change their ways. In its main plot, then, *The Tender Husband* proposes that women seduced by luxury imports can also be seduced by luxurious men. Clerimont's main anxiety, however, is her spending, as he sets out to entrap his wife through seduction because he has no such complaint to lay at her doorstep and that would allow him to berate her.

Luxury spending also becomes a major theme in Colley Cibber's comedy, *Love's Last Shift*. First produced in 1696, Cibber's play appeared 2 years before Jeremy Collier's diatribe and is sometimes credited with being the first sentimental comedy (Fone 1968: 33). Whether or not it merits that title, the play clearly responds to

[6] *The Lying Lover* 5.1.61-64 in Kenny 1971. Future references to Steele's plays from this edition and cited in the text.

reforming impulses and calls attention to its distinctions from and continuities with Restoration drama. In the main plot, Lovelace had abandoned his wife Amanda after only a few months of marriage to pursue rakish delights on the continent. When the play begins he has returned; he thinks his wife is dead. She is alive, however, and a friend informs her of her husband's reappearance. The friend and Amanda plot to have Amanda seduce Lovelace as a mistress in order reform him back into marriage. Lovelace has not only left his wife; he also left his country and spent money extravagantly on other women – and foreign women at that. Thus Lovelace epitomizes the eroticized luxury spender. In the character that Cibber wrote for himself, however, we find another luxury spender, although this one comically proposes the very argument that Mandeville would make in all seriousness. Lovelace knows that he has not been virtuous and feels some shame around that, which makes him able ultimately to be reformed. Sir Novelty Fashion, however, takes great pride in his spending and considers it a positive economic act:

> I must confess, Madam, I am for doing good to my Country: For you see this Suit, Madam... I suppose you are not ignorant what a hard time the Ribbon-Weavers have had since the late Mourning: Now my design is to set the poor Rouges up again, by recommending this sort of Trimming: The Fancy is pretty well for second Mourning... By the way, Madam, I had fifteen hundred Guineas laid in my Hand, as a Gratuity, to encourage it: But, I'gad, I refus'd'em, being too well acquainted with the Consequences of taking a Bribe in a national Concern![7]

Cibber's Sir Novelty Fashion makes a joke about corrupt politicians, but he also proposes that he turned down payment for an early version of product placement. As a patriot, he instead parades in the ribbons at no charge for the economic benefit of his country. Not only has he made the purchase, but his own indulgence creates the fashion for similar spending. Thus while indulgence in luxury nearly destroys Lovelace and casts sexual suspicion on Mrs. Clerimont, Sir Novelty Fashion makes the case for the economic usefulness of luxury consumption.

While Sir Novelty insists on his own economic contributions, however, the play clearly casts him as a fool. He tries to woo Narcissa away from Young Worthy while avoiding his long-time mistress, Mrs. Flareit. He parallels Lovelace in his spending, although their money goes to different luxuries; Novelty Fashion can afford his indulges, whereas Lovelace cannot. Nevertheless, Novelty Fashion's own spending, as he himself suggests, encourages the spread of luxury indulgence through his own example. Lovelace would be on his way to becoming a version of Novelty Fashion, except that he runs out of money. Cibber casts Novelty Fashion as a comically entertaining but also dangerous example. Novelty Fashion hopes to seduce Narcissa, but declares that he will resort to force if she does not comply. In this play the fop, like the excessively fashionable ladies on stage (such as Mrs. Clerimont), embody anxieties about excess spending and must be brought to some kind of fiscal virtue. Mrs. Clerimont ultimately complies, but Sir Novelty Fashion remains dangerous in his spending, ultimately pledging 300 pounds a year to Mrs. Flareit to renounce all

[7] *Love's Last Shift* in Viator and Burling 2000: 2.1.49-57.

claims to him, a deal that the Elder Worthy describes as "extravagant" (5.3.486-7) but that Novelty Fashion defends as a proto-Mandevillian strategy to keep the price of such women high.

The central themes of the early forays of the English stage into sentimentalism, then, include the dangers of dueling, of female spending, and of foppish masculinity. Mandeville singles all of these out for attention in *The Fable,* explicitly defending the first two and implicitly defending the last. Mandeville must go out his way to discuss dueling, for it digresses from the central economic concerns of *The Fable*. He works in this topic in an elaboration on a line in the poem that describes the newly honorable behavior of the bees, who now pay for everything outright. This observation leads to a digression on a variety of honorable and dishonorable behaviors. Mandeville defends dueling as having the positive social effect of keeping high the level of fear. For Mandeville dueling does not defy honor; instead, it *defines* honor, even if honor itself is not a product of virtue but instead a form of social manipulation. It is a necessary one for Mandeville, though, since the fear of being challenged supports civilized behavior (1: 158). While Mandeville pities those who end up crossing swords, alleviating society of this threat would ruin all conversation, for if there were no possibility of being challenged, then no one would take enough care to avoid offending others. Dueling thus remains necessary; the handful of men killed by this practice is more than compensated by the way its threat supports a "Politeness of Manners". For Mandeville, fear is necessary for social polish. Mandeville does not mention who in particular has been "rail[ing]" at the custom of dueling; certainly, it was a topic for Christian reformers. But as we have seen, an exploration of a calamity caused by dueling furnishes the central dynamic of Steele's early foray into sentimentalism on stage. It would later return as a crucial issue in his *Conscious Lovers* as well.

While the exploration of dueling comes up as an elaboration of another issue, the problem of luxury spending lies at the heart of both "The Grumbling Hive" and *The Fable of the Bees.* In note S of *The Fable*, Mandeville elaborates on the wheedling wife who manages to get money out of her husband that she spends on things beyond her station. The wheedling wife became a stock figure of early eighteenth-century comedy: Susanna Centlivre gives us just such a wife in her *Basset Table*, first performed in 1705. Mrs. Sago is married to a tradesman but socializes (and, unfortunately, gambles) with the elite, flattering her husband into supplying her with cash and gifts. She even, as Mandeville suggest of many such women (1:162), acquires funds through a kind of domestic embezzling. Steele's Clerimont married Mrs. Clerimont because she had money in the first place, but has become unable to control her spending or her luxurious taste for foreign imports. Mandeville acknowledges that such practices could ruin families, but he also suggests that they benefit the nation by lubricating the economy. For Mandeville, *The Tender Husband*, more than a diatribe from Christian moralists, would represent the dangerous lesson in how to courtesy, so to speak. Not only does the play persuade the audience of the hazards of Mrs. Clerimont's spending, but it flatters them with their own ability to reform, and to become happier in this reformation. It is this sentimentalism, born in the theater, that would later be taken up by philosophical rebuttals of Mandeville in works by Francis Hutcheson and others. Much of Mandeville's attack, then, may have been prompted by the newly sentimental stage.

The play most often credited with introducing sentimentalism to the English stage – Colley Cibber's *Love's Last Shift* – strangely, as noted, anticipates Mandeville's own argument and puts it in the mouth of the fop. In his extended consideration of both novelty and fashion, as well as in his self-consciousness about their new significant in the social world, Sir Novelty Fashion makes the argument his own luxurious consumption supports the economy by employing tradesmen who otherwise would not have work. The central lesson in politeness, however, address the profligate rake Lovelace and how the love of his wife can persuade him to give up spending money on whores. Mandeville, of course, see prostitutes as beneficial to the economy and would also approve of Novelty Fashion's generosity to Mrs. Flareit, who will turn around and spend the money on luxuries. Novelty Fashion does not reform, and ends the play uncoupled (although happily so). Loveless, however, must learn not just sexual but also financial constraint. When he appears in the first act, he has become so bedraggled with poverty that his friends do not recognize him. By the end of the play, however, he has recovered his proper attire, and, by turning away from prostitutes and toward his wife, has recovered his estate as well.[8] Thus whether Cibber achieves his ends through sentiment or not, he clearly suggests the economic more saliently than the moral dangers of the libertine attraction to prostitutes.

Richard Steele returns to Mandevillian problems of luxury, reform, desire, and exchange in his most famous play, *The Conscious Lovers* (1722), written after the publication of *The Fable of the Bees. The Conscious Lovers* revisits the kinds of sentimental figures that he and Cibber had previously created, and that Mandeville had attacked as misleading lessons in manipulation. Mandeville singled out Steele for his misrepresentations of human nature; Steele, perhaps, offers a response in *The Conscious Lovers.* He declares that he wrote the whole play "for the sake of the Scene of the Fourth Act, wherein Mr. *Bevil* evades the Quarrel with his Friend" (299). Steele's own declaration exaggerates the importance of the dueling scene; as Nicole Horejsi points out, audiences responded more readily to the Indiana plot and the recognition scenes, which constitute the structural resolution of the play (Horejsi 2003: 12). Perhaps in emphasizing the importance of the dueling scene, Steele was highlighting his engagement with Mandeville. In *The Lying Lover*, as we have seen, Steele had dramatized the terrible potential consequences of crossing swords; in *The Conscious Lover*, Steele displays Bevil's internal process and negotiation with his friend Myrtle that lead to restraint. He also gives us a Cibberian fop in the character of Cimberton, the coxcomb selected by Mrs. Sealand to marry Lucinda, who is in love with Myrtle and the immediate cause for the averted duel (but due to a misunderstanding). Cimberton, however, possesses considerably less charm than Sir Novelty Fashion, and becomes disturbingly threatening when he enumerates Lucinda's physical features in objectifying detail in a way that recalls Mandeville.

[8]On the importance of Loveless's wardrobe throughout this play, see Gollapudi 2004.

Mandeville notes that while a few men have a genuine passion for their wives, most just take pride in them out of mere possession, "as a Coxcomb does in a fine horse" (1:162). Steele echoes this language when he has Lucinda object to Cimberton as a "Monster! There's no bearing it. The hideous Sot! – there's no enduring it, to be thus survey'd like a Steed at Sale" (3. 293-4). Mrs. Sealand herself is exactly the kind of status-conscious wife who Mandeville celebrates and sentimental and reformist dramatists excoriate.

All of these elements suggest Steele's engagement with not just early sentimental comedy, but with Mandeville's critique of sentimental optimism as an economically dangerous and politically manipulative ideology. But there is nevertheless an additional element in this play that goes to the heart of the Mandevillian critique and that makes the case for the sentimental vision that would have the most enduring impact and that brings us back to the original question of this essay. In Mandeville, sexuality is a significant motivator of human behavior; romantic love, however, plays practically no role. People marry for status, money, and sexual satisfaction, but romantic love does not appear to be a factor. Mandeville considers the passions of pride and envy as significant motivators of human action, but love is a little like virtue or courage, in that those who express it usually have another agenda. Steele, however, offers a vision of coupling emphatically distinct from Mandeville's and perhaps also different from the vision offered by earlier reform comedy. Men seek women, according to Mandeville, to feed their pride. Sexuality exists for Mandeville as a physical impulse and little more:

> [B]y Love we understand a strong Inclination, in its Nature distinct from all other Affections of Friendship, Gratitude, and Consanguinity, that Persons of different Sexes, after liking, bear to one another: It is in this Signification that Love enters into the Compound of *Jealousy*, and is the Effect as well as happy Disguise of that Passion that prompts us to labour for the Preservation of our Species. This latter Appetite is innate both in Men and Women, who are not defective in their Formation, as much as Hunger or Thirst, tho' they are seldom affected with it before the Years of Puberty. Could we undress Nature, and pry into her deepest Recesses, we should discover the Seeds of this Passion before it exerts itself, as plainly as we see the Teeth in an Embryo, before the Gums are form'd. There are few healthy People of either Sex, whom it has made no Impression upon before Twenty: Yet, as the Peace and Happiness of the Civil Society require that this should be kept a Secret, never to be talk'd of in Publick; so among well-bred People it is counted highly Criminal to mention before Company any thing in plain Words, that is relating to this Mystery of Succession: By which Means the very Name of the Appetite, tho' the most necessary for the Continuance of Mankind, is become odious, and the proper Epithets commonly join'd to Lust are *Filthy* and *Abominable* (1:142–3).

Mandeville does not condemn this impulse; nevertheless, as this passage makes explicit, love has its roots in lust. Later he briefly considers the possibility of love without connection to lust. It might be possible to reach this unnatural state, given "the force of Education, and a Habit of thinking as we are taught." But there are

> many more whose Pretences to those refin'd Notions are only upheld by Art and Dissimulation. Those, who are really such Platonick Lovers are commonly the pale-faced weakly People of cold and phlegmatick Constitutions in either Sex; the hale and robust of bilious Temperament and a sanguine Complexion never entertain any Love so Spiritual as to exclude all Thoughts and Wishes that relate to the Body. But if the most Seraphick Lovers

would know the Original of their Inclination, let them but suppose that another should have the Corporal Enjoyment of the Person beloved, and by the Tortures they'll suffer from that Reflexion they will soon discover the Nature of their Passions (1:144).

Steele's *Conscious Lovers* stands out in the comedies of the period not just for its self-conscious embrace of the sentimental, but for the author's extreme attempt to disentangle romantic love from sexual desire and personal gain, the fusion of which constitutes perhaps the most memorable gauntlet thrown down by Mandeville. In this sense *The Conscious Lovers* differs significantly from Cibber's *Love's Last Shift:* Loveless only returns to Amanda when, through her disguise trick, she persuades him that the marriage will include exciting sex. Further, financial security provides another powerful argument in favor of marriage, a motivation that Cibber does not represent as dishonorable. Yet in *The Conscious Lovers,* Steele calls attention to his play's departure from this model by setting up a situation that *appears* Mandevillian: to some of the other characters, Bevil seems to be keeping Indiana by paying her expenses presumably in exchange for sex. Indiana had been born in England, but captured by pirates on her way to join her father in the Indies and raised in France by her captor but later left vulnerable to his unscrupulous brother. In the play, Bevil has fallen in love with the mysterious Indiana, whom he rescued from prison and poverty in France. Bevil Sr, however, has concluded a deal with the merchant Sealand for his son to marry Sealand's daughter Lucinda. But Lucinda is already in love with Bevil's best friend Myrtle, who returns her affection. Instead of scheming to trick or entrap the older generation, as the characters would have done had they been in a Restoration comedy, Bevil Jr tries to figure out a way to avoid marriage to Lucinda and unite instead with Indiana, but without offending his father. He tries to explain his position to Lucinda in a letter, but this correspondence raises the jealousy of Myrtle, who challenges him to a duel. Bevil, after initially accepting, changes his mind and talks his friend out of this rash and violent act. Sealand finds out about the relationship with Indiana, and takes it upon himself to figure out whether or not Bevile is a rake, as the situation with Indiana suggests.

The significance for the main plot in the context of Mandeville's powerful critique lies in the way Steele sets up the relationship between Bevil and Indiana, which defies Mandeville not just for its chastity, but for its chastity within a specific economic context. Bevil not only loves Indiana, but he also supports her financially. She has no family (that she knows of) and no means of support; Bevil has set her up in a household with her aunt, visits her regularly, leaves tickets for her to attend the opera, and stops by to discuss the various performances to which he has treated her. She is a stranger in England, and entirely dependent on him. His actions mystify Indiana as much as they do Sealand, who worries that he has betrothed his daughter to a rake. Indiana does not know whether or not he loves her. She doesn't ask, as that would violate both female propriety and also the laws of gratitude. Her aunt makes the obvious point that most men offering money in that situation would expect something in return, and simply accepting the money would in many contexts compromise the girl. But Indiana, against the better judgment of her aunt, nevertheless accepts the money and Bevil's attention. Thus, Steele sets up a kind of anti-prostitution, anti-Mandevillian scenario: Bevil gives money to a beautiful woman, but expects nothing in return. She does not spend it extravagantly, but lives quietly with her aunt. Even her affections, Bevil insists, must remain free of influence by his offering. Mandeville defines human relations through exchange; in *The Conscious Lovers*, however, the expected circuit of exchange remains incomplete in that Bevil asks Indiana for nothing. Thus the relationship between Bevil and Indiana *looks* like prostitution, or at least keeping. Even Bevil's friend Myrtle suggests Indiana's objectification, as Horejsi observes, but also her lack of consequence when he call her Bevil's "Foreign trinket" (Horejsi 2003: 14). Bevil's sexual and romantic passivity confuses Indiana herself. Bevil is not simply a non-rake. He has brought a virginal foreign girl to England with (apparently) no family to protect her; there would be no real consequences to Bevil for discreetly seducing her. Steele thus systematically eliminates all of the Mandevillian motives that Bevil might have for restraining his sexual desires. The relationship remains a secret and could easily have continued as such, so Bevil would not lose, as Mandeville would suggest, his public reputation by seducing her. In *Love's Last Shift*, by contrast, Amanda brings Lovelace back into the respectability of marriage through sexual desire and great gains in wealth. Thus, Amanda tempts her husband back with the two Mandevillian motives for coupling. Bevil, however, cannot be motivated by money, because Indiana has none and in fact costs him money; he cannot be motivated by status, since Indiana (as far as he knows) has no status in England; and finally and most dramatically, he cannot be motivated by sexual desire because he has no intention of asking for anything in return for his money. Indiana's extreme vulnerability and total dependence on Bevil serve to dramatize the significance of his lack of sexual motivation or expectation. Unlike Mandeville and even Cibber, then, Steele deliberately separates personal attachment from sexual desire, distinguishing the sentimental from the reformist.

It could be a coincidence that Mandeville published his *Modest Defence of the Public Stews* (1724) not long after *The Conscious Lovers* (1722) made such a splash; the closeness of the dates, however, is suggestive. Several characters in the play suspect Bevil of keeping Indiana as his mistress; with these suspicions, Steele highlights the structural expectation that *The Conscious Lovers* sets up of prostitution. Bevil's greatest heroic action of the play lies in choosing not to request sex in return for all of his funding. In a sense, Bevil's financial support of Indiana is superfluous to the plot: he could have raised suspicions without actually sending her money, and there certainly could have been different barriers to their union. Steele, however, uses it to emphasize Bevil's anti-Mandevillian motives.

There is a way of looking at the *Defence*, then, as a form of resistance to the representation of human feelings in sentimentalism in general and perhaps even Steele's *Conscious Lovers* in particular. With little biographical evidence of Mandeville's theatergoing habits, this possibility must remain speculative. Nevertheless, reading Mandeville in the context of sentimental and reformist drama suggests an important dialogue. Both Mandeville and the reformers, it is worth noting, take as their goal the preservation of virtuous women. The disturbing part of the *Defence* lies in the full embrace of commodified sex as a strategy for securing gendered conventions. Mandeville differs, however, in the extremity of his identification of sexuality as a commodity – so much so that critics have continued to speculate on whether or not the essay is parodic. There are two sides, however, to the shock of commodified sex: one is the placing in the commercial marketplace the emblem of what was emerging as the most personal form of identity, but the other, which has attracted less attention, is the way Mandeville's model does not allow for any kind of emotional connection attached to sex. This is actually true whether he is discussing a prostitute or a virgin's management of her desire, which is always strategic, or a man's quest to find an appropriate wife, which is for Mandeville equally strategic and based on getting the optimal combination of sexual fulfillment and money. Brothels would, he argues, allow these men to make the best decisions about marriage because they would not be acting impulsively out of pressing sexual desire. In all cases, individuals face tensions between physical longing and personal benefit in the form of either money or status. In Mandeville's scheme, men and women have *passions* but they do not have emotions that motivate coupling.

Finally what this dialogue reveals is the function of sentimentalism in smoothing the transition into a commercial market society. Mandeville succeeded in exposing the contradictions that change ushered in. He saw sentimentalism as an effective strategy for social manipulation, potentially surpassing the power of religion itself. Sentimentalism also provided the most powerful and enduring answer to his philosophical assertions. Certainly religion continued to hold sway, but as FB Kaye pointed out long ago, Mandeville is effective in his arguments in part because he takes up the terms laid out by Christian moralists, and then points out the social and economic benefits of vice (1: xlviii–liv). Although sentimentalism has its roots in the same moral foundation, it attempts to move away from the categories with which Mandeville grapples. Thus, Mandeville points out that Pope Sixtus the Fifth, "who was so strictly Severe in the Execution of Justice," nevertheless tolerated brothels (Mandeville 1724: 7–8), as some polluted member of the community can be expelled to preserve the purity of others. Traditional Christianity, then, is not inconsistent with commodified sex. Mandeville only differs in suggesting that the polluted ones do not differ fundamentally, and that society in general benefits from their polluted state. In Steele's sentimental vision, however, a mysterious orphan in an Indian mantle, who could be readily cast as polluted, might in fact turn out to be someone's daughter.[9] Perhaps just as significantly, she might also inspire warm feelings unrelated to money or sexual fulfillment. In order to highlight those feelings, however, the relationship between Indiana and Bevil must be stripped of both sexual and financial desire, while constantly being shadowed by these Mandevillian possibilities. Cibber's reformed rake can declare that love "is a tender plant that can't live out of a warm bed" (5.3.426-7); post-*Fable* representations of sentimental coupling, however, can go to remarkable lengths to sheltering their seedlings from any trace of Mandevillian desires.

[9]See Horejsi's excellent reading on this point, especially 29–30.

References

Fone, B.R.S. 1968. Colley Cibber's *Love's last shift* and sentimental comedy. *Restoration and 18th Century Theatre Research* 7(1): 33–43.

Gollapudi, Aparna. 2004. Seeing is believing: Performing reform in Colley Cibber's *Love's last shift. Restoration and 18th Century Theatre Research* 19(1): 1–21.

Gollapudi, Aparna. 2011. Why did Steele's *The lying lover* fail? Or, the dangers of sentimentalism in the comic reform scene. *Comparative Drama* 45(3): 185–211.

Horejsi, Nicole. 2003. (Re)valuing the 'Foreign Trinket': Sentimentalizing the language of economics in Steele's *Conscious lover. Restoration and 18th Century Theatre Research* 18(2): 11–36.

Hume, Robert D. 1976. *The development of English drama in the late seventeenth century*. Oxford: Clarendon.

Hundert, E.J. 1994. *The enlightenment's fable: Bernard Mandeville and the discovery of society*. Cambridge: Cambridge University Press.

Hynes, Peter. 2004. Richard Steele and the genealogy of sentimental drama: A reading of *The conscious lovers. Papers on Language and Literature: A Journal for Scholars and Critics of Language and Literature* 40(2): 142–166.

Kenny, Shirley Strum (ed.). 1971. *The plays of Richard Steele*. Oxford: Oxford University Press.

Mandell, Laura. 1992. Bawds and merchants: Engendering capitalist desires. *ELH* 59(1): 107–123.

Mandeville, Bernard. 1724. *A modest defence of publick stews: Or, an essay upon whoring, as it is now practis'd in these kingdoms. Written by a Layman*. London: A. moore.

Mandeville, Bernard. 1988. *The fable of the bees or private vices, publick benefits*, 2 vols. With a commentary critical, historical, and explanatory by F.B. Kaye. Indianapolis: Liberty Fund.

McGirr, Elaine M. 2013. Rethinking reform comedies: Colley Cibber's desiring women. *Eighteenth-Century Studies* 46(3): 385–397.

Pollock, Anthony. 2009. *Gender and the fictions of the public sphere, 1690–1755*. New York: Routledge UP.

Samet, E.D. 2003. Spectacular history and the politics of theater: Sympathetic arts in the shadow of the bastille. *PMLA* 118(5): 1305–1319.

Viator, Timothy J., and William Burling. 2000. *The plays of Colley Cibber*. Teaneck: Fairleigh Dickinson University Press.

Chapter 5
Humorism *A Posteriori*: Fables and Dialogues as a Method in Mandeville's Thought

Alessandro Chiessi

Abstract In this essay, I would show how Mandeville's literary experimentalism seems to be the consequence of an epistemological aim: the objectification of the case study. Taking seriously Mandeville's statements about the "Reader's Diversion" in a "Good Humor'd manner", it is possible to see an explicit aim with an explicit method, which brings to the estrangement from the object analysed. Here came out a particular outline of empiricism that, besides essays and treatises, uses also fables and dialogues to explain Human Nature and its expressions in society. Specifically fables and burlesque poems are genres used for the estrangement and the objectification of their literary subjects. At the same time, Mandeville achieves similar effects in dialogues through the play of points of view. Why 'Humorism *a posteriori*'? First of all, because good humour, diversion and amusement are the result of Mandeville's stylistic choices; secondly, because Humorism is the effect of figures of speech and literary genres. In this perspective, Humorism can be considered an experimental method adopted for the empirical description of Human Nature.

Keywords Mandeville • Method • Empiricism • Fables • Poems • Burleque poems • Dialogues • Irony • Satire • Humorism *a posteriori*

5.1 Introduction

Can anyone adopt words like *Humorism* and *a posteriori* associated with Mandeville's writings? Probably yes. Here, beyond considerations about literary genres, I would like to put together two different notions: the first is related to emotional effects and their conceptualization; the second is effectively connected to a method. Nevertheless, discussing about *Humorism*, in the English language, is discussing about a neologism. Usually it is possible to find the noun *Humour* and not

A. Chiessi (✉)
Department of Philosophy and Communication, Alma Mater Studiorum - University of Bologna, Bologna, Italy
e-mail: alessandro.chiessi@unibo.it

E. Balsemão Pires, J. Braga (eds.), *Bernard de Mandeville's Tropology of Paradoxes*, Studies in History and Philosophy of Science 40, DOI 10.1007/978-3-319-19381-6_5

Humorism, but the last syllable, the final *clausula* – this "*-ism*" – can be considered the label of a conceptualization. A conceptualization linked to the unusual empirical method of Bernard Mandeville: the same method through which he considers Nature and Human Nature a connection of causes and effects. Not only one cause and one effect, but possibly a lot of causes under one effect, as in a chain there is not only one link, but there are a lot of links.[1]

Searching into Mandeville's writings, one can find some purposes and a manner – or more than one manner – to realize these purposes. Reading what he writes in the *Preface* of *The Fable of the Bees*, appears his consciousness about explicit purposes and results. As Mandeville claims: "If you ask me, why I have done all this, *cui bono*? and what Good these Notions will produce? truly, besides the Reader's Diversion, I believe none at all" (Mandeville 1924a: 8). Beyond this clear assertion, he confirms the same position in *The Vindication of the Book* – after the impeachment by the Grand Jury of Middlesex and the *Abusive Letter to Lord C.* – where he states: "The *Fable of the Bees* was designed for the Entertainment of People of Knowledge and Education, when they have an idle Hour which they know not how to spend better" (Mandeville 1924a: 404).

Mandeville's purposes in *The Fable of the Bees* – but, from my point of view, not only here – seem to be "Diversion" and "Entertainment" of the reader. At the same time, the manner he composes his books – as it is reported in *A Letter to Dion* – is "an open good-humour'd Manner" (Mandeville 1954: 25). Once again, here, the reference is to *The Fable of the Bess* and the related defence against its critics, but probably it is possible to extend this outlook to other writings and try to define what I would like to circumscribe into the paradigm of 'Humorism *a posteriori*'. If explicit purposes are "Diversion" and "Entertainment", and the explicit manner is – one way or another – "good-humour'd", here there is a method; a method could be conceptualized into *Humorism*. This method, however, has clear aims and achievements, which open a path to other results. First of all the analysis of Human Nature and the unmasking of primary principles: passions, which leading are "Pride" with "Self-love" and "Self-liking".

In this paper, I would like to show three aspects of Mandeville's conceptualized method, i.e. *Humorism*: (a) how *Humorism* is displayed and obtained in Mandeville's writings, focusing particular attention to fables, poems and dialogues; (b) in which way *Humorism* creates an effect of estrangement and so an objectification of the observed case-study; and from these premises (c) how this particular *Humorism* is part of Mandeville's empiricism, and so of the knowledge *a posteriori*. To better understand these three characteristics of *Humorism* in Mandeville, I would like to analyse some fables, poems and dialogues – not forgetting *The Fable of the Bees* – written during his literary career.[2] Here it is important to remind that effectively *The Fable of the Bees* is a book composed of writings of different genres and

[1]This is a similitude of *The Fable of the Bees* (cf. Mandeville 1924a: 91, 402–404).

[2]The suggestion on *Humorism* and its conceptual implications comes from Luigi Pirandello and his essay of the beginning of the Twentieth Century (cf. Pirandello 1908).

styles, where *Humorism* becomes a peculiar part of Mandeville's method, of his epistemology and so of his philosophical project. Actually in *The Fable of the Bees* there are the fable, series of notes, essays and – in its second part – dialogues.

5.2 Satire, Irony or Literary Genres Experimentalism: *The Fable of the Bees*

A lot of scholars tried to define the literary style of Bernard Mandeville, focusing their attention especially on *The Fable of the Bees*. One of the most prominent questions is related to the classification of his literary genre or genres. Does Mandeville write satire or he adopt other solutions, for example irony and further categories? Surely, ranging over his different writings, it is quite clear he tested a lot of genres, experimenting unusual stylistic approach, achieving sometimes dull aesthetic results.

A new season of studies begins during the first part of the last Century when Frederick Benjamin Kugelman – mostly known as Kaye – publishes a critical edition of *The Fable of the Bees*. His important *Introduction* focused on the life of Mandeville, history of the text, thought and background, spends few words about rhetorical and genres results. Beyond some considerations on the "freshness of his style" (Kaye 1924: xxxviii), mostly in the first volume of the *Fable*, "abounding in wit and humour, rich yet clear, equally adapted to speculation and to narrative" (Kaye 1924: xxxviii), Kaye does not deepen questions related to a designed literary genre.[3] In 1953, Professor Jacob Viner, after years of study and reflections on Mandeville's thought, publishes an important critical edition of *A Letter to Dion*. Here Viner emphasizes the importance of satire, which is revealed in ethical "rigorism" as a way to conceal his libertinism.[4] Few years later, after this significant survey, Friedrich August von Hayek lectures on a "Master Mind" and presents Mandeville as a keynote thinker of his historical period. Showing his achievements in social theories from the description of human nature, Hayek recognizes his initial aim in the "satire on the conceits of a rationalist age" (Hayek 1966: 126; 1978: 250).

It is possible to reflect on how scholars changed their focus, passing from a consideration of manner, or literary *style*, to consideration of shape, or literary *genre*, probably without a clear consciousness of this shift. Near this distinction, quoting Ronald Paulson, it is not wrong to remind what he writes in the 60s: "It is not at all the same thing to say that [someone] wrote satire and to say that he wrote *a* satire. The latter introduces ideas of form and convention as well as of tone; but a novel, or a play, or a poem, can be satire without being *a* satire, and the adjective "satiric" is much the most popular form of the world" (Paulson 1967: 4). So the *style* of a book can be satiric, but the same book could be not ascribed into the *genre* of satire. At the end of the 50s and during the 60s, there is however a flourishing debate about

[3]Kaye considers satire only *The Grumbling Hive* (cf. Kaye 1924: lxxii).

[4]Cf. arguments of Professor Viner (1953: 1–15; 1958: 332–342).

literary genres in the theory of literature in which the most important contributions come from Northrop Frye[5] and from the just mentioned Ronald Paulson.

In this renewing and renewed cultural context George Hind, influenced by the theoretical indications of Frye about Swift's *Gulliver's Travels*, associates *The Fable of the Bees* with the Menippean satire. So he considers this compound book an example of a designed genre (Hind 1968). One year later, Phillip Harth publishes an essay where he rethinks Kaye's interpretation of Mandeville's thought. Here Harth problematizes the idea of a "*reductio ad absurdum*" rising by the collision of "rigoristic" (for private) and "utilitarian" (for public) standards in judgement.[6] From Harth point of view, Mandeville's satire is directed against one specific aim, i.e. "rigorism" (Harth 1969). But in this paper there is not a clear theoretical distinction between *style* and *genre*, and it is not easy to understand the difference between "satire" and "satiric" (reminding Paulson's claims).

During the 70s, Irwin Primer publishes a significant collection of essays, devoting one section of that work to *Style, Satire and Paradox* (cf. Primer 1975: 157–211). One of these papers – Robert Adolph's paper – detects the effects on style of different tensions and paradoxes in Mandeville's thought. Even if its style is "correct" and can be contextualized (with some particularity) into Restoration age, *The Fable of the Bees* can be furthermore categorized into the "paradoxical satire". For this reason, Adolph talks about an external "*persona*" not involved in conceptual contradictions, coming out the plot, and he quotes Frye and his idea of the "low normal" and "high normal" satiric point of view of a speaker outside the narration. Mandeville, in Adolph reconstruction, is between these two categories and embraces paradox as a stylistic device. His use of burlesque, lampoon, parody, exaggerations and so on, are examples of paradox: one of the most important characteristics of his satire. But, at the end, Mandeville creates "satire by writing within the Augustan canons of uniformity, objectivity, precision and clarity" (Adolph 1975: 167).

Robert H. Hopkins, into the same section of *Mandeville Studies*, considers *The Fable of the Bees* a "comic satire". In this reading, Hopkins refers to Chaisson's "pressure of the social scene" (1970: 504) and interprets stylistic choices – like paradox – or selections of literary genres as instruments for social criticism. The optimistic view of human nature proposed at the begging of the Eighteenth Century by some philosophers (i.e. Shaftesbury) struggles with the consequences of "possessive individualism"[7] both in politics and in economics. In this context, Mandeville would have criticised those positive thoughts, directing his satire against them and adopting paradoxes as stylistic devices (cf. Hopkins 1975).

A different interpretation is provided by Philip Pinkus. His point of view is that Mandeville does not write satire – as Swift does – because he does not criticize

[5]Frye's book was also an important touchstone for the theorizations of Paulson (cf. Frye 1957).

[6]Or – as Kaye writes – there is a "rigoristic criterion to motive" actions and an "utilitarian criterion to conduct" them (cf. Kaye 1924: xlix).

[7]The reference is to Crawford Brough Macpherson and his milestone book about early modern Political Philosophy (cf. Macpherson 1963).

evil in society. There is no evidence Mandeville judges *The Fable of the Bees* a satire, so Pinkus explicitly asks: "if the *Fable* is indeed satire, than what is being satirized?" (1975: 201). He argues – after denying any possible objects of satire – that Mandeville satirized nothing; insofar it is not possible to find any targets. Here it is significant to remind the importance of "hypocrisy" and "self-deceit" in Mandeville's *Fable of the Bees*, which can problematize the lack of an object of satire. From these premises, Pinkus states: "every satiric passage in the *Fable* is subordinated to the paradox that private vices make public benefits" (1975: 203). So the only possible literary genre is paradox; but in the place of genre, Pinkus softens his claims introducing the notion of "*mode*" as a "predominant attitude" or a "guiding principle" (1975: 205).

The unusual reading of Pinkus, compared to others, shows how scholars try to assign *The Fable of the Bees* into different categories, with dissimilar results. What appears quite clearly is a vague use of taxonomies once regarding *style*, once again literary *genres* and, sometimes, *logical framework* of argumentations. Surely it is not easy to understand and follow the numerous suggestions of this particular writing, but to define previous theoretical distinctions can probably help to discover more coherence than discordance.

Recently Frank Palmeri, in his survey on narrative forms in early modern literature, places *The Fable of the Bees* into the paradigm of "paradoxical satire" (cf. Palmeri 2003: 123–125). Palmeri seems close to Adolph's analysis and reasonably critical to Pinkus' inquiry. In this perspective, the issue is related to possible literary genres and follows the path traced by Professor Viner and subsequent scholars.

Mauro Simonazzi, in one of his last monographs on Mandeville, traces an original – and from some point of view controversial – interpretation of *The Fable of the Bees*. Simonazzi, recalling Frye's separation between satire and irony, considers the latter an independent literary genre (Simonazzi 2008: 92–96). "The chief distinction between irony and satire is that satire is militant irony: its moral norms are relatively clear, and it assumes standards against which the grotesque and absurd are measured", writes Frye (1957: 162). Embracing this difference into literary categories, Simonazzi associates *The Fable of the Bees* to the genre of irony.[8] Although Frye wants to trace in his essays this variation, I think it is necessary to contextualize these statements into the "mythical patterns" or "structural principles" he wants to discover alongside the history of literature. He directly asks: "are there narrative categories of literature broader than, or logically prior, the ordinary literary genres?" (Frye 1957: 162). His aim seems to be a renovation of conventional genres, attempting to discern common patterns. For this reason he introduces four categories, associated with four seasonal myths (images of relational and cyclical movement, from spring to summer, and then from autumn to winter): the comic, the romantic, the tragic, and the ironic or satiric. These categories are also defined

[8] In particular, Simonazzi traces a comparison between Mandeville and Swift in their different use of hyperbole to reveal their contrastive choice in literary genres: Mandeville irony and Swift satire (cf. Simonazzi 2008: 94–96).

by Frye "narrative pregeneric elements of literature", which are called "*mythoi* or generic plots" (Frye 1957: 162). These aspects of "pregenre", myth and plot, are the common traits of archetypical structural principles of literature. From these premises, reckoning the wide interpretative Frye's project into Western literature, I would prefer to be prudent and to preserve the distinction between satire – as a literary genre – and irony – as a figure of speech.[9]

Mandeville, in *The Fable of the Bees*, wants to write a satire, a paradox or an irony (considered as a genre)? Assuming a distinction in *style*, *genre* and *logical framework* of argumentations, probably it is possible to better understand its peculiarities: uniformities or unconformity, and coherences or incoherencies. Undoubtedly *The Fable of the Bees* presents sometimes ironic, some other times satiric and, occasionally, paradoxical elements in its *style*. About its genre, categorizing a gradually evolved writing into satire, can be too hazardous: the absence of a previous design and unity, but also the expansion from *The Grumbling Hive* till the six dialogues in the second part of the *Fable*, attest a lot of literary genres chosen to satisfy Mandeville's own necessities and taste. This gradual development, sometimes to unexpected directions, seems to characterize – from my point of view – the whole *Fable of the Bees* an expression of *genre experimentalism*, the same experimentalism it is possible to find in other writings of Mandeville.[10]

The further development of this essay concerns the possible discovery of a *logical framework* beneath Mandeville's argumentations, or a method gathered from his writings (not only from *The Fable of the Bees*, although this two volume book is a paradigmatic example of his genres experimentalism), i.e. *Humorism.*[11]

---

[9]In this perspective cf. N. Frye (1957: 131–140). Actually Frye claims: "The aim is to give a rational account of some of the structural principles of Western literature in the context of its Classical and Christian heritage" (1957: 133).

[10]The controversial definition of its literary genre can not find evidence in the same *Fable*. Pinkus, for example on the one hand, writes there are no explicit statement about the design of a satire. He is, in this, quite confirmed by the *Preface* of the *Fable*; Mandeville, reflecting on *The Grumbling Hive*, writes: "I am in reality puzzled what Name to give them [to the verses]; for they are neither Heroick nor Pastoral, Satyr, Burlesque nor Heroi-comick; to be a Tale they want Probability, and the whole is rather too long for a Fable" (Mandeville 1924a: 5). Harth, on the other hand underlines, some years before Pinkus, the reference to satire and, quoting Cleomenes (Mandeville's spokesman), emphasizes the purpose of the author of the first part of *The Fable of the Bees* (the same Mandeville) for writing satire (Mandeville 1924b: 105; 1954: 8, 36). Although, at the end of his literary parable, Mandeville can consider the first part of *The Fable of the Bees* a satire, the second part (in dialogues) problematizes this ascription if it is assumed portion of the same work (cf. Harth 1969; Pinkus 1975).

[11]The problem related to the genre of *The Fable of the Bees* is really close to the problem of its argumentative unity: being a two volume book, scholars try to define if they are separate elements, or the second is a kind of elaborate defence of the first part of the book. For an extensive study on the historical context where *The Fable of the Bees* appears, see the monograph of Mikko Tolonen, who recreates an history of the text and arguments for a separation of the first and the second volume. Rightly Tolonen detect an intellectual development in Mandeville's thought, and the publication of two different parts is a sign of this development. I think my paradigm

5.3 Fables and (Burlesque) Poems

Mandeville first English publications are translations of Jean de la Fontaine's fables, collected and edited during the first years of the Eighteenth Century. The reference is to *Some Fables after the Easie and Familiar Method of Monsieur de la Fontaine* and *Aesop Dress'd*. These books are interesting because they are results of a cultural atmosphere and signs of Bernard Mandeville's literary experimentalism. In this context, on one hand, John Dennis composes *Miscellanies in Verse and Prose* (published in 1693) where it is possible to find translations in burlesque style of poems or fables, taken from Horace, Juvenal, Boileau and de la Fontaine. And on the other hand, Mandeville writes and published in *Some Fables* and *Aesop Dress'd* two original fables: *The Carp* and *The Nightingale and Owl*. The main difference between Dennis' and Mandeville's translations is given by style: *Miscellanies in Verse and Prose* is an example of burlesque poetry, which seems more an improper manner than a creative interpretation, applied to verses and authors with different aesthetic purposes. *Some Fables* and *Aesop Dress'd*, on the other side, want to recreate the atmosphere of Jean de la Fontaine fables.[12]

In Mandeville's collections of fables, as in *The Grumbling Hive*, it is possible to discover some aspects of the "good-humour'd Manner" and so some characteristics of his *Humorism*. The easy and familiar method (of de la Fontaine) can be referred to stylistic and linguistic solutions of his fables. In this perspective, a simple language is necessary in order to describe the subject of the fable – traditionally an animal – for typifying his moral behaviour. For example the carp is defined in this manner: "A Handsome Carp genteely bred, /In fresh and running Water fed, /Puff'd up with Pride and Vanity, /Forsook the *Thames* and went to Sea" (Mandeville 1703: 24; 1966: 24). These opening verses of the first stanza give immediately, with few simple and onomatopoeic words, an idea of the character of the carp. This moral typification at the beginning of the tale is obtained both with a humble subject and a neutral language. So the stylistic solution, at the same time, is an essential part of the final teaching of *The Moral*, where Mandeville writes: "Some Fops that visit France and Rome, /Before they know what's done at home, /Look like our Carp when come again. /Strange Countries may improve a Man, /That knew the World before he went; /But he, that sees out ignorant, /Whom only Vanity intices, /Brings Nothing from'em, but their Vices" (Mandeville 1703: 27; 1966: 27).

Similarly in *The Grumbling Hive* there is identification between men and animals – fishes, birds or bees doesn't matter – and likewise it is possible to find a neutral language: a language displayed for typifying moral characters. Why do not remind all kinds of "knaves"? Are they bees or men? "Sharpers, Parasites, Pimps, Players, /Pick-pockets, Coiners, Quacks, South-sayers" (Mandeville 1924a: 19). But

of "genre experimentalism" corroborates the interpretation of a gradually-evolved thinking (cf. Tolonen 2013: 103–146).

[12]To problematize the dabate are important *English burlesque poetry* (Bond 1964) and the *Introduction* to Mandeville's *Aesop Dress'd* (Shea 1966).

ironically at the same level it is possible to find "Lawyers", "Physicians", "Priests", "Soldiers", "Ministers", because the first class of bees "were call'd Knaves, but bar the Name, /The grave Industrious were the same" (Mandeville 1924a: 19). The description of behaviours, with an explicit moral evaluation, is aimed to detect the natural roots of their actions, but this aim is not obtained directly, only through irony (considered as figure of speech) and sarcasm (i.e. the effect of stylistic choices). As Mandeville states: "every Part was full of Vice, /Yet the whole Mass a Paradise" (Mandeville 1924a: 24). This is the logical premise of Jupiter moral reformation and, consequently, the premise for a new beehive: a poor beehive. The correspondence of animals and men, and the ironical/sarcastic development of the fable, neutrally described, using a neutral language, creates a sort of emotional parallelism till the teaching of *The Moral*. I utilize the term parallelism because there is a complete and irreversible correspondence between two sides of morals and economics: the vicious beehive (α) implies wealth (β), as the honest beehive (α^*) implies poverty (β^*). From this logical framework seems impossible the homology between vicious beehive (α) and poverty (β^*) or honest beehive (α^*) and wealth (β). So at the end of the fable Mandeville states: "Then leave Complaints: Fools only strive/(X) To make a Great an Honest Hive/(Y) T' enjoy the World's Conveniencies, /Be fam'd in War, yet live in Ease, /Without great Vices, is a vain/EUTOPIA seated in the Brain. /Fraud, Luxury and Pride must live, /While we the Benefits receive [. . .] /Bare Virtue can't make Nations live/In Splendor; they, that would revive/A Golden Age, must be as free, /For Acorns, as for Honesty" (Mandeville 1924a: 36–37).

From these few examples, I would like to clarify how *Humorism* is revealed in Mandeville's fables. There is a humble subject, namely animals (corresponding to men); at the same time, there is a moral behaviour description through neutral language. Using these literary solutions, associated with irony and sarcasm – the "good-humour'd Manner" – and with amusement, it is possible to obtain the estrangement from the same subject of the fable. This estrangement creates an objectification of the subject so an objectification of the moral character. Consequently, in Mandeville's philosophy, *Humorism* with the related estrangement produces the objectification of human passions.

A similar effect with different literary solutions it is traced in Mandeville's burlesque poems, which are *Typhon* (edited in 1704) and *Wishes to a Godson* (published in 1712). Beyond the just mentioned John Dennis' *Miscellanies in Verse and Prose*, Samuel Butler's *Hudibras* (in three books: 1663, 1664, 1678 with the one volume edition published in 1684) and Paul Scarron's *Virgile travesti* (1648–1653) are archetypes of burlesque poetry. Indeed burlesque poetry is characterized by a discrepancy between the subjects of the poem – usually haughty and fastuous – and the narration style, generally vulgar and trivial.[13] I think Mandeville experimented and adopted this poetical manner, creating another kind of *Humorism*. For example in *Typhon Preface* Mandeville explicitly put together godly, human and animal nature: "There you [reader] shall see Gods no wiser than some of us turn'd into

[13]Cf. R.P. Bond (1964) and J.S. Shea (1966: ii–v).

Bears, Bees, Storks, and such like Creatures resembling one another, which is as much to say, as, That the best is to come" (Mandeville 1704: The Preface). Here amusement results from the discrepancy between the subjects of *Typhon* – that are Gods and Giants – and the indecorous description of their behaviour, through an adequate language (a language adequate to inappropriate behaviours and not to subjects).

Another example of "Diversion" and "Entertainment" in a "good-humour'd Manner" is *Leander's excuse to Cloris* printed in *Wishes to a Godson*. Here, the classical love poem becomes a sort of cheating justification. Love declaration passes through the admission of other love affairs and so through betrayals. The discrepancy is again between the subjects of the poem – Leander's reiterated cheating – and the narration style, referred to love declarations. In the second stanza, Mandeville writes: "When wanton Passion leaves my Breast, /Of Womankind, I love you best; /Tho' I've been catch'd in *Celia*'s Arms, /And Conquer'd by *Bellinda*'s Charms; /When Lust has led astray your Swain, /'Twas Love that call'd him back again; /And you my Dear, may still with justice boast, /Where're I play, that you I love the most" (Mandeville 1712: 13).

I think the logical framework of *Humorism* in burlesque poems is similar – but not the same – to fables. In these writings there are haughty subjects (Gods similar to men and animals); and there is a moral character description through vulgar and trivial language, which produces an internal discrepancy. This discrepancy is part of irony and sarcasm, used to create amusement. This amusement is essential for the estrangement from the subject, and so for the objectification of the subject. That is an objectification of moral characters or, in other words, an objectification of human passions into Mandeville's philosophy.

Aside fables and (burlesque) poems, dialogues give another nuance of *Humorism* in Bernard Mandeville's writings.

5.4 Dialogues

The first examples of dialogues seem to be the discussions of Lucinda and Artesia in the *Famale Tatler* (1709–1710). As Maurice Marks Goldsmith underlined, in these papers it is possible to detect a lot of themes and argumentations, Mandeville adopts and develops during the different editions of *The Fable of the Bees* (Goldsmith 1999: 11–72). In the same period, Mandeville publishes a whole dialogic book, namely *The Virgin Unmask'd*, where the aunt Lucinda and her niece Antonia discuss about manners and talk about love stories (cf. Mandeville 1709). Few years later Mandeville publishes the first edition of his unusual medical treatise, *The Treatise of Hypochondriack and Hysterick Passion* (edited in 1711). Here three characters – a physician, a patient and his wife – examine mental diseases, or, as Cheyene defined them, examine causes and effects of "The English Malady". These characters are respectively Philopirio, Misomedon and Polytheca. In the *Preface* of this book, there is an important statement can help to clarify the role of dialogues in Mandeville philosophy.

> In these Dialogues, I have done the same as *Seneca* did in his *Octavia*, and brought my self upon the Stage; with this difference, that he kept his own Name, and I changed mine for that of *Philopirio*, a Lover of Experience, which I shall always profess to be: Wherefore I desire my Reader to take whatever is spoke by the Person I named last, as said by my self. (Mandeville 1711: xi)

Although the declaration is clear and is directed in favour of empiricism that is the "love of experience", I would like to underline – for the aim of this paper – the explicit correspondence of Mandeville's point of view and Philopirio's assertions. In this textual passage there is more than an empirical announcement; there is a clear identification between the author and one character of the book: in this case Mandeville and Philopirio. This is not a new solution, because during all the history of philosophy a lot of thinkers, writing dialogues, embraced the point of view of one character to teach principles. I recall, first of all (and only), the Socratic irony in Plato's dialogues. But, in Mandeville, the struggle of points of view creates a different irony – producing sarcasm and then amusement – as results of explicit purposes – that are "Diversion" and "Entertainment" – through a peculiar manner: the "good-humour'd Manner". Here the objectification of the arguments of discussion is part of what I would define *Humorism*.

Some years later, although it's not clear if Mandeville is or is not the author of this pamphlet, appeared *The Mischiefs that Ought Justly to be Apprehended from a Whig-Government*. Kaye considers this text doubtful, but during the 60s, Dickinson assigns it to Mandeville's writing and so, some years later, Goldsmith.[14] Here the different point of view of Loveright, supporter of Whigs, and Tantivy, exponent of Tory Party, shows how dialogues, into political matters, can be a dialogue between deaf citizens. Probably, this can be considered another example of irony and, from my point of view, of *Humorism*. Significant is the final discussion and farewell.

> *Tant*. Now I have let you run your length without contradicting you, do you imagine you have any ways convinced me?
>
> *Love*. No, I am persuaded Tories are not to be convinc'd, or else the bare reflection on their Actions would be sufficient to shew them their Folly; for how can a Man more egregiously contradict his Principles than by openly shewing himself a Malecontent at the same time he defends the Doctrine of Passive Obedience.
>
> *Tant*. You are an incorrigible Whig, and so fare youwell
>
> *Love*. Remember Passive Obedience and then fare you well likewise. (Mandeville 1975: 40)

Beyond the historical references to Passive Obedience, here I would highlight in which way the communication in political arguments can be profoundly dogmatic and – if it is possible to use this expression – appears a sort of incommunicability.

But the most important example of dialogues in Mandeville's writings is displayed in the second part of *The Fable of the Bees*, and in *An Enquiry into the Origin of the Honour, and the Usefulness of Christianity in War*. In these books two characters – Cleomenes and Horatio – talk about and deepen topics examinated in

[14]For a survey about the philological matter it is a right thing to compare what write F.B. Kaye (1924: xxxi), H.T. Dickinson (1975: i–xiii) and M.M. Goldsmith (1985: 91–92).

the first part of *The Fable of the Bees.*[15] Properly in the second part of *The Fable of the Bees* there is a sort of clear declaration about the use of dialogues in philosophy.

> When partial Men have a mind to demolish an Adversary, and triumph over him with little Expence, it has long been a frequent Practice to attack him with Dialogues, in which the Champion, who is to lose the Battel, appears at the very beginning of the Engagement, to be the Victim, that is to be sacrificed, and seldom makes a better Figure, than Cocks on *Shrove-Tuesday*, that receive Blows, but return none, and are visibly set up on purpose to be knock'd down [. . .]. But it is as true, that there is no other manner of writing, by which greater Reputation has been obtain'd. Those, who have most excell'd all others in it were the two most famous Authors of all Antiquity, *Plato* and *Cicero* [. . .]. It is evident then, that the Fault of those, who have not succeeded in Dialogues, was in the Management, and not in the manner of Writing. (Mandeville 1924b: 8)

It is the "management" of dialogue that creates a condition to make philosophy. Mandeville again stated: "But tho' the Names I have chosen [Cleomenes and Horatio] are feign'd, and the Circumstances of the Persons fictitious, the Characters themselves are real, and as faithfully copied from Nature, as I have been able to take them" (Mandeville 1924b: 10). From this perspective, the opposition of point of view is just an opposition of characters, and so of principles of actions, and – at the end – of passions. Here there is another shade of *Humorism*; here ironically the estrangement comes up from the discussion. In other words, once again, it is possible to verify the objectification of the argumentation that aiming to "Diversion" and "Entertainment" in a "good-humour'd Manner", creates the estrangement through irony, sarcasm or, at least, amusement.

5.5 Conclusions

In this essay, I tried to show how *Humorism* is created in fables, burlesque poems and dialogues. In the first literary genre, Mandeville chooses a humble subject, namely animals (corresponding to men); at the same time, he describes a moral behaviour through neutral language. These literary solutions produce irony and sarcasm – in other words are the "good-humour'd Manner" – and consequently amusement. The same amusement is cause of estrangement from the subject of the fable. This estrangement creates an objectification of the subject, and at the end, an objectification of its moral character until its roots, which are human passions. Instead, in burlesque poems, there are haughty subjects (for example Gods similar to men); there is a moral character description through vulgar and trivial language, which produces an internal discrepancy. This discrepancy is part of irony and sarcasm, in creating amusement. This amusement is essential for the estrangement from the subject, and so for the objectification of the subject: that is an objectification of moral characters or, again, an objectification of human passions. In

[15] I remind that for few pages, in the second part of *The Fable of the Bees*, there is another speaker: Fulvia. Cf. the *I Dialogue* (Mandeville 1924b: 32–41).

dialogues the estrangement results from discussions – possibly ironic discussions – and is aimed at the discovery of Human Nature rules. Here the objectification is part of the literary style.

So, the last question I must answer is: why do I define this conceptualization – the paradigm of *Humorism – a posteriori*? Because it is the result of the management of literary styles and genres (fables, burlesque poems and dialogues), and because is the results of some purposes – "Diversion" and "Entertainment" – through an explicit manner: the "good-humour'd Manner". But there is another aspect: the estrangement and the objectification of literary subjects or, at the end, of passions – I think – could be considered part of the empirical approach into Bernard Mandeville's epistemology. Not the only characteristic, but one of some characteristics of his analysis method.

References

Primary Sources

Mandeville, Bernard. 1703. *Some fables after the easie and familiar method of Monsieur de la Fontaine*. London: Printed for Richard Wellington at the Dolphin and Crown at the West-End of St. Paul's Church-Yard.

Mandeville, Bernard. 1704. *Typhon: Or the wars between the gods and the giants: A burlesque poem in imitation of the comical mons. Scarron*. London: Printed for J. Pero, at the Swan, and S. Illidge, at the Rose and Crown in Little-Britain, and Sold by J. Nutt near Stationers-Hall.

Mandeville, Bernard. 1709. *The virgin unmask'd: Or female dialogues betwixt an elderly maiden lady and her niece on several diverting discourses on love, marriage, memoirs and morals of the times*. London: Printed, and are to be Sold by J. Morphew, near Stationers-Hall, and J. Woodward in Thread-needle-street.

Mandeville, Bernard. 1711. *A treatise of the hypochondriack and hysterick passions*. London: Printed and Sold by Dryden Leach, in Elliot's Court, in the Little-Old-Baily, and W. Taylor, at the Ship in Pater-Noster-Row.

Mandeville, Bernard. 1712. *Wishes to a godson, with other miscellany poems, by B. M.* London: Printed for J. Baker, at the Black-Boy, in Pater-Noster-Row.

Mandeville, Bernard. 1924a. *The fable of the bees or private vices, publick benefits*, vol. 1, ed. F.B. Kaye. [1732]. Oxford: Clarendon.

Mandeville, Bernard. 1924b. *The fable of the bees or private vices, publick benefits*, vol. 2, ed. F.B. Kaye. [1729]. Oxford: Clarendon.

Mandeville, Bernard. 1954. *A letter to Dion*, ed. B. Dobrée. [1732]. Liverpool: University Press of Liverpool.

Mandeville, Bernard. 1966. *Aesop dress'd*. [1704]. Los Angeles: William Andrews Clark Memorial Library.

Mandeville, Bernard. 1975. *The mischiefs that ought justly to be apprehended from a Whig-government*. [1714]. Los Angeles: William Andrews Clark Memorial Library.

Secondary Sources

Adolph, Robert. 1975. "What pierces or strikes": Prose style in the fable of the bees. In *Mandeville studies*, ed. I. Primer, 157–167. The Hague: Martinus Nijhoff.

Bond, Richmond P. 1964. *English burlesque poetry: 1700–1750*, 2nd ed. New York: Russel & Russell.

Chiasson, Elias J. 1970. Bernard Mandeville: A reappraisal. *Philological Quarterly* 49: 489–519.

Dickinson, Harry T. 1975. Introduction. In Mandeville, Bernard. *The mischiefs that ought justly to be apprehended from a Whig-government*, i–xiii. Los Angeles: William Andrews Clark Memorial Library.

Frye, Northrop. 1957. *Anatomy of criticism*. Princeton: Princeton University Press.

Goldsmith, Maurice Marks. 1985. *Private vices, public benefits*. Cambridge: Cambridge University Press.

Goldsmith, Maurice Marks. 1999. Introduction. In Mandeville, Bernard. *By a society of ladies*, 7–74. Bristol: Thoemmes.

Harth, Phillip. 1969. The satiric purpose of the fable of the bees. *Eighteenth-Century Studies* 2: 321–340.

Hayek, Friedrich August. 1966. Dr. Bernard Mandeville. Lecture on a master mind. *Proceedings of the British Academy* 52: 125–141.

Hayek, Friedrich August. 1978. Dr. Bernard Mandeville. Lecture on a master mind. In Hayek, Friedrich August. *New studies in philosophy politics, economics and the history of ideas*, 249–266. London: Routledge.

Hind, George. 1968. Mandeville's fable of the bees as Menippean satire. *Genre* 1: 307–315.

Hopkins, Robert H. 1975. The cant of social compromise: Some observations on Mandeville's satire. In *Mandeville studies*, ed. I. Primer, 168–192. The Hague: Martinus Nijhoff.

Kaye, Frederick Benjamin. 1924. Introduction. In Mandeville, Bernard. *The fable of the bees*, vii–cxlvi. Oxford: Clarendon.

Macpherson, Crawford Brough. 1963. *The political theory of possessive individualism*. Oxford: Clarendon.

Palmeri, Frank. 2003. *Satire, history, novel: Narrative forms, 1665–1815*. Newark: University of Delaware Press.

Paulson, Ronald. 1967. *The fictions of satire*. Baltimore: Johns Hopkins Press.

Pinkus, Philip. 1975. Mandeville's paradox. In *Mandeville studies*, ed. I. Primer, 193–211. The Hague: Martinus Nijhoff.

Pirandello, Luigi. 1908. *L'umorismo: Saggio*. Lanciano: Carabba.

Primer, Irwin (ed.). 1975. *Mandeville studies: New explorations in the art and thought of Dr. Bernard Mandeville (1670–1733)*. The Hague: Martinus Nijhoff.

Shea, John Stephen. 1966. Introduction. In Mandeville, Bernard. *Aesop dress'd*, i–xv. Los Angeles: William Andrews Clark Memorial Library.

Simonazzi, Mauro. 2008. *Le favole della filosofia*. Milano: Franco Angeli.

Tolonen, Mikko. 2013. *Mandeville and Hume: Anatomists of civil society*. Oxford: Voltaire Foundation.

Viner, Jacob. 1953. Introduction. In Mandeville, Bernard. *A letter to Dion*, 1–15. Berkeley: William Andrews Clark Memorial Library.

Viner, Jacob. 1958. Introduction to Bernard Mandeville a letter to Dion. In *View and the short*, ed. The Long, 332–342. Glencoe: Free Press.

Chapter 6
Mandeville, Pope, and Apocalypse

Peter Knox-Shaw

Abstract Some years before the Scriblerians brought a comic realism to bear on the themes of prophecy and apocalypse, Mandeville gave millenarians a taste of their own medicine by showing – in the conclusion to *The Grumbling Hive* – that a land free of the offences decried by the pious would indeed prove to be ruinous. In so doing he inaugurated a tradition of secularised apocalypse that finds one of its most famous expressions in the *Dunciad*. Both Pope and Mandeville make use of the millenarian motifs of Elkanah Settle's pageants for Lord Mayor's day, and though the Williamite politics of *The Fable of the Bees* was deeply inimical to the Tory wits (as appears in the satires of the Scriblerians), time has exposed a paradoxical congruence between Pope and Mandeville that underlies their official enmity.

Keywords Millenarianism • Reform societies • Lord Mayor's day • Elkanah Settle • The Scriblerians • Secularisation • *The Dunciad* • *The Fable of the Bees*

The opening decades of the eighteenth century were haunted, like its last, by spectres of apocalypse, but apocalypse is a two-sided affair, and its applications to history are diverse. Because it is in essence the divine plan for replacing a corrupt and evil order with an unblemished one, apocalypse involves both destruction and renewal. But while to initiates the dawning of the new age is presaged by a series of pre-ordained signs, such interpretation inevitably depends on circumstance. So Coleridge, writing in the aftermath of great upheaval, would summarise the last part of his poem, *Religious Musings* (1794): 'French Revolution, Millennium, Universal Redemption' (Coleridge 1974: 64). In Mandeville's era apocalypse tended to show its more threatening face. At one extreme, many Huguenot refugees (some 50,000 after the revocation of the Edict of Nantes) were inspired by the millenarian ideas of Pierre Jurieu who advocated war against a Papal or Bourbon anti-Christ; and sharing this militancy were a number of radical English sects surviving from the civil war. But the main stream of apocalyptic thought in Britain was conformist, comprehensively Anglican, and took the form of envisaging a divine retribution that

P. Knox-Shaw (✉)
Department of English, University of Cape Town, Cape Town, South Africa
e-mail: Peter.Knox-Shaw@uct.ac.za

E. Balsemão Pires, J. Braga (eds.), *Bernard de Mandeville's Tropology of Paradoxes*, Studies in History and Philosophy of Science 40, DOI 10.1007/978-3-319-19381-6_6

was imminent.[1] Periods of comparative stability seem often to stimulate fantasies of disaster. After a succession of political beginnings, the Stuart Restoration of 1660 and the Glorious Revolution of 1689 – both of which were initially compared to that milder classical counterpart of the millennium, the return of Astraea and the Golden Age – the Augustan peace was continually punctured by clerical warnings of annihilation.

No devotional tract of the period could make a greater claim to centrality than *The Causes of the Decay of Christian Piety* (1667), a jeremiad by the author of *The Whole Duty of Man* to which in many printings it provided a second part. On the title page of the fourteenth edition which appeared a year before Mandeville published *The Grumbling Hive* (1705) there is a plate that shows the old St Paul's consumed by flames in the great fire of 1666, with the caption e*tiam periere ruinae* (even its ruins have perished). In the Preface the fire is graphically ascribed to the revenge of a wrathful God on 'the Scenes of our Luxury, or our Fraud', more generally on a infamously secularised religion. Indeed, in exception to the findings of later analysts like Weber and Tawney, the *Causes of the Decay* identifies the spirit of commerce as deeply incompatible with Christianity, while predicting that a creed constantly sapped by avarice and self-seeking will result in a 'last Scene' to which the Fire and Plague are a mere curtain-raiser (Allestree 1704a, Preface, np).

Many years ago W. A. Speck made a powerful case for the relevance to *The Grumbling Hive* of the numerous societies for the Reformation of Manners that sprang up at the turn of the century: these embodied, he argued, precisely the sort of repressive attitudes to a flourishing economy that Mandeville was out to combat (Speck 1975). But missing from his account is the link, so conspicuous in the period, between calls for reformation and apocalyptic belief, a conjunction particularly clear in *Causes of the Decay.* And to this text Mandeville indirectly refers in the preface to his *Modest Defence of Publick Stews* (1724) when – after mentioning *en passant* the strong Stuart sympathies of the author of the *Whole Duty of Man* – he cites the postscript of a celebratory sermon preached by Bishop Gibson to a gathering of the Reform societies (Mandeville 1724: Preface np). Here the Bishop attributes to reformist intervention over 86,000 prosecutions for 'debauchery and profaneness', boasts of the distribution of some 400,000 tracts, and quotes a passage on the need for corporate action from the *Causes of the Decay* which he rightly acclaims as the founding document of the Reform Societies.[2] But there is good reason to

[1]See, for an excellent synoptic account, Paul J. Korshin, 'Queuing and Waiting: the Apocalypse in England, 1660–1750', in *The Apocalypse in English Renaissance thought and literature*, ed. C. A. Patrides and Joseph Wittreich (Ithaca, 1984), pp. 240–265. For the continental background, *Millenarianism and Messianismin in early Modern European Culture*, vol. 4 ed. J. C. Lauresen and R. H. Popkin (Dordrecht, 2001). And on the early impact of the Huguenot millenarians, Hillel Schwartz, *The French Prophets: the History of a Millenarian group in Eighteenth-Century England* (Berkeley, 1980).

[2]Gibson in his Sermon of 1723 quotes from the *Causes*: 'The Scandal brought upon Religion, as it was not constructed by the irregularities of one or two persons, but by associated and common crimes, so neither will it be removed by a few single and private Reformations. There must be

suppose that Mandeville was conscious of the *Causes* before he set to work on *The Grumbling Hive*, three editions of it having appeared between 1703 and 1704 alone.[3] The poem's grumbler-in-chief who threatens that 'the land must sink . . . For all its Fraud', is clearly a fictive and composite figure, and a somewhat comical one at that: a well-heeled millenarian who complains that the world is going to pot because he's been cheated by a shopkeeper:

And whom d'ye think
The Sermonizing Rascal chid?
A Glover that sold Lamb for Kid.[4]

The Causes of the Decay makes a litany of the claim, 'the Christian religion above all things excludes fraud and falshood', on the grounds that to lie is to defer the 'new Jerusalem' (1704a: 28, 25), but when its author (probably the royalist Richard Allestree) descends to specific transgressions he first enlists the Cromwellian sequestration of Church and Royalist property, and then immediately adds,

> But since those Leviathans are withdrawn, the lesser Devourers supply their Place: Fraud succeeds to Violence, and in all Places, all Occasion of Commerce, we still meet with Sequestrators. The adulterated Wares and false Measures in *Shops* . . . are too irrefragable Proofs thereof. (Allestree 1704a: 226)

Small fry, indeed, to merit the blaze of a national conflagration!

While *The Grumbling Hive* has often been read as an answer to the sermonizing reformists, no attention appears to have been paid to its parody of apocalypse, which typically proceeds by means of mimicry and inversion. The author of the *Causes of the Decay* who insists that there can be no cure while 'the arrow [of Avarice] remains in the flesh' (1704, 361) is comically mirrored by the Grumbler whose bluster eventually provokes a fit of divine pique:

But *Jove* with Indignation mov'd,
At last in Anger swore, *He'd rid*
The bawling Hive of Fraud; and did. (Mandeville 1924: I, 27)

The very absence of the causes of complaint proves then to be the cause of ruin:

But, Oh ye Gods! What Consternation,
How vast and sudden was th' Alteration! (Mandeville 1924: I, 27)

Jove's retribution, moreover, not only takes the form of giving the godly what they want, but of delivering a millennium that proves to be a punishment indeed. The last

---

Combinations and publick Confederacies in Virtue' (Gibson 1723: Postscript, 2). See also *A Letter from a residing member of the Society for Promoting Christian Knowledge* (London, 1714), p. 47.

[3]In addition to the separate edition of 1704, see *The Whole Duty of Man* (1703), pp. 213–468; and *The Works of the author of the Whole Duty of Man* (Oxford, 1704), pp. 215–464.

[4]Mandeville 1924: I, 27. All references to the *Fable* are to the edition by F. B. Kaye.

point is clinched in the conclusion to the main narrative and to its Moral in both of which the much dwindled hive is equated with a return to the Golden Age.[5]

Mandeville had far too much confidence in the resilience of the society he describes to suppose that history would follow this course. The catastrophe of *The Grumbling Hive* is tongue-in-cheek, but that did not stop it from becoming a sore point when the *Fable* came under fierce attack in the 1720s. George Blewitt complained bitterly of the poem's 'dreadful Account of Ruin and Desolation' (Blewitt 1715: 14), and William Law vented his fury (Law 1724: 92–3) upon a further burlesque of apocalypse provided by Mandeville in the *Remarks*, the end state of which is passive obedience to the King.[6] In hindsight it seems that Mandeville's parodic gambit marked out the territory that would occupy him for the rest of his career. The regression of a thriving mercantile society to an original natural state, however ironically pictured, posed questions about the reverse process. What were the progressive stages that lay between? And the contrast between the values associated with the golden age and those that actually obtained in contemporary society suggested the idea that different kinds of ethos were appropriate to different kinds of social dispensation, while challenging the practicality of a creed that elevated avarice to the status of the cardinal sin.

Mandeville's *Fable* shows that while avarice may be the 'root of evil' in communities where property is held in common (I, 25), justice is the virtue that takes precedence in a society where property is a matter of *meum* or *tuum*. From *The Grumbling Hive* alone it is clear that his reputation as an amoralist is ill-founded, for the poem not only attaches supreme importance to the operation of law (vice requires to be 'lopt and bound'), but repeatedly satirises the failure of the legal system to deliver justice as traditionally and morally conceived: hence the remarks on the iniquity of sentencing that favours the 'Rich and Great' at the expense of the poor (I, 37, 24). The bustling hive is a 'paradise' only in name, but its ruinous reformation is most graphically realised through the vignette of an institution that begins to crumble through lack of clients and competitiveness. Elsewhere Mandeville remarked on the contrast between the symbolic pomp of the law and the grim realities of its instrumentation, but a judiciary in collapse reveals depths still uglier:

JUSTICE hang'd some, set others free;
And after Goal delivery . . .
First march'd some Smiths with Locks and Grates,
Fetters, and Doors with Iron Plates:
Next Goalers, Turnkeys and Assistants:
Before the Goddess, at some distance,
Her chief and faithful Minister,

[5]See Ovid's description of the Golden Age, *Amores*, III. 8.

[6]For Mandeville's further parody of apocalypse, see I, 231. It is worth noting that the doctrine of passive obedience ranks high in the creed of the Goddess of Dullness in the *Dunciad*: see *The Dunciad*, ed. James Sutherland, Twickenham ed., vol. 5 (London, 1965): 'The RIGHT DIVINE of Kings to govern Wrong', (B) IV, l. 188, p. 360; and see especially the note to (B) IV, l. 453, p. 385.

'Squire CATCH, the Law's great Finisher,
Bore not th' imaginary Sword,
But his own Tools, an Ax and Cord:
Then on a Cloud the Hood-wink'd Fair,
JUSTICE herself was push'd by Air:
About her Chariot, and behind,
Were Serjeants, Bums of every kind,
Tip-staffs, and all those Officers,
That squeeze a Living out of Tears.[7]

Even without the detailed commentary on the Lord Mayor's show under *Remark O* in the *Fable*,[8] Mandeville's readers would have recognised in this passage a travesty of the famous London pageant. But his mention there of the Lord Mayor's 'great two-handed Sword' highlights his pointed substitution of the gruesome figure of the state executioner for the Sword-bearer who traditionally led the return procession from Westminster clad in a gown of black damask.[9] The 'imaginary Sword', the symbol of justice eclipsed by the axe, was central to this set of mildly millenarian events, the invocation of which contributed significantly to the original cultural reference of the poem.

At the turn of the century the once annual show fell into a decline so marked that a player in the last grand pageant of 1702 referred nostalgically to the 'splendour which formerly shined forth on this solemn city festival, now almost dropt into oblivion' (Fairholt 1843: 118). This decline, as well as one of its most probable causes – the disapproval of the Reform Societies[10] – was a circumstance well suited to the air of dereliction that the poem's straggling procession of Justice and her out-of-work retinue is meant to convey. Mandeville, who was undismayed by what he calls the habitual 'Gluttony and Drunkenness' of Lord Mayor's Day', and who was an ardent fan of pageantry *per se* (viz. his rapturous remarks on a canonization in Rome),[11] could rely on his readers' memory of the great spectacles staged by

[7]I, 28–9. At the close of the seventeenth century Celia Fiennes described the 'pageants' – or 'floats' as they were later called - in her diary as follows: 'a sort of Stages Covered and Carryed by men and on ye top many men and boys acting ye respective trades or Employ for Each Company, some in shipps for ye Merchts . . . '. Quoted by Robert Withington in his *English Pageantry: An Historical Outline*, 2 vols. (London, 1920), II, 67.

[8]I, 163–5. Mandeville argues in these paragraphs on the Lord Mayor's show that the chief function of the pageantry of justice is 'to animate not to deter'.

[9]For the positioning of the Sword-bearer, see the detailed description of Lord Mayor's Day in Guy Miège, *The New State of England, under our present monarch King William III* (London, 1701), pp. 177–8; for the Sword-bearer's 'gown of black damask', see John Gough Nichols on Lord Mayor's Day 1697, *London Pageants* (London, 1837), p. 82. This gown seems consistent with the one worn by the processing Sword-bearer shown in Plate 7 (opposite p. 178) of the 5th edition of Swift's *A Tale of a Tub* (London, 1710).

[10]Indicative here is Allestree's aversion to 'Shews and Pageantries of Justice', and his belief that a 'pageant like piety' was a defilement that 'required no slighter purgation than that of FIRE', see *Causes*, p. 207, and Preface, also pp. 28, 147.

[11]See *An Enquiry into the Origin of Honour*, pp. 94–6; also Kaye, I, xix.

Elkanah Settle during William III's reign.[12] In keeping with tradition Settle never once failed to give Justice a place of honour in a float, but in his most lavish show of 1698, entitled 'Glory's Resurrection', Justice is borne through the streets in a 'triumphant chariot of gold',[13] from which she descends in the person of Astraea to acclaim William as the restorer of the Golden Age (a trope touched on by Mandeville also in his elegy on the King)[14]:

> Justice of old by long Oppression driven,
> Left the Tyrannick World and flew to Heav'n,
> But when Great NASSAW, Albion's Sceptre bore,
> Our Laws and Rights sent kindly to restore,
> She visited the *Alien World* once more. (Settle 1698: 4)

Of a piece with the Golden age, the golden chariot and scales were especially appropriate to the sponsorship of the Goldsmith's company, but the customary celebration of London as the hub of world trade – hence all the sideshows of exotic places and goods – grew particularly pronounced in the 1690s. Justice was implicated, too, in the original rationale of the day (sanction of the city's choice of a new Mayor by the justices at Westminster), but increasingly assumed a larger symbolic role as the watchdog of commerce. Settle's Astraea solemnizes, moreover, the conclusion of William's long campaign against the French, a war that gave rise to the Bank of England, to national debt and public credit, to what has been termed the 'fiscal military state',[15] and on the economic front to capitalism, though in a form partially controlled by a comparatively democratic government at Westminster.

Elkanah Settle is known now chiefly for his role as a master of ceremonies in Pope's *Dunciad*, a poem set on Lord Mayor's day which provides many variations on the theme of apocalypse,[16] culminating of course in its famous final lights-out, a culturally induced – and pointedly demythologised – *Götterdämmerung*. Since in the last version of this work Mandeville is listed among the dunces, it might seem that any similarities between it and *The Grumbling Hive* are merely the product of the symmetry that arises from total opposition. But placed in the context of Pope's long, cumulative composition (1719?–1744),[17] the poem tells a more complex story which conforms only in its last stages to that unhappy stereotype of Mandevillean

[12]For some account of these see Frank Clyde Brown, *Elkanah Settle:His Life and Works* (Chicago, 1910), pp. 28–32.

[13]*Glory's Resurrection being the Triumphs of London Revived for the Inauguration of the Right Honourable Sir Francis Child* (London, 1698). The Chariot of Justice is depicted in Plate III.

[14]*The Pamphleteers: A Satyr* (London, 1703): see, for example, 'Such was poor Albion's case, when William came, /Rescue'd our Isle from the devouring Flame'.

[15]See the analysis by John Brewer in his *The Sinews of Power: War, Money, and the English State, 1688–1783* (London, 1989), particularly pp. 140–5.

[16]These include, for example, the 'second birth' of the swallowed gold coins that Annius recovers with the help of a 'soft, obstetric hand', see *The Dunciad*, ed. James Sutherland, Twickenham ed., vol. 5 (London, 1965), (B) IV, ll. 386, 394, pp. 379–80.

[17]See Maynard Mack's Introduction, pp. xiii–xiv.

influence, the writer who bites the hand that fed. Satire demands latitude, however, and Mandeville, who had so successfully paid back millenarians in their own kind with *The Grumbling Hive*, would perhaps not have been unduly perturbed to find himself subject to his own tactics, and upheld as a case of the biter bit. However marked their differences, it seems clear that Pope and his fellow Augustans were in Mandeville's debt not only for a range of far-reaching ideas but for a tenacious literary form also.

Mandeville figures as a butt in two satires by John Gay, who drew on the *Fable of the Bees*, nonetheless, both in his city georgic *Trivia* (witness the tell-tale linkage of opulence and dirt)[18] and in *The Beggars Opera* (witness Peachum's cheerful cynicism),[19] as has been widely recognized. But 'The Degenerate Bees', in his second volume of *Fables* (1738), was intended as an excoriating rebuttal. Here a stubbornly altruistic bee who resists the blandishments of a swarm addicted to luxury and acquisitiveness, warns the hive that disaster will soon ensue:

> Know, that in selfish ends pursuing,
> You scramble for the public ruin. (Gay 1974: II, 416)

In these later fables Gay simultaneously upholds civic virtue and extolls the supreme value of trade and industry ('Be commerce then thy sole design'), leaving the impression that the benevolence of the butcher, the brewer, or the baker are indeed enough to make a dinner (Gay 1974: II, 407). A scramble for self-preservation, however, is the theme of Gay's earlier end-of-the world satire, 'A True and Faithful Narrative of the Recent Consternation' (1716), a send-up of William Whiston's prediction that a comet would incinerate the earth prior to the coming of the New Jerusalem. And here Dr Mandeville is pictured abjectly penitent in the face of apocalypse (Gay 1974: II, 471). Thus was the doctor treated to a dose of his own medicine after a pattern that derives from Swift's Bickerstaff papers, written soon after *The Grumbling Hive.* These visited on the astrologer Partridge – who had forecast the imminent death of a fellow Londoner – not only a precise prediction of his own demise, but an account of his illness and burial.

Whiston's warnings of the coming cometary catastrophe (subject to much revision on his part) provided a favourite topic for the Scriblerians. Pope wrote a ballad on Mary Toft who was reported to have given birth to (variously specified) numbers of rabbits, an event the former Lucasian Professor took as a fulfilment of that portent foretold in the Book of Esdras (2.5.8), 'women shall bring forth monsters'.[20] In his 'Mr Joanidion Fielding' (1716) Arbuthnot presents a visionary who speculates on the manner in which a comet over Constantinople will precipitate the fall of the Ottoman empire, before it turns its bewigged face towards the west.

---

[18]Stephen Copley and Ian Haywood, 'Luxury Refuse and Poetry: John Gay's *Trivia*' in *John Gay and the Scriblerians*, ed. Peter Lewis and Nigel Wood (London, 1988), pp. 68–70, 72, 80.

[19]David Nokes, *John Gay: A Profession of Friendship* (Oxford, 1995), pp. 440–1.

[20]'The Discovery' (1727), *Alexander Pope: Minor Poems*, ed. Norman Ault (London, 1964a), pp. 259–62, 262n.

And Pope in his 'Annus Mirabilis' (1722) imagines a millennium in the form of a universal sex change which – although it obliges the Pope to 'undergo a new groping' – leaves the world a better place (Pope 1744: 78).

But in the most searching satire of this group, another short prose piece by Pope entitled 'God's Revenge against Punning' (1716), a fiercely protestant narrator, well versed in the rhetoric of the Reform societies, opens with a dry account of the chastisements inflicted by heaven on his sinful nation, before singling out the vice which has drawn down an epidemic of punning:

> Scarce had this Unhappy Nation recover'd these funest Disasters, when it pleased God to suffer the Abomination of Play-houses to rise up in this Land; From hence hath an Inundation of Obscenity flow'd from the Court, and overspread the Kingdom: Even Infants disfigured the Walls of holy Temples with exorbitant Representations of the Members of Generation; nay, no sooner had they learnt to spell, but they had Wickedness enough to Write the Names thereof in large Capitals; an Enormity, observ'd by Travellers, to be found in no Country but *England*.[21]

Hereafter the narrator's logic becomes increasingly perplexed. Punning is the punishment that God has devised for the evils that have sprung from the play-houses he has been pleased to 'suffer', but for punning he is obliged to mete out further punishment, so punishing Londoners for his own punishment of *pun*-ishment, for God, it seems, is not above punning himself.

Like Mandeville, Pope enjoyed clearing away the theological cobwebs that obscured the workings of history, and we can be sure that he, in company with Montesquieu,[22] took particular note of that passage from the *Fable* in which Mandeville insists that cultural decline is a matter of secondary causes:

> But here I expect a full-mouth'd Cry against me; What! has God never punish'd and destroy'd great Nations for their Sins? ... of all the famous States and Empires the World has had to boast of hitherto, none ever came to Ruin whose Destruction was not principally owing to the bad Politicks, Neglects, or Mismanagements of the Rulers. (Mandeville 1924: I, 117)

When Pope in the *Dunciad* – by way of stark contrast to the millenarian fantasies of his dunces – provides a global review of great vanished cultures, he underlines the ideological causes that fuelled the burning of libraries from east to west ('one

[21] *The Prose Works of Alexander Pope*, ed. Norman Ault, 2 vols. (Oxford, 1936), I, 270–1.

[22] Compare Montesquieu's remarks on the aetiology of decline: 'Fortune never interposes in the Government of this World; and we may be convinced of this truth by the Romans, who enjoyed a continual Series of Prosperity when they regulated their Conduct by one invariable Plan, but suffered an uninterrupted Train of Calamities, when they acted upon different principles. There are a set of general Causes, either Moral or Physical, which operate in every Monarchy, and either raise or maintain it, or else involve it in Ruin. All accidental Conjunctures are subordinate to these Causes; and if the Hazard of a Battle, which, in other Words, is no more than a particular Cause, has been destructive of a State, some General Cause presided, and made a single Battle be the inevitable Ruin of the State. In a Word, the Tendency of the main Principle draws after it all the particular incidents.' Montesquieu, Charles de Secondat, *Reflections on the causes of the Grandeur and Declension of the Romans*, trans. from the French (London, 1734), pp. 193–4.

bright blaze turns learning into air'), with a reminder of the ever present bigotry that keeps the 'beams of Science' at risk.[23] It was an essentially modern outlook that inspired Pope's vision of cultural decline, and there can be no reasonable doubt that Mandeville contributed significantly to it.

Pope once blamed ecclesiastics for their very free hand with the 'thunderbolts of God', and he returned to this theme when accounting for the rise of superstition in *An Essay on Man*,[24] a poem which reveals – as I have recently argued elsewhere – numerous debts to the *Fable* (Knox-Shaw 2014). These range from the idea that vice is often actively beneficial (a draft MS actually contained the line, 'And public good extracts from private vice')[25] to the understanding that social evolution is both non-contractual and gradualist. And though Pope drew as well on Bayle and La Rochefoucauld for his account of the primacy of the emotional life and of the ego ('So drives Self-love, thro' just and thro' unjust', Ep. III, l. 269), Mandeville left a distinct imprint, particularly through his comparison of the makings of sociability to the maturation of wine from the grape. Pope's notion of a natural balance obtaining between miserliness and prodigality as, too, his wider sense of an economy that is self-equilibrating derives substantially from the *Fable* also.

Though Pope parts company with his sometime mentor in the final book of the *Dunciad*, which presents among other things a critique of capitalism, the poem shares with *The Grumbling Hive*– quite apart from its explicit setting – an all-pervading structural feature. It opens with the prophecy that Dunce the Second, heir to the city poet Elkanah Settle, whose crowning proceeds to take place on Lord Mayor's day 1720, will restore the Golden Age. This millennium proves, however, like Mandeville's starving hive in the hollow oak, to be the ultimate terminus of cultural decline: 'a new Saturnian age' only in the sense that Saturn is the alchemical sign for lead.[26] Elkanah Settle's millenarian hopes for his successor are orchestrated in a spectacle created by impresario Rich but so like his own that he cries out, 'What pow'r ... what pow'r these wonders wrought?'[27]:

Hell rises, Heav'n descends, and dance on Earth,
Gods, imps, and monsters, music, rage, and mirth,
A fire, a jig, a battle, and a ball,
Till one wide Conflagration swallows all.

---

[23]*The Dunciad*, (A) III, ll. 59–114, pp 155–60; (A) III, l. 70, p. 156. The 'A' text is based on the quarto edition of 1729.

[24]To John Caryll, 19 July 1711, *The Correspondence of Alexander Pope*, 5 vols. (Oxford, 1956), I, 126; *An Essay on Man*, ed. Maynard Mack, Twickenham ed. (London, 1964b), III, ll. 241–68, 267-8n, pp. 116–20.

[25]Pope first altered this line to 'That draws some Virtue out of ev'ry Vice', and finally to 'That disappoints th'effect of ev'ry vice' (II, 240). The original is scored out on the last page of Epistle II in the MS of the *Essay* in the Pierpoint Morgan Library; see *Alexander Pope*, An Essay on Man:*Reproductions of the Manuscripts in the Pierpoint Morgan Library and the Houghton Library*, introduced by Maynard Mack (Oxford, 1962), np.

[26]See *Dunciad*, (A) I, l. 26, and 26n, p, 63; also (A) III, l.318, p. 186.

[27]See (A) III, l. 246, p. 178; also (A) III, ll. 273–4, p. 183 and note.

Thence a new world, to Nature's laws unknown,
Breaks out refulgent, with a heav'n its own:
Another Cynthia her new journey runs,
And other planets circle other suns:
The forests dance, the rivers upward rise,
Whales sport in woods, and dolphins in the skies,
And last, to give the whole creation grace,
Lo! one vast Egg produces human race ((A) III, ll. 233–44, p. 177).

Refracted through the lens of pantomime the new heaven and earth of Isaiah (65.17) reappear as a catalogue of impossibilities, but more than the popular genre of theatrical extravaganza is at stake. When accounting for the origin of political societies in *An Essay on Man* Pope epitomises the reign of a personified Superstition (which he brackets with Tyranny) in imagery that is parallel:

She from the rending earth and bursting skies,
Saw Gods descend, and fiends infernal rise ...
Gods partial, changeful, passionate, unjust,
Whose attributes were Rage, Revenge, or Lust.[28]

This critique of apocalypse in the third epistle of the *Essay* is integral to its Mandeville-like portrayal of the slow, erratic rise of civility through the piecemeal aggregation of separate families and groups, as also through a growing recognition of the rewards of a socially directed self-love.[29] A similar dismissal of supernatural intervention marks the closing 'vision' of the *Dunciad* of 1728 which – spoken by Settle – is intended, in part, to satirise a 'wild, ungrounded, and fictitious' millenarianism.[30] In the final *Dunciad* this element of literary burlesque falls away, however, when Pope not only extends the famous lines but gives them to the Muse to speak. Gone, too, is the assurance (which matches Mandeville's of 1729)[31] that the inroads of dullness on culture are ultimately limited, and in its place there stands the dire fulfilment of the Goddess's boast to make one mighty Dunciad of the land.[32] But Pope's enactment of the *fin du globe* differs markedly from the traditional versions of annihilation. Retribution by an offended and fiery God yields to a dynamic of depletion, to an autotelic process of irreversible decay. And though natural theology is among the chief casualties of Dullness, Pope's end of all endings remains father to a long line of more fully secularised versions of apocalypse, a line that goes back to Mandeville's ironic parable of the devastated hive.

[28]*An Essay on Man*, III, ll. 253–4. 257–8, p. 118.

[29]See especially III, ll. 122–130, 199–210, 269–282, pp. 104–5, 113, 120–1; and Kaye, II, 132–3. Pope's account of patriarchal authority invites comparison also with Mandeville's discussion of 'reverence', see *Essay* III, ll. 215–34, pp. 114–5, and Kaye II, 280–1.

[30]See Scriblerus's note to *Dunciad*, (A) III, ll. 337, p. 192.

[31]Mandeville in his Sixth Dialogue concludes his satire on scholarly myopia and the vagaries of the learned with the admission that learning is soundly established nonetheless, see Kaye, II, 342–4. In the same Dialogue he argues that the unmeritorious are serviceable to society even in the highest office, see II, 324–9.

[32]See second paragraph of Scriblerus's note to (A) III, l. 337, p. 192; and (B) IV, l. 604, p. 403.

Whereas Settle's 'Glory's Resurrection' provided a perfect foil for Mandeville's picture of social collapse, for Pope such shows are symptomatic of the collapse itself. As a devout if unorthodox Catholic, Pope had reason to be critical of the Williamite pageantry of his youth, but when he takes note in the *Dunciad* of the infamous Pope-burnings that first won Settle fame, it is to indicate that his distaste transcends the old divide between Whig and Tory.[33] What was it then that Pope found so obnoxious in the culture that he hypostasizes in his dunces, and how are we to understand his position in relation to the cultural bearings of our own day? In keeping with some of the more virulent strains of post-modernism, Pope's dunces scoff at the empirical (Settle dismisses Bacon, Newton and Locke in Book III),[34] choose 'the high Priori Road',[35] regard language as something entirely divorced from the real ('Words we teach alone' ... 'on Words is still our whole debate'),[36] and believe, above all, that the past can be exorcized and a new order instituted with the mere wave of a wand. In the case of these maladies Pope's satire continues to perform its work of healing. A realist to the core, Mandeville would surely have welcomed this outcome of his legacy.

References

Primary Sources

Allestree, Richard (?). 1703. *The whole duty of man.* London: Edward Paulet.
Allestree, Richard (?). 1704a. *The causes of the decay of Christian piety.* London: Edward Paulet.
Allestree, Richard (?). 1704b. *The works of the author of the whole duty of man.* Oxford: Edward Paulet.
Anon. 1714. *A letter from a residing member of the Society for Promoting Christian Knowledge.* London: J. Downing.
Arbuthnot, John. 1716. *Mr Joanidon Fielding: His true and faithful account of the comet.* London: E. Berington.
Blewitt, George. 1715. *An enquiry whether a general practice of virtue tends to the wealth or poverty, benefit or disadvantage of a people?* London: R. Wilkin.
Coleridge, Samuel Taylor. 1974. *Coleridge: Poems*, ed. John Beer. London: J. M. Dent and Sons.
Gay, John. 1716. *A true and faithful narrative of the recent consternation.* Poetry and prose. vol. 2.
Gay, John. 1738. '*The Degenerate Bees*', in his second volume of *Fables.* Poetry and prose, vol. 2.
Gay, John. 1974. *Poetry and prose*, 2 vols, ed. Vinton A. Dearing. Oxford: Oxford University Press.
Gibson, Edmund. 1723. *A Sermon preach's to the Societies for Reformation of Manners.* London: John Wyat.
Law, William. 1724. *Remarks upon a late book entituled, The fable of the bees.* London: William and John Innys.
Mandeville, Bernard. 1703. *The pamphleteers: A satyr.* London.

[33]*Dunciad*, (A) III, ll. 282–4, p. 183.

[34]See (A) III, ll. 211–14, p. 175.

[35]*Dunciad*, (B) IV, l. 271, p. 386.

[36]See (B) IV, ll. 150, 219, pp. 356, 364.

Mandeville, Bernard. 1724. *A modest defence of publick stews*. London: A. Moore.
Mandeville, Bernard. 1729. *An enquiry into the origin of honour.* London: John Brotherton.
Mandeville, Bernard. 1924. *The fable of the bees*, 2 vols, ed. F.B. Kaye. Oxford: Clarendon Press.
Miège, Guy. 1701. *The new state of England, under our present monarch King William III.* London: H. Mortlock.
Montesquieu, Charles de Secondat. 1734. *Reflections on the Causes of the Grandeur and Declension of the Romans*, trans. from the French. London: William Innys.
Pope, Alexander. 1716. *God's revenge against punning*. London: J. Roberts.
Pope, Alexander. 1727. *The discovery.* London: A. Campbell.
Pope, Alexander. 1744. Annus Mirabilis. In *Miscellanies in verse and prose*. London: John Thomas.
Pope, Alexander. 1936. *The prose works of Alexander Pope*, 2 vols, ed. Norman Ault. Oxford: Basil Blackwell.
Pope, Alexander. 1956. *The correspondence of Alexander Pope*, 5 vols, ed. George Sherburn. Oxford: Clarendon Press.
Pope, Alexander. 1964a. *Minor poems*, vol. 6, ed. Twickenham Norman Ault. London: Methuen.
Pope, Alexander. 1964b. *An essay on man*, vol. 3i, ed. Twickenham Maynard Mack. London: Methuen.
Pope, Alexander. 1965. *The dunciad*, vol. 5, ed. Twickenham James Sutherland. London: Methuen.
Settle, Elkanah. 1698. *Glory's resurrection being the triumphs of London revived for the inauguration of the Right Honourable Sir Francis Child.* London: R. Barnbam.
Swift Jonathan. 1710. *A tale of a tub.* London: John Nate.

Secondary Sources

Brewer, John. 1989. *The sinews of power: War, money, and the English state, 1688–1783*. London: Unwin Hyman.
Brown, Frank Clyde. 1910. *Elkanah settle: His life and works*. Chicago: University of Chicago Press.
Copley, Stephen, and Ian Haywood. 1988. Luxury refuse and poetry: John Gay's *Trivia*'. In *John Gay and the Scriblerians*, ed. Lewis Peter and Wood Nigel. London: Vision Press.
Fairholt, F.W. 1843. *Lord mayor's pageants*. London: the Percy Society.
Knox-Shaw, Peter. 2014. Pope, Mandeville, and an essay on man. *Notes and Queries* 61(3): 398–400.
Korshin, Paul J. 1984. Queuing and waiting: The Apocalypse in England, 1660–1750. In *The Apocalypse in English renaissance thought and literature*, ed. C.A. Patrides and Wittreich Joseph. New York: Cornell University Press.
Laursen, J.C., and R.H. Popkin (eds.). 2001. *Millenarianism and messianism in early modern European culture*, vol. 4. Dordrecht: Kluwer.
Mack, Maynard. 1962. *Alexander Pope,* an essay on man: *Reproductions of the manuscripts in the Pierpoint Morgan Library and the Houghton Library*. Oxford: The Roxburghe Club.
Nichols, John Gough. 1837. *Lord Mayor's day 1697*. London: London Pageants.
Nokes, David. 1995. *John Gay: A profession of friendship*. Oxford: Oxford University Press.
Schwartz, Hillel. 1980. *The French prophets: The history of a millenarian group in eighteenth-century England*. Berkeley: University of California Press.
Speck, W.A. 1975. Mandeville and the eutopia seated in the brain. In *Mandeville studies*, ed. Primer Irwin. The Hague: Martinus Nijhoff.
Withington, Robert. 1920. *English pageantry: An historical outline*, 2 vols. Cambridge MA: Harvard University Press.

Chapter 7
The Fable of the Bees: *proles sine matre*?

Béatrice Guion

Abstract From a literary point of view, *The Fable of the Bees* is a mixture of genres. The premise of this paper is that formal choices are not only formal, but say something about Mandeville's intents, and are not without relevance about the signification of the *Fable*. The first version, *The Grumbling Hive*, appears as a poem that can be inscribed within the satirical verse tradition, which was well represented in England at the turn of the seventeenth and the eighteenth centuries. The second version, whose title explicitly refers the genre of the fable, adds to the social and political dimension a concern for moral unmasking, much in the manner both of La Fontaine's *Fables*, which Mandeville translated into English, and of the seventeenth century French Augustinian Moralists. It adds prose remarks, which can recall the remarks found in Bayle's *Dictionary*. As for the second part of the *Fable*, it takes on the form of a philosophical dialogue, with a revendicated French patronage.

Keywords Fable • French moralists • Irony • Libertinism • Paradox • Philosophical dialogue • Satire

The motto chosen by Montesquieu in his *captatio benevolentiæ* for *De l'esprit des lois* could easily be applied to Mandeville's *Fable of the Bees*: from a literary point of view, his work is a mixture of genres. My premise is that formal choices are not only formal, but may tell something about Mandeville's intents, and are not without relevance about the (much discussed) signification of the *Fable*.

The first version, *The Grumbling Hive*, reads as a poem derived from the satirical verse tradition, which was well represented in England at the turn of the seventeenth and the eighteenth centuries, and which assumes an aim of political, moral and/or religious criticism. The second version, whose title explicitly refers to the genre of the fable, reveals, beyond the social and political dimension, a concern for moral unmasking, much in the manner both of La Fontaine's *Fables*, which Mandeville translated into English, and of the seventeenth century French Moralists. Such an addition of prose remarks is reminiscent of the remarks in Bayle's *Dictionary* – first

B. Guion (✉)
Faculté des lettres, Université de Strasbourg, Strasbourg, France
e-mail: beatrice.guion@unistra.fr

E. Balsemão Pires, J. Braga (eds.), *Bernard de Mandeville's Tropology of Paradoxes*, Studies in History and Philosophy of Science 40, DOI 10.1007/978-3-319-19381-6_7

of all because of the ironical tone, but also because of the discrepancy between the original text and the remarks, which are much longer. As for what is usually called the second part of the *Fable*, it takes on the form of a philosophical dialogue – a genre much used in England at this time, and even more so in England than in France: it is all the more significant that Mandeville should claim that a French philosopher, the neo-Epicurean Pierre Gassendi, be his formal model.

So from the first version in 1705 to 1728 Mandeville chose three different literary genres. The issue raised here is not that of the unity of the work, nor that of the evolution of his thinking between the first and the second part of the *Fable*,[1] but the meaning of his formal choices. As Hundert pointed out, Mandeville is a self-conscious member of the Republic of Letters[2]: his formal choices, – which are at the same time generic, rhetorical and stylistic choices –, and their grounding in well known traditions of writing are not devoid of significance. Both the English and the French background of these traditions should be taken into account.

7.1 *The Fable of the Bees* as a Satire

The nucleus of *The Fable*, *The Grumbling Hive*, appeared as a verse satire, as were Mandeville's first texts: he played an active part in the writing of verse lampoons at the time of the Costerman Riot,[3] and his first known English publication, *The Pamphleteers*, was a political verse satire. The verse satire, which is at the beginning of the eighteenth century commonly held as a Roman invention, is defined both by its form, and by its purpose: it criticizes contemporary mores, with a view to moral reformation. For the humanist Casaubon, for example, the essence of latin verse satire is the scourging of vices and the exhortation to virtue («*vitiorum insectatio, & ad virtutem cohortatio*»[4]). Dryden in his *Discourse concerning the Original and Progress of Satire* (1693) recalls the aim of Roman satire:

> Only as *Dacier* has observ'd before me, we may take notice, That the word Satire is of a more general signification in *Latin*, than in *French*, or *English*. For amongst the *Romans* it was not only us'd for those Discourses which decry'd Vice, or expos'd Folly; but for others also, where Virtue was recommended.[5]

He condemns the evolution of modern satire: «For in *English*, to say Satire, is to mean Reflection, as we use that word in the worst Sense; or as the *French*

---

[1]The two most important books lately published on Mandeville lay the emphasis on the difference between the two parts: see Simonazzi 2008 and Tolonen 2013. I would like to thank Denis Lagae-Devoldere for his reading of the present paper and his translation suggestions.

[2]Hundert 1996: 14.

[3]See Dekker 1992.

[4]I. Casaubon 1605: f. ã2 r°. I use Dryden's translation, but I restore the plural of «*vitiorum*» (Dryden 1974 [1693]: 55).

[5]Dryden 1974 [1693]: 48.

call it, more properly, *Medisance*.»[6] Shaftesbury also complains that «[o]ur Satir [...] is scurrilous, buffooning, and without Morals or Instruction, which is the Majesty and Life of this kind of writing.»[7] As a matter of fact, the only object of verse satire in early modern Europe is not just morals. It also aims at politics and religion; in England, where it became a major genre after the Restoration, the Augustan age is considered as its heyday.[8] Hundert assumes that *The Grumbling Hive* imitates Samuel Butler's manner in *Hudibras*, which was one of Mandeville's favourite works[9]: if *Hudibras* was at first a political satire directed against puritans, its condemnation of hypocrisy led to a more general painting of folly, which could remind the reader of Erasmus's *Encomium Moriæ*.[10]

When Mandeville in the 1714 Preface claims that his verse are not «Satyr, Burlesque nor Heroi-comick»,[11] his words are not to be taken at face value – all the less so since in the next page he speaks of a «satyr»:

> The Satyr therefore to be met with in the following Lines upon the several Professions and Callings, and almost every Degree and Station of People, was not made to injure and point to particular Persons, but only to shew the Vileness of the Ingredients that all together compose the wholesome Mixture of a well-order'd Society [...].[12]

That satire be directed towards vice in general, and not towards «particular Persons» is a traditional claim of humanist theoreticians of the satire genre; it is, also, what separates satire from lampoon.[13] It is not always the case, nor in ancient, nor in modern verse satire; it was not the case in Mandeville's first writings.

If *The Grumbling Hive* belongs to the genre of verse satire, it is another question to know whether the *Fable* in prose is a satire or not. The issue, much discussed by critics, is all the more difficult since the satire is both a genre – the verse satire – and a tone[14]: it is, of course, more difficult to define and to characterize the satire as a tone than as a literary genre with clear-cut formal features. However, it seems that the word, at least in the common parlance of the early eighteenth century, was not only applied to a peculiar literary form, but also used with a loose meaning, denoting the spirit of a writing as soon as its contains criticism.[15] In this regard, the text of 1714, whose title explicitly refers to the genre of the fable, could still

---

[6]Dryden 1974 [1693]: 48. See also 28.

[7]Shaftesbury 2001a [1710] (*Soliloquy*, part II, section 3): t. I, 126 [266] (the first reference is to the page in the modern edition, the second to the page in the 1732 edition). On the moral purpose attributed to satire see Elkin 1973: 71–89.

[8]See Wheeler 1992: 312.

[9]See Hundert 1996: 4.

[10]See Jack 1964 [1952]: 15–42.

[11]Mandeville 1988a: 5.

[12]Mandeville 1988a: 6.

[13]See Jack 1964: 43; Elkin 1973: 20–24.

[14]See Debailly 2012.

[15]See Elkin 1973: 11.

be considered as a satire: fable and satire are not mutually exclusive. One should bear in mind that the fable was widely used in England at the turn of the seventeenth and the eighteenth centuries for political satire,[16] and that Mandeville translated into English some of La Fontaine's *Fables*, which contain sharp criticisms not only of human vices, but also of the French society and politics of his time. Mandeville himself in *A Letter to Dion* claims that his intent was to ridicule vices.[17]

Cleomenes in part II of the *Fable* speaks of satire as opposed to panegyrick, thus inscribing satire within the rhetorical genre of epideixis: «There is, generally speaking, less Truth in Panegyricks than there is in Satyrs.»[18] Mandeville seems to refer to satire in order to characterize the first part of the *Fable* (the prose remarks) in the third dialogue of the second part. Horatio blames the author of *The Fable of the Bees* for mingling serious matter with laughing matter: «But who knows, what to make of a Man, who recommends a thing very seriously in one Page, and ridicules it in the next ?»[19] Cleomenes answers that «he shews the Folly of Vice and Pleasure, the Vanity of Wordly Greatness»[20]: one could think of Democritus, who laughs at men's folly, one is also reminded of Erasmus's *Praise of Folly*. One can all the more think of Erasmus as *Typhon*, published in 1704, was dedicated to the «Numerous Society of F.» – that is to say Fools – : it is easy to see here a passing wink at Erasmus, who is one of the few Dutch writers whom Mandeville praised, and to whom he is greatly indebted.[21] He may be remembering *The Praise of Folly* when he writes that he targets vices, and not specific persons: this is the very claim of Erasmus, who in his dedication to Thomas More stresses that he gives no name, unlike saint Jérôme, who was «not sparing sometimes, mens very name».[22] In his well known letter to Martin Dorp, in which he answers the criticism Dorp, and more generally the clergy, had directed at the book, he repeats, as a defence, that he did not name the persons he criticized.[23] When Mandeville writes that he did not intend «to injure and point to particular Persons»,[24] his argument smacks of Erasmus's defence of the satire.

The Praise of Folly is both a satire and a paradoxical eulogy: *The Fable of the Bees* could also be read as a paradoxical encomium of vices, as Henry Knight Miller

[16]See Noel 1975: 32, 36–37; Simonazzi 2008: 69–70.

[17]«[. . .] it is manifest, that, unless I was a Fool, or a Madman, I could have no Design to encourage or promote the Vices of the Age. It will be difficult to shew me an Author, that has exposed and ridicul'd them more openly.» (Mandeville 1954 [1732]: 32).

[18]Mandeville 1988d. The First Dialogue: 59.

[19]Mandeville 1988d. The Third Dialogue: 102.

[20]Mandeville 1988d. The Third Dialogue: 102.

[21]The most complete and useful study about Mandeville's connections with Erasmus remains Irwin Primer's (Primer 1993: 313–335). This question surely deserves more scholarly investigation.

[22]Erasmus 1668. Epistle to Sir Thomas More: f. A4 r°.

[23]«nullius omnino nomen perstringitur [. . .] An non vides me toto opere sic a nominibus hominum temperasse [. . .] ?» (Erasmus to Martin Dorp, 1515, in Erasmus 1910: 95).

[24]Mandeville 1988a. Preface: 6.

suggested more than 50 years ago.[25] The paradoxical encomium, associated in the Renaissance to scepticism, still widely used in the Augustan age, has a double aim, comic and satirical, in a critical purpose: its aim coalesces with that of satire. One should recall that the eulogy of drunkenness is a *topos* in the *Paradoxes* published in the sixteenth and seventeenth centuries, as is the eulogy of poverty: several texts are entitled «For Drinkers», or «For Povertie». More specifically, one also finds eulogies of beer in English *Paradoxes*.

If the critics often spoke of the satirical intent of the *Fable of the Bees* in prose,[26] some substantiate their claim further: George Hind, referring to Frye's *Anatomy of Criticism*, characterised it as Menippean satire,[27] which appears to be neither certain nor useful. First of all, Frye says very little about Menippean satire in the pages devoted to «satire and irony»; it is, moreover, quite difficult to define. The editor of one of the best known Menippean satires in the modern age, the French *Satire Ménippée*, writes in 1595 that not everybody understands the title, and knows what a Menippean satire is; he defines it as a slanderous writing whose aim is moral censorship of vices, and, moreover, mingles verse with prose.[28] The mingling of prose and verse has long been held as a formal characteristic of the Menippean satire, and a defining one at that: for Dryden, «[t]his sort of Satire was not only compos'd of several sorts of Verse, [...] but was also mix'd with Prose [...].»[29] In that respect, the addition of prose remarks to the poem does not result in a genuine prosimetrum: the poem and the remarks can be read separately; nor are the quotations of poetry in the remarks enough to speak of Menippean satire, as George Hind does. It is no more certain that «the often extensive use of dialogue» be «[a]nother distinguishing trait of Menippean satire»,[30] so that the second part of the *Fable* could be considered as a satire from that perspective.

The Fable of the Bees has been classified as a satire for so long that it has become a commonplace assumption, challenged by Philip Pinkus in the *Mandeville Studies* in 1975,[31] and by Mauro Simonazzi. Mauro Simonazzi, also referring to Frye's *Anatomy of criticism*, draws a different conclusion: he relies on the sharp distinction

---

[25]Miller 1956: 145–178.

[26]See for instance Jacob Viner: «if Mandeville's rigorism were sincere, the whole satirical structure of his argument [...] would be incomprehensible, and there would be manifest inconsistency between his satirical purposes and his procedures as a writer.» (Viner 1991 [1953]: 179).

[27]Hind 1968: 307–315. Irene E. Gorak also speaks of a Menippean satire (Gorak 1990: 5).

[28]«car tous ceux qui sont nourris aux lettres savent bien que le mot de Satyre ne signifie pas seulement un poème de médisance, pour reprendre les vices publiques, ou particuliers de quelqu'un: comme celles de Lucilius, Horace, Juvénal, et Perse: mais aussi toutes sortes d'écrits, remplis de diverses choses et de divers arguments, meslez de proses, et de vers entrelardés [...].» (*Satyre ménippée* 2007: 160). The Flemish scholar Justus Lipsius is named among the modern writers of Menippean satire, which maybe is not devoid of significance for the Dutch Bernard Mandeville (*Satyre ménippée* 2007: 161–162).

[29]Dryden 1974 [1693]: 46.

[30]Hind 1968: 310.

[31]Pinkus 1975: 193–211.

Frye makes between satire and irony to conclude that the *Fable* is an ironical, and not a satirical work.[32] Frye asserts that satire is «militant irony» founded on «relatively clear» moral norms, while in ironical writing «the reader is not so sure», either about «the author's attitude», or about what he is supposed to think.[33] Pinkus, for one, claims that *The Fable* is a paradox and not a satire; but satire and paradox are not mutually exclusive. One can even wonder if the text could not be read in the light of the paradoxical eloquence of the Renaissance – along the lines of Erasmus's *Praise of Folly*.

That satire and paradox are not mutually exclusive is confirmed by the writings of the French seventeenth century moralists. The word «satire» has sometimes been used by the early readers of La Rochefoucauld and of La Bruyère to characterize their works. One of them sees in the *Maxims* both a satire, and a paradoxical writing: «it is a very strong and a very ingenious satire of the corruption of the nature by the original sin, of self-love and of pride, and of the malignity of the human mind [...] although it is full of paradoxes, these paradoxes are nevertheless true, as long as one speaks of natural reason and natural virtue, without grace.»[34] As a matter of fact, one can find in La Rochefoucauld's *Maxims* a mixture of satire, irony and paradox.

7.2 Paradoxical Writing: The Inheritance of French Moral Tradition

Much has been written about Mandeville's knowledge of the French tradition of moral thought, and how much he borrowed from it – both the Augustinian moralists and the libertines, which he could read in French and in English translations. Many critics agree with Kaye: «The great source of Mandeville's psychology was France».[35] Despite opposite premisses, Augustinian Moralists and libertines share a pessimistic view of man (of fallen man for the Augustinians), driven by self-love and passions instead of reason.

Mandeville's debt towards the French moralists was noticed – often to be denounced – by the first readers (and censors) of the *Fable of the Bees*: Bluet for example, in *An Enquiry Whether a general Practice of Virtue tends to the Wealth or Poverty* (1725) blames Mandeville for having plundered Montaigne, Bayle, Jacques Esprit and La Rochefoucauld, without ever acknowledging his borrowings. Adam

[32]Simonazzi 2008: 92–96.

[33]Frye 1969 [1957]: 223.

[34]«[...] c'est une satire très forte et très ingénieuse de la corruption de la nature par le péché originel, de l'amour-propre et de l'orgueil, et de la malignité de l'esprit humain [...] quoiqu'il y ait partout des paradoxes, ces paradoxes sont pourtant très véritables pourvu qu'on demeure toujours dans les termes de la vertu morale et de la raison naturelle, sans la grâce.» (lettre d'auteur inconnu à Mme de Schonberg, 1663, in La Rochefoucauld 1992: 568–569; my translation).

[35]Mandeville 1988: xciv. See Horne 1978: 19–31; James 1975: 43.

Smith in the *Theory of Moral Sentiments* (1759) associates La Rochefoucauld and Mandeville, in the chapter devoted to «licentious systems» (part VII, section 2, chapter 4).

The affinities between Mandeville's portrayal of human nature and that of French moralists are well known, especially the prominent role given to self-love and the denunciation of false virtues it entails. What I would like to consider here is not so much the commun themes as the inscription in a specific kind of writing: the point I would like to enhance is that Mandeville's debt toward the French seventeenth century writers appears also in his manner of writing, and the use of characteristic formal devices.

It is not without reason that the same questions were raised about Mandeville and about both La Rochefoucauld and Bayle; and that such a wide range of contradictory interpretations has been given of these authors: all of them have been accused of being immoral and cynical, all of them have also been read (mostly by modern critics) as rigorist Augustinians. Such questions never arose either for Jacques Esprit or for Pierre Nicole, although their conception of human nature was similar. That is why the manner of writing is of the utmost importance: Mandeville in the *Fable* dismisses the theological background, just as La Rochefoucauld did, when Pascal, Nicole, Jacques Esprit do not hide either their theological premises or their apologetic intent. In the *Enquiry into the origin of moral virtue* he claims he will speak only of «Man in his State of Nature and Ignorance of the true Deity».[36] He writes again in Remark O that he is speaking of «all Men *in Nature*», excepting «Devout Christians».[37] One can read the exact same declaration in the «Preface to the reader» of La Rochefoucauld's *Maxims* (this preface, which appears in the second French edition, in 1666, was translated only in the English 1694 edition):

> the judicious person who made them, only considers Mankind in the present Deplorable State of Nature, as 'tis over-run with Ignorance and corrupted by Sin; and therefore whatever he says of that infinite number of defects that are to be found in their apparent Vertues, does not in the least concern those Happy but few favourites whom Heaven is pleased to preserve from them by a particular Grace.[38]

This statement is presented as the grounding of the work: «and' tis in a manner the Foundation of all these Reflections».[39] One can wonder if such words are to be taken seriously, or not: they could be read as a way of circling around the risks of religious censorship – La Rochefoucauld knew, as early as 1663, *i.e.* before the publication, that some readers found his maxims dangerous, and viewed them as

[36]Mandeville 1988: 50.

[37]Mandeville 1988a: 166.

[38]La Rochefoucauld 1694. The Preface to the Reader: f. A5 v°. The translation is on the whole faithful to the French text; there are nevertheless some significant additions: «*the judicious person*», «*over-run with Ignorance*, those *Happy but few favourites* whom Heaven *is pleased* to preserve from them by a particular Grace». See La Rochefoucauld 1992: 5.

[39]La Rochefoucauld 1694. The Preface to the Reader: f. A5 v°.

threatening to Christian morals[40] –. He was accused of attacking not only vices, but also virtues: the same reproach was made to Mandeville. Horatio lashes out at the *Fable* for laughing not only at vices, but also at virtues: «Vice is expos'd in it, and laugh'd at; but it ridicules War and martial Courage, as well as Honour and every thing else».[41]

Not only does Mandeville dismiss the theological background, but he also shares with La Rochefoucauld the refusal of a systematic thinking, and therefore of writing; he also shares with him a manner of writing which is deliberately ironical, paradoxical and leading to aporia – so that the reader is not sure of what the author means (the same could be said about Pierre Bayle).

The use of paradox is an organic part of both La Rochefoucauld's *Maxims* and the writings of the French free thinkers, such as Gassendi's *Exercitationes paradoxicae adversus Aristoteleos* or La Mothe Le Vayer. It should also be recalled that in the sixteenth century the genre of paradox was closely associated with scepticism, as the works of Thomas More show. La Mothe Le Vayer, who is both a libertine and a sceptic, explains his frequent use of paradox by presenting it as a remedy against stoic confidence as well as against the errors and follies of vulgar opinion.[42] Gassendi, who denounces both «the weakness and the arrogance of dogmatic philosophers»,[43] justifies the title of his *Exercitationes paradoxicæ* by saying they «contain paradoxes, or opinions beyond the vulgar's understanding».[44] The word «paradox» is to be understood in its etymological sense – what is against the *doxa*, the common opinion. Mandeville is quite aware of this meaning, as is made clear in *A Letter to Dion*: «Many Things are true, which the Vulgar think Paradoxes.»[45]

7.3 Philosophical Dialogue: Gassendi as a Model

The choice of philosophical dialogue, in the second part of the *Fable*, is not devoid of significance. No doubt it had a meaning in the English context: much used in England between 1650 and 1750, it played a dominant role in eighteenth-century moral philosophy, and it became, for this very reason, the object of critical

[40]See the letters sent to Madame de Sablé in 1663 about the *Maxims* (La Rochefoucauld 1992: 565, 576, 577).

[41]Mandeville 1988d. The Third Dialogue: 103.

[42]La Mothe Le Vayer 1988. Lettre de l'auteur: 14. One of these dialogues, «De la vie privée», was translated into English in 1678 (*The Great Prerogative of A Private Life: By way of Dialogue*, London).

[43]«Miserescere proinde, ac pudere cœpit me levitatis, et arrogantiæ Dogmaticorum Philosophorum [. . .].» (Gassendi 1959. Præfatio: 7; my translation).

[44]«Hinc et Exercitationes inscripsi Paradoxicas, quod Paradoxa contineant, seu opiniones præter vulgi captum.» (Gassendi 1959. Præfatio: 11; my translation).

[45]Mandeville 1954 [1732]: 49.

reflection.[46] It was also extensively used in theological and political polemics, as Mandeville himself noticed in the Preface of the second part:

> When partial Men have a mind to demolish an Adversary, and triumph over him with little Expence, it has long been a frequent Practice to attack him with Dialogues, in which the Champion, who is to lose the Battel, appears at the very beginning of the Engagement, to be the Victim, that is to be sacrifised [...].[47]

Mandeville does not fail to name the antic patterns: Plato, Cicero, Lucian – the same (classical) names La Mothe Le Vayer gave in the Introduction to his own *Dialogues*, to justify why he chose this particular form.[48] But he significantly adds the name of a modern philosopher:

> I would never have ventur'd upon it [this manner of writing], if I had not liked it in the famous *Gassendus*, who by the help of several Dialogues and a Friend, who is the chief Personage in them, has not only explain'd and illustrated his System, but likewise refuted his Adversaries: Him I have followed [...].[49]

These lines suggest a definition of the philosophical dialogue, intended as a tool in intellectual polemics (to explain one's system, and to refute one's adversaries). It is significant that Mandeville should claim a French writer, the neo-Epicurean Gassendi, as a formal model. It is nethertheless puzzling, because if Gassendi advocated a manner of writing which was «neither ciceronian nor scolastic», and which included the bitings of satire,[50] he never wrote dialogues. John Robertson suggests that Mandeville may have been thinking of the *Disquisitio metaphysica seu dubitationes et instantiæ adversus Renati Cartesii Metaphysicam et responsa* (1644)[51]: nevertheless this text is not a real dialogue; following what the title heralds, it exposes Gassendi's initial question (*dubitatio*) for each difficult point, Descartes's answer (*responsio*) and Gassendi's new objections to it (*instantia*). It has also been said that «Gassendus» was an error for Galileo[52]: if admittedly philosophical dialogue has been practised in the sixteenth and the seventeenth centuries by the most unorthodox Italian writers as Bruno, Galileo, Campanella and Vanini, it was also much used by the French free thinkers, especially La Mothe Le Vayer, and later Fontenelle.

One could wonder if Mandeville, who is well acquainted with the libertine tradition, is not thinking of seventeenth century French writers rather than of Italian

[46] See Prince 1996: 12–13.

[47] Mandeville 1988d: 8. On the vogue of pamphlets in dialogic form after the Revolution, see Hirzel 1963 [1895]: t. II, 399–400.

[48] La Mothe Le Vayer 1988: 12.

[49] Mandeville 1988d. Preface: 21.

[50] «neque Ciceronianus sum, neque Scholasticus omnino. [...] Quod stylus porro videri possit interdum paulo mordacior: materies sane id exigit. Hac enim præcipue in parte difficile est satyram non scribere.» (Gassendi 1959. Præfatio: 17).

[51] Robertson 2005: 271.

[52] See Simonazzi 2008: 71.

naturalists: Malcom Jack writes that «the essayist style of the *libertin érudit* may be recognized in all his work».[53] Moreover Gassendi is an important figure for him, as Grégoire[54] and Hundert pointed out; one ought to remember that in his mature age he embraced his conception of the animal. More generally, he is close to Gassendi's neo-Epicurianism. Hundert adds that he may have read Bernier, to whom he was probably led via La Mothe Le Vayer.[55] Still Gassendi never practised philosophical dialogue. One would rather expect La Mothe Le Vayer's name: he wrote dialogues which are characterized by their scepticism and their anti-dogmatic turn. As for Fontenelle's *Dialogues des Morts* (1683), they focus on predominance of passions in man, at the expense of reason, and they question the existence of a genuine virtue. Mandeville shares with the French libertines a pessimistic conception of human nature, the refusal of dogmatic thinking, and also a distrust towards common opinion. He ought to be taken earnestly, when he writes in the remark T: «I don't expect the Approbation of the Multitude. I write not to many, nor seek for any Well-wishers, but among the few that can think abstractly, and have their Minds elevated above the Vulgar.»[56] One is bound to think of La Mothe Le Vayer, and more generally of the French *libertins érudits* of the first half of the seventeenth century, who made exactly the same claim.

When he names Gassendi as a precedent which inclined him to choose the form of the dialogue, Mandeville is not only naming a philosopher of the modern age, but also a French author, which may be significant. Undoubtedly he has a continental training, and is indebted to the French moral tradition; but one might wonder if naming a French writer could be a way to avoid speaking of Shaftesbury. It is difficult to imagine that Mandeville did not read and meditate what Shaftesbury wrote about philosophical dialogue, in *The Moralists* (1709) and in the *Soliloquy*, first published in 1710. Shaftesbury notes the dislike of modern writers for dialogue:

> This brings to my mind a Reason I have often sought for; why we Moderns, who abound so much in *Treatrises* and *Essays*, are so sparing in the way of *Dialogue*; which heretofore was found the politest and best way of managing even the graver Subjects.[57]
>
> I have formerly wonder'd indeed why a *Manner*, which was familiarly us'd in Treatrises upon most Subjects, with so much Success among the Antients, shou'd be so insipid and of little esteem with us Moderns.[58]

[53]Jack 1987. Preface: n.p.

[54]Grégoire 1947: 164–165.

[55]Hundert 1996: 39, 45 *sq*, 95-96. The (deceptive) thesis of Rolf W. Puster concentrates on atomism, and does not take the form of Gassendi's writings into account (Puster 1991).

[56]Mandeville 1988a: 231.

[57]Shaftesbury 2001b [1709] (*The Moralists, A Philosophical Rhapsody*, part I, section 1): t. II, 90 [187].

[58]Shaftesbury 2001a [1710] (*Soliloquy*, part I, section 3): t. I, 99 [198-199]. La Mothe Le Vayer already noticed that this kind of writing was despised in his time (La Mothe Le Vayer 1988: 12).

The form of the dialogue constitutes for him a perfect device for «moral Painting» or «Philosophical portraiture».[59] It also represents a manner of thinking opposed to dogmatism: he sees in it «a certain way of Questioning and Doubting, which no-way sutes the Genius of our Age» because «Men love to take party instantly».[60]

One can agree with Irwin Primer, who noticed that Mandeville's debt to Shaftesbury is more extensive and various than is generally said.[61] One could wonder if Mandeville, when he chose the form of the dialogue to defend his ideas, and to refute Shaftesbury – explicitly attacked in the second part of the *Fable* –, did not decide to compete with him not only with respect to ideas, but also in the use of the genre: it would be as if Mandeville used the very form of *The Moralists* in order to convince that Shaftesbury's doctrine is false. It can be added that in several occasions Mandeville speaks of the *Fable* as a rhapsody, which is the very word used by Shaftesbury in the title of the *Moralists: A Philosophical Rhapsody*. Mandeville underlines that the *Fable* is written «in an open good-humour'd Manner, free, I dare say, from Pedantry and Sourness».[62] He insists on this: «The whole is a Rhapsody void of Order or Method, but no Part of it has any thing in it that is sour or pedantick».[63] The same refusal of pedantism is found in Shaftesbury: in his *Miscellaneous Reflections*, he writes about the dialogue entitled the *Moralists* that «[i]t conceals what is *scholastical*, under the appearance of a polite Work»,[64] and that the author «bear[s] himself out, against the appearance of *Pedantry*».[65]

* * *

Shaftesbury has led a vivid reflection on literary genres and styles[66]: if he is denounced as a philosophical adversary in *A Search into the Nature of Society* and in the second part of *The Fable of the Bees*, Mandevilles nevertheless shares with him a concern about the most efficient kind of writing for moral philosophy, as well as the taste for fable and dialogue. When he chose the dialogue for the second part of the *Fable*, he may have decided to counter Shaftesbury's ideas with one of his favourite forms.

---

[59] «We need not wonder, therefore, that the sort of moral Painting, by way of *Dialogue*, is so much out of fashion; and that we see no more of these philosophical Portraitures now-a-days. For where are *the Originals*?» (Shaftesbury 2001b [1709], *The Moralists, A Philosophical Rhapsody*, part I, section 1: t. II, 91 [188]).

[60] Shaftesbury 2001b [1709], *The Moralists, A Philosophical Rhapsody*, part I, section 1: t. II, 91 [189]).

[61] Primer 1975: 140–141.

[62] Mandeville 1954 [1732]: 25.

[63] Mandeville 1988c. *A Vindication of the book*: 405. The word «rhapsody» is used again in the Preface of the second part (Mandeville 1988d: 5), and in *A Letter to Dion* (Mandeville 1954 [1732]: 30). Mandeville also applies it to *An Essay on Charity, and Charity-Schools* (Mandeville 1988b: 322).

[64] Shaftesbury 2001c [1711] (*Miscellaneous Reflections*): t. III, 119 [285].

[65] Shaftesbury 2001c [1711] (*Miscellaneous Reflections*): t. III, 120 [287].

[66] See Wolff 1960; Jaffro 1998.

Mandeville refuses methodical and systematic writing, because he refuses systematic thinking, *i.e.* dogmatism. He shares this refusal with Shaftesbury, but also with French free thinkers and some French moralists: hence his dilection for ironical and paradoxical writing, in the manner of both La Mothe Le Vayer and La Rochefoucauld.

If he is well aware of the English issues, the importance of the French background should not be forgotten: from the early works onwards, the adaptation of Scarron and the translations of La Fontaine's *Fables*, to the ostentatious patronage of Gassendi in the Preface of the second part of the *Fable*, Mandeville's thinking and writing is definitely marked by his acquaintance with continental thought.

References

Primary Sources

Casaubon, Isaac. 1605. «In Persium Prolegomena». In *Auli Persi Flacci Satirarum liber. Isaacus Casaubonus recensuit, & Commentario libro illustravit.* Parisiis: apud Ambrosium & Hieronymum Drouart.

Dryden, John. 1693. *Discourse concerning the original and progress of Satire.* In 1974 *The works of John Dryden*, H.T. Swedenberg dir., vol. IV. Berkeley/Los Angeles/London: University of California Press.

Erasmus, Desiderius. 1910. *Opus epistolarum Des. Erasmi Roterodami* t. II, ed. P.S. Allen. Oxonii: in typographeo Clarendoniano.

Erasmus, Desiderius. 1668. *Moriæ Encomium; or, The Praise of Folly, written Originally in Latine, by Des. Erasmus of Rotterdam; and Translated into English, by John Wilson.* London: printed for William Leak.

Gassendi, Pierre. 1959. *Dissertations en forme de paradoxes contre les aristotéliciens. Exercitationes Paradoxicæ Adversus Aristoteleos*, livres I et II, ed. Bernard Rochot. Paris: Vrin.

La Mothe Le Vayer (François de). 1988. *Dialogues faits à l'imitation des Anciens.* Paris: Fayard.

La Rochefoucauld (François de). 1694. *Moral maxims and reflections.* London: Gillyflower in Westminster Hall, S. Sare at Grays Inn-Gate and J. Everingham in Ludgate street.

La Rochefoucauld (François de). 1992. *Maximes*, ed. Jacques Truchet. Paris: Bordas.

Mandeville, Bernard. 1954 [1732]. *A letter to Dion*, ed. Bonamy Dobrée. University Press of Liverpool.

Mandeville, Bernard. 1988a [1924]. *The fable of the bees*, vol. 1, ed. F.B. Kaye. Indianapolis: Liberty Fund.

Mandeville, Bernard. 1988b [1924]. *An essay on charity and charity-schools.* In *The fable of the bees*, ed. F.B. Kaye, t. I, 253–322. Indianapolis: Liberty Fund.

Mandeville, Bernard. 1988c [1924]. *A vindication of the book.* In *The fable of the bees*, ed. F.B. Kaye, t. I, 381–412. Indianapolis: Liberty Fund.

Mandeville, Bernard. 1988d [1924]. *The fable of the bees*, vol. 2, ed. F.B. Kaye. Indianapolis: Liberty Fund.

Martial, Martin (ed.). 2007. *Satyre ménippée.* Paris: Champion.

Shaftesbury, Anthony Ashley Cooper. 2001 [1709]. *The moralists, a philosophical rhapsody.* In *Characteristicks of men, manners, opinions, times* t. II, ed. Douglas Den Uyl. Indianapolis: Liberty Fund.

Shaftesbury, Anthony Ashley Cooper. 2001 [1710]. *Soliloquy.* In *Characteristicks of men, manners, opinions, times* t. I, ed. Douglas Den Uyl. Indianapolis: Liberty Fund.

Shaftesbury, Anthony Ashley Cooper. 2001 [1711]. *Miscellaneous reflections*. In *Characteristicks of men, manners, opinions, times* t. III, ed. Douglas Den Uyl. Indianapolis: Liberty Fund.

Secondary Sources

Debailly, Pascal. 2012. *La Muse indignée. Tome I. La satire en France au XVIe siècle*. Paris: Classiques Garnier.

Dekker, Rudolph. 1992. 'Private vices, public virtues' revisited: The Dutch background of Bernard Mandeville. *History of European Ideas* 14(4): 481–498.

Elkin, P.K. 1973. *The Augustan defence of satire*. Oxford: Clarendon Press.

Frye, Northrop. 1969 [1957]. The mythos of winter: Irony and satire. In *Anatomy of criticism. Four essays*, 223–239. New York: Atheneum.

Gorak, Irene E. 1990. The satirist as producer: Mandeville's *The Fable of the Bees Part Two*. *Genre* XXIII: 1–14.

Grégoire, F. 1947. *Bernard de Mandeville et la 'Fable des abeilles'*. Nancy: Imprimerie Georges Thomas.

Hind, George. 1968. Mandeville's *Fable of the bees* as Menippean satire. *Genre* I: 307–315.

Hirzel, Rudolph. 1963 [1895]. *Der Dialog. Ein literarhistorischer Versuch*, Hildesheim, Olms, 2 vol.

Horne, Thomas A. 1978. Mandeville and the French moral tradition. In *The social thought of Bernard Mandeville. Virtue and commerce in early eighteenth-century England*, 19–31. London: MacMillan.

Hundert, E.J. 1996 [1994]. *The enlightenment's* Fable. *Bernard Mandeville and the discovery of society*. Cambridge: Cambridge University Press.

Jack, Ian. 1964 [1952]. *Augustan satire. Intention and idiom in English poetry 1660–1750*. Oxford: Clarendon Press.

Jack, Malcolm. 1987. *The social and political thought of Bernard Mandeville*. New York/London: Garland.

Jaffro, Laurent. 1998. *Éthique de la communication et art d'écrire. Shaftesbury et les Lumières anglaises*. Paris: PUF.

James, E.D. 1975. Faith, sincerity and morality: Mandeville and Bayle. In *Mandeville studies. New explorations in the art and thought of Dr. Bernard Mandeville (1670–1733)*, ed. Irwin Primer, 43–65. The Hague: Martinus Nijhoff.

Miller, Henry Knight. 1956. The paradoxical encomium with special reference to its vogue in England, 1600–1800. *Modern Philology* LIII(3): 145–178.

Noel, Thomas. 1975. *Theories of the fable in the eighteenth century*. New York/London: Columbia University Press.

Pinkus, Philip. 1975. Mandeville's paradox. In *Mandeville studies. New explorations in the art and thought of Dr. Bernard Mandeville (1670–1733)*, ed. Irwin Primer, 193–211. The Hague: Martinus Nijhoff.

Primer, Irwin. 1975. Mandeville and Shaftesbury: Some facts and problems. In *Mandeville studies. New explorations in the art and thought of Dr. Bernard Mandeville (1670–1733)*, ed. Irwin Primer, 126–141. The Hague: Martinus Nijhoff.

Primer, Irwin. 1993. Erasmus and Bernard Mandeville: A reconsideration. *Philological Quarterly* 72: 313–335.

Prince, Michael. 1996. *Philosophical dialogue in the British enlightenment. Theology, aesthetics, and the novel*. Cambridge: Cambridge University Press.

Puster, Rolf W. 1991. *Britische Gassendi-Rezeption am Beispiel John Lockes*. Stuttgart: Frommann-Holzboog.

Robertson, John. 2005. *The case for the enlightenment. Scotland and Naples 1680–1760*. Cambridge: Cambridge University Press.

Simonazzi, Mauro. 2008. *Le favole della filosofia. Saggio su Bernard Mandeville*. Milano: FrancoAngeli.

Tolonen, Mikko. 2013. *Mandeville and Hume: Anatomists of civil society*. Oxford: Voltaire Foundation.

Viner, Jacob. 1991 [1953]. Introduction to Bernard Mandeville, *A letter to Dion* (1732). In *Essays on the intellectual history of economics*, ed. Douglas A. Irwin, 176–188. Princeton: Princeton University Press.

Wheeler, Angela J. 1992. *English verse satire from Donne to Dryden. Imitation of classical models*. Heidelberg: Carl Winter/Universitätsverlag.

Wolff, Erwin. 1960. *Shaftesbury und seine Bedeutung für die Englische Literatur des 18. Jahrhunderts. Der Moralist und die literarische Form*. Tübingen: Max Niemeyer.

Chapter 8
Mandeville as a Sceptical and Medical Philosopher

Rui Bertrand Romão

Abstract Bernard Mandeville, in his double condition as physician and as moral and political philosopher of the early eighteenth century in Great Britain was clearly linked by many of his contemporaries to the sceptics, and not only to his contemporary sceptics but also to Ancient Pyrrhonians. However, many of his most relevant theories do not explicitly show this link. We then must ask what kind of scepticism was his if any? In order to characterize it, this paper suggests there is a need for a reassessment of some of the most celebrated Mandevillean theories and themes (self-liking and self-love, criticism of self-denial, rigorism, the relation of individuals with society, the role of luxury, economic conceptions, and so on) focusing his intertwined inheritance of a medical outlook stressing empiricism and of a sceptical tradition of thought applied to moral and political issues. Thus our paper sustains that Mandeville's own peculiar combination of individualism, empiricism, conservatism and "rigorism" mixes up with a sort of mitigated scepticism.

Keywords Scepticism • Pyrrhonian tradition • Self-interest • Unintended consequences • Paradox • Rigorism

The understanding of such a work as Mandeville's and of the originalities of his thought has greatly benefitted from being adequately placed in the eclectic context of the intersection of several Seventeenth Century currents of thought. One of the areas light has been shed on recently corresponds to the relation between Mandeville's medical thought and his political philosophy.[1]

Yet perhaps one dimension of this relation that still remains hitherto relatively undervalued and that demands more attention than the one that seems to have

[1]The most complete treatment of this subject is nowadays Phillip Hilton's (cf. Hilton 2010).

R.B. Romão (✉)
Institute of Philosophy, University of Porto, Porto, Portugal

E. Balsemão Pires, J. Braga (eds.), *Bernard de Mandeville's Tropology of Paradoxes*, Studies in History and Philosophy of Science 40, DOI 10.1007/978-3-319-19381-6_8

been bestowed to it from a philosophically updated point of view consists of its connection to the varied and manifold important tradition of scepticism in the Early Modern Thought.

As to the medical branch of the tradition, Francisco Sanches's naturalistic sceptical empiricism marks it, establishing somehow a bridge between ancient and early modern trends of medical thought. Mandeville in a rather idiosyncratic way outstandingly illustrates in the early eighteenth century the relation between scepticism and medical empiricism established at least since Sextus Empiricus.

In the *editio princeps* of A *Treatise of the Hypochondriack and Hysterick Passions [...]*[2], Mandeville, showing solid knowledge of the history of Ancient medicine, calls the character that represents his own views Phylopirio. This naming clearly announces adherence to some sort of empiricism related to the one followed by the Empirical school: "In these Dialogues I have done the same as Seneca did in his *Octavia*, and brought my self upon the Stage; with this difference, That he kept his own Name, and I changed mine for that of *Philopirio*, a Lover of Experience, which I shall always profess to be" (Mandeville 1715: xi). Such an endorsement of the "absolute Necessity of Experience" (Mandeville 1730: vii) goes along with Mandeville's anti-rationalistic stance exhibited in all editions of this dialogical treatise. In an important passage of the first edition, after declaring that he does not "make use of any [hypothesis]" (Mandeville 1715: 47), he shows his agreement with the Sect of the "*Empyricks*" (Mandeville 1715: 48) as to conferring the paramount importance in clinical practice to observation and experience, while disparaging the therapeutical effectiveness of the use of deductive methods in Medicine: "They were of Opinion, that the Art of Physick consisted in down-right Observation, and a world of Experience; and that all manner of reasoning about the Causes of Distempers, and being Witty in deducing the Symptoms from 'em, were very insignificant in curing people that were Sick: /So far I am an *Empyrick*" (Mandeville 1715: 48–49). However Mandeville disagrees with what their "professed enemies" such as Galen say of them, concerning their alleged refusal to repeat experiments and to make predictions based on experience (cf. Mandeville 1715: 49–50).[3]

Empiricism and anti-rationalism correspond to two relevant features essential to the understanding of Mandeville's attitude as also shown in his non-medical writings, including *The Fable of the Bees* in its several metamorphoses.

---

[2]The complete title of the first edition of this treatise in three dialogues is: *A Treatise of the Hypochondriack and Hysterick Passions, Vulgarly call'd the Hypo in Men and the Vapours in Women; In which the Symptoms, Causes and Cure of those Diseases are set forth after a Method intirely new. The whole interpreted, with Instructive Discourses on the Real Art of Physick itself, And Entertaining Remarks on the Modern Practice of Physicians and Apothecaries: Very useful to all, that the Misfortune to stand in need of either*, reprinted in the 1715 issue (cf. Mandeville 1715: i). The 1715 reprint of this edition is called in the frontispiece second edition. However, the true new edition, significant is the 1730 one, of which the revised and enlarged the complete title becomes abridged: *A Treatise of the Hypochondriack and Hysterick Diseases. In Three Dialogues*, London, J. Tonson, 1730.

[3]See also, e.g., Mandeville 1715: 67–68, 105–106, 140; Mandeville 1730: 55–60, 126–130, 227.

As for Mandeville's possible ties with Sanches, we will not claim here that there is any clear evidence linking the two philosophers nor that the author of *Quod Nihil Scitur* and of the posthumous *Opera Medica* has somehow directly influenced the writer of *A Treatise of the Hypochondriack and Hysterick Passions*. I just want to hint here they seem to share some traits, among which we should highlight an enquiring approach on nature and on man that integrates many features of empiricism, the jointly understanding of medicine as an art and as a science, as partly overlapping with the study of nature, an emphasis put on the paramount importance of the combined role of judgment and experience on scientific and medical topics and an extensive criticism of speculative theorizing applied to medicine.

If this linking of Mandeville to his fellow physician and philosopher has been practically ignored in spite of a brief mention of Sanches' name by Kaye in a list of examples of the skeptical method of dealing with the "relation of private judgment and traditional religion" (Kaye 1924: 42), one cannot say the same happened with his relation with Montaigne as well as with the philosophers he inspired including the libertine ones and the Jansenistic current of thought that mingled with that sceptical tradition. It is famous the linking Kant in the *Critique of Practical Reason* made between the two philosophers as to their conceptions of the subjective external "Practical Material Principles of Determination of the Foundation of Morality" (cf. Kant 1898: 129). In his groundbreaking edition of the *Fable of the Bees*, both in the introduction and in the notes to the text, Kaye had not only identified some key passages of Montaigne's *Essays* quoted or paraphrased by Mandeville but also clearly shown much common ground shared by the two philosophers. The kinship thus highlighted led Kaye to even envisage Montaigne as a forerunner of Mandeville concerning essential features of the Dutch author's thought such as his anti-rationalistic account of the consequences of man's subjection to his passions and self-imposture: "Mandeville's anti-rationalism is developed with such/literary inventiveness that it gives the effect of great originality. It was however, merely, the most brilliant handling of a conception which, from the time of Montaigne, had been common in French thought, and which, besides, had been profoundly stated by Spinoza" (Kaye 1924: lxxix–lxxx). Notwithstanding this evidence, including the circumstance that Mandeville twice in *The Fable of the Bees* explicitly mentioned the author of the *Essays* as a major influence – the one claiming in the "Preface" the French essayist as his model of critic of human faults and frailties ("'T was said of *Montagne*, that he was pretty well vers'd in the defects of Man-kind, but unacquainted with the Excellencies of human Nature: If I fare no worse, I shall think my self well used" – Mandeville 1924a: 5), the other one dealing with self-liking and a critical view of anthropocentrism (cf. Mandeville 1924b: 131) – his philosophical affinities with Montaigne have been relatively understated. It would be unfair, of course, to here omit mentioning many scholars, like for instance Donati, Simonazzi, Primer or Hundert, just to mention a few names, who did not fail to acknowledge Montaigne as playing an important role as one of the foremost thinkers in the early modern tradition presenting conceptions strongly influential on Mandeville's philosophy (cf., e.g., Donati 2011: 69; Simonazzi 2008: 201, 211; Primer 2006: 43; Hundert 2005: 32, 39; see also Romão 2015: 129–148). However, these learned references to Montaigne do not necessarily involve a really in-depth comparative analysis of Mandeville and Montaigne's philosophies.

Though this study is not the appropriate place to pursue that aim, we have here to stress that Mandeville's explicit reference to Montaigne as an example of dealing with misery of man instead of insisting in exploring its dignity means more than a mere acknowledgment of an influence or than the recognizance of affinities between their words: it shows that his consideration of the "vilest and most hateful qualities" of men sits at the at the core of the project of scrutinizing Human Nature, "abstract from Art and Education", revealing the central paradox expressed in *The Fable of the Bees*.

At the beginning of this "Preface" of *The Fable of the Bees*, a relatively conventional prelude to a *sui generis* work, Mandeville, using the not uncommon metaphor of the "Political Bodies of Civil Societies", immediately draws the attention of the readers to the motivation of man in two related ways. The first one consists of stressing the importance of looking for small apparently insignificant trifles. The second one corresponds to the brutal gesture of a surgeon cutting through skin and flesh in order to find the overlooked elements that are supposed to act as springs of the motion of the human machine.

The examination of natural man or in Mandeville's exact terms the examination "into the Nature of Man, abstract from Art and Education" leads him to the consideration of sociability as somehow and paradoxically derived from repulsive individual attitudes and traits: "[. . .] may observe that what renders him a sociable animal, consists not in his desire of Company, Goodnature, Pity, Affability, and other Graces of a fair Outside; but that his vilest and most hateful Qualities are the most necessary Accomplishments to fit him for the largest, and according to the World, the happiest and most flourishing Societies" (Mandeville 1924a: 4). This apparently means that, at least concerning worldly criteria, sociability turns out to reside in what could be called man's bad nature, for the hateful qualities alluded to are shown to be as effectual in fitting a thriving society as the loveable and kind human attributes should have been and fail to be. We should not of course despise the circumstance stressed by Mandeville that his analysis and satire concerns a particular kind of society: an especially large and prosperous one corresponding to the model of England. In more than one passage he explores the contrast between economic prosperity and the generalized adoption of virtue within a community. However, sometimes he also seems to tend to universalize the paradox he summed up so brilliantly in the famous under title he contrived, extending it all societies.

This duality reflects the coexistence of two of the most controversial sides of his philosophy: an unabated rigorist attitude consistently sustained throughout his work, in every field it touched upon, and the relentless pursuit of the full consequences of the assumption of egoism as the motivation grounding human behaviour. The conciliation between these two apparently incompatible claims always formed a problem that his adversaries as well as his followers tried to somehow solve. The easiest way seems to dissociate the two opposed terms, opting for just one of them and rejecting the other, on moral, religious, political or philosophical assumptions.

The Mandevillean unremitted insistence on the intertwining of those two *leitmotiven* of his readings of human nature tends to be *sui generis* for, unlike most of his inimical contemporaries interpreted, it does not really shun a religious attitude: instead, it raises human conduct when religiously led to the extremely high standards of few extraordinary human beings, such as some mystics and ascetics, if not to those of people like catholic saints. Nevertheless, one must emphasize that this conception of virtue as truly and thoroughly self-denying and virtuous in the full sense of the word does not mean rejecting it completely as a model or ideal. Mandeville shows us that self-denial is an ideal falsely adopted and in truth extremely difficult and unrewarding to pursue, but he does not really present it as erroneous when taken by itself. The problem lies in the lack of harmony between theory and practice, not in the philosophical adherence to high principles. That is why he so severely attacks stoical conception of fortitude.

Though his writings are full of ironical remarks and observations about the behaviour of pretended pious men, the levels of irony possibly used when he speaks of piety do not necessarily enclose dissimulation or collides. The case is that his rigorism intertwines closely with his realistic point of view and with his apology of self-examination. The fact that he sustains realistic positions does not entail that he glorifies power or that he extols the pure exercise of authority. Mandeville constant decrying of hypocritical abuse of power complemented somehow by his notion of "dextrous management", also linked to self-examination, serves to show that his realistic attitude does not mean any reverence for authoritarianism: "it is the Work of Ages to find out the true Use of the Passions, and to raise a Politician, that can make every Frailty of the Members add Strength to the whole Body, and by dextrous Management turn *private Vices into publick Benefits*." (Mandeville 1924b: 319).

Thus, one must not deem his attitude to piety as insincere for then one would paradoxically envisage what he disparages as true and consider the defence of the idea itself of a truthful and unstained virtue as not only unattainable but also somehow misleading. When Mandeville depicts the trading system he so well describes he shows both its grandeurs and its miseries.

The circumstance that the eighteenth century reception of Mandeville focused mainly on The *Fable of the Bees* and secondarily on *The Free Thoughts* should not drive us to neglect that his chief medical work, *A Treatise of the Hypochondriack and Hysterick Passions*, met considerable popular success, if we take into account the fact that it had two editions, any of them having been twice printed.

On his account, man cannot become sociable but through unsociability. Human will and human reason are deemed incapable of achieving what they want, mere chance revealing itself far more potent than both of them.

But his pessimism about human nature bleak though it is, often strikes an optimistic note. Evil for him may naturally beget good. This may happen with unbecoming individuals who are the unwilling and unforeseen benefactors of other individuals. But its most important instance he likes to stress is the good and above all necessary effect of faulty and even criminal actions and conducts in the accruing of richness and splendour to the nation where they were performed.

Mandevillian philosophy seems to contain ambivalent elements, as many of his readers and critics have stressed since the eighteenth century. This ambivalence can be related to a general use of *antilogies* and to a clear fondness for paradoxes. However, ambivalence does not here mean insincerity (cf. Munro 1975: 178–267). It rather corresponds to the expression of a sort of dialectical thinking that reflects the tensions observed in things. It also derives from a *Weltanschauung* similar to that of Montaigne, affirming the inseparability of good and evil: "We must learn to endure what we cannot avoid. Our life is composed, like the harmony of the world, of contrary things, also of different tones, sweet and harsh, sharp and flat, soft and loud. If a musician liked only one kind, what would he have to say? He must know how to use them together and blend them. And so must we do with good and evil, which are consubstantial with our life. Our existence is impossible without its mixture, and one element is no less necessary than the other." (Montaigne 2003: 1018 Frame).

Mandeville, adopting in the wake of early Modern sceptical philosophers but also following other political philosophers he has been related to a viewpoint rather descriptive than normative and full of extremely subtle psychological analyses, develops a sophisticated theory of passions grounded on the extreme complexity and contradictory character of human nature: "What strange contradictions Man is made of!" (Mandeville 1924a: 66).[4]

Joining together self-examination with the study of passions as springs of action, Mandeville seems to conceive the former as absolutely necessary in order, as we have said, to reduce to the utmost unforeseen consequences of human behaviour. However, the valorisation of the role of unintended effects of individual acting in an organised community has as its counterpart the rejection of an atomistic conception of society combined with what we may call a deprecation of self. This is patent in, for instance, Mandeville's description in *Remark (N)* of that "strong Habit of Hypocrisy, by the Help of which, we have learned from our cradle to hide even from ourselves the vast Extent of Self-Love, and all its different Branches" (Mandeville 1924a: 135).

Mandeville's vision of human nature, presenting us the most serious indictment of hypocrisy and the most rigorous analysis of the mechanisms of the concealment of passions ever produced in philosophical writings, is not reducible to the observation of the asymmetry of vice and virtue leading to a uniform conception of a benign society composed of malign individuals. The core of his complex philosophy exploring paradoxes and reflecting the dark hues of a pitiless vision of mankind, approached from an empirical point of view, seems to reside in an uncommon ability to find ways of seeing as conciliated what at first sight would seem irreconcilable: self-interest and public good, corruption and salvation, the singled-mindedness of man's behaviour and the contradictory character of human nature, the affirmation of both the strength and the weakness of reason, the defence of the misery and of the dignity of man, the acceptance of progress and the need for conservation. No coin has but one side. In real life, unlike what rulers may pretend, qualities are

[4]Cf. Mandeville 1924a: 168; Mandeville 1924b: 100; Mandeville 1987:90.

mixed up. To everything corresponds a counterpart. Vices joined together do not necessarily entail virtuous consequences. On the other hand, the author can with clear-sightedness calculate that an assembly of virtuous elements most likely results in a stagnant community.

Mandeville has a voice of his own, whose beautiful convoluted long phrases, sophisticated arguments, and extremely refined and subtle psychological and social analyses go deep into each of his readers and still fascinate us. His constant research about "what it really is to believe", his attempt to understand "public interest" independently from what appears mingled with, his unique mixture of realism, empiricism, conservatism, moderate scepticism, sheer common sense and "rigorism" join with his talent for understanding the labyrinths of the interplay between human beings and with his genius to imaginatively examine the intricacies of social mechanisms.

References

Primary Sources

de Montaigne, Michel. 2003. *Complete works*, translated by Donald M. Frame. New York/London/Toronto: Everyman's Library.

Kant, I. 1898. *Critique of practical reason*, Translated by T.K. Abbott London/New York Bombay.

Mandeville, Bernard. 1715. *A treatise of the hypochondriack and hysterick passions, vulgarly call'd the hypo in men and the vapours in women* [. . .], 2nd ed. London: David Leach (digital scan by Gale Ecco Print Editions).

Mandeville, Bernard. 1924a. *The fable of the bees: Or private vices, publick benefits*, ed. F.B. Kaye, vol. I. Oxford: Clarendon Press (facsimile by Liberty Fund, Indiana).

Mandeville, Bernard. 1924b. *The fable of the bees: Or private vices, publick benefits*, ed. F.B. Kaye, vol. II. Oxford: Clarendon Press (facsimile by Liberty Fund, Indiana).

Mandeville, Bernard. 1981. *A treatise of the hypochondriack and hysterick diseases (1730)*, vol. II of the *Collected Works* of Bernard Mandeville, facsimile editions prepared by Bernhard Fabian and Irwin Primer. Hildesheim/New York: Georg Olms.

Mandeville, Bernard. 1987. *Free thoughts on religion, the church, and national happiness (1720)*, vol. V of the *Collected Works* of Bernard Mandeville, facsimile editions prepared by Bernhard Fabian and Irwin Primer. Hildesheim/Zurich/New York: Georg Olms.

Secondary Sources

Donati, Roberto. 2011. *Le Ragioni di un pessimista: Mandeville nella cultura dei Lumi*. Firenze: Edizioni ETS.

Hilton, Phillip. 2010. *Bitter honey: Recuperating the medical and scientific context of Bernard Mandeville*. Bern: Peter Lang.

Hundert, E.J. 2005. *The enlightenment's fable. Bernard Mandeville and the discovery of society*. Cambridge: Cambridge University Press [1994].

Kaye, F.B. 1924. "Introduction" in Mandeville 1924: xvii–cxlvi.

Munro, Hector. 1975. *The ambivalence of Bernard Mandeville*. Oxford: Oxford University Press.

Primer, Irwin. 2006. *Bernard Mandeville's "a modest defence of publick stews". Prostitution and its discontents in early Georgians England.* New York/London: Palgrave Macmillan.
Romão, Rui Bertrand. 2015. Bernard Mandeville's skeptical political philosophy. In *John Christian Laursen and Gianni Paganini. Skepticism and political thought in the seventeenth and eighteenth centuries.* Toronto: University of Toronto Press.
Simonazzi, Mauro. 2008. *Le favole della Filosofia. Saggio su Bernard Mandeville.* Milano: FrancoAngeli.

Chapter 9
Is Adam Smith Heir of Bernard Mandeville?

Işıl Çeşmeli

Abstract Sociability, sympathy, moral sense, benevolence and self-love constituted the main themes of the eighteenth century moral discourse. Scottish philosophers of the era principally focused on the fact that moral sentiments could not be offspring of self-interest. Hutcheson's theory of moral sense, Hume's and Smith's theories of sympathetic identification lay at the heart of their moral theory. Conversely, Bernard Mandeville annihilated social and moral principles by asserting self-love as a primary motivation of man. His notoriety among eighteenth century moralists was due to his famous motto "private vices cause public benefits", his undaunted confession of self-love as a basic motive of human nature and its beneficial effects for a prosperous society. Many philosophers and moral theorists concentrated on his selfish hypothesis by offering counter arguments and justifications about origin of human nature, character of moral virtue and moral judgments. In his earlier work, *The Theory of Moral Sentiments (TMS)*, Adam Smith defends the mechanism of sympathy as a ground of moral sentiments and draws attention to altruistic character of human nature. Although Smith designates Mandeville's thesis as "licentious system" in *TMS* by propounding sympathy at the basis of morality, he uses self-interest principle in *The Wealth of Nations* (*WN*) as a basic motive of human beings. This article aims to analyze whether Smith's moral theory is successful to overcome Mandeville's selfish principle or he successfully reconciles sympathy and self-love.

Keywords Self-love • Sympathy • Virtue • Vice • Impartial spectator • Self-command

9.1 Introduction

Sociability, sympathy, benevolence, moral sense and self-love constituted the main themes of the eighteenth century moral discourse. Scottish philosophers of the era principally focused on the fact that moral sentiments could not be offspring

I. Çeşmeli (✉)
Department of Philosophy, Middle East Technical University, Ankara, Turkey
e-mail: cisil@metu.edu.tr

E. Balsemão Pires, J. Braga (eds.), *Bernard de Mandeville's Tropology of Paradoxes*, Studies in History and Philosophy of Science 40, DOI 10.1007/978-3-319-19381-6_9

of self-interest. Hutcheson's theory of moral sense, Hume's and Smith's theories of sympathetic identification lay at the heart of their moral theories. Conversely, Bernard Mandeville annihilated social and moral values and principles by asserting self-love as a primary motivation of man. He investigated selfish moral conduct of human species in *The Fable of the Bees*. In his famous poem "Grumbling Hive" and *The Fable* Mandeville's narration of selfish basis of human nature and origin of moral virtue and his undaunted confession regarding beneficial effects of vices for a flourishing society affected later moral thinking. Although his "doggerel poem" (Mandeville 1988a: 5) at first attracted little attention, when final form of *The Fable* with inclusion of remarks in prose, "An Essay on Charity and Charity-Schools and "A Search into the Nature of Society" appeared Mandeville became the target of harsh criticisms. Many philosophers and moral theorists concentrated on his selfish hypothesis and his theory regarding artificial origin of moral virtue and vice by offering counter arguments and justifications about origin of human nature, character of moral virtue and moral judgments. Hutcheson, as an eighteenth century moral philosopher, rejected the theory that all human actions were solely motivated by self-interest. In all his writings on moral philosophy he employs moral sense or feeling which is incommensurable with psychological egoism. According to him, the only way to retain the doctrine of virtue is to give up egoistic theory of motivation. In opposition to Mandeville, he defends sociability as a part of human nature and he proposes benevolence which has a strong influence on human beings' motives and disinterested judgments. Adam Smith, disciple of Hutcheson, propounded the idea that sympathy was at the basis of morality. In *The Theory of Moral Sentiments (TMS)* he argues against the definition of vanity, origin of moral virtue and vice portrayed by Mandeville in *The Fable.*

9.2 Human Nature and Origin of Morality in the Fable of the Bees

In *The Fable*, Mandeville depicts dark side of the human nature and posits mankind driven solely by selfish passions and motives. He underlines the significance of selfishness in the second volume of it: "Man centers everything in himself and neither loves nor hates, but for his own Sake. Every individual is a little World by itself, and all Creatures, as far as their Understanding and Abilities will let them, endeavour to make that Self happy: This in all of them is the continual Labour, and seems to be the whole Design of Life" (Mandeville 1988b: 178). According to him, either educated or uneducated, man acts in accordance with his selfish instincts. While helping others who are in need seems to be an altruistic behavior, in fact we help others in order to relieve our compassion and pity. Similarly, acts of an educated man also spring from selfishness because his good acts are either from love of praise or fear of blame. Therefore, it seems that all apparent acts are done either in order to satisfy natural selfish impulses, or of the selfish passion of pride. In the second volume of *The Fable*, Mandeville distinguishes self-love from self-liking. Although these two are innate in human beings, self-love is an instinct for self-preservation and self-liking arises from our desire to be esteemed.

After revealing the selfish nature of man, Mandeville points out the possibility to govern selfish instincts and inclinations in "An Enquiry into the Origin of Moral Virtue". He states that although human beings are selfish it is possible to make them tractable by a dominant power; therefore they may be easily subjugated in this way. He asserts that it was considered as the task of lawgivers to convince mankind to believe that "it was more beneficial for everybody to conquer than indulge his Appetites and much better to mind the public than what seemed his private interest" (Mandeville 1988a: 42). As a result of circumspection, careful and intimate examination of man's nature and also with the help of eulogies, lawgivers and moralists succeeded to enchant human beings. This mechanism worked as follows: "Making use of this bewitching Engine [Flattery], they extoll'd the Excellency of our Nature above other Animals, and setting forth with unbounded Praises the Wonders of our Sagacity and Vastness of Understanding, bestow'd a thousand Encomiums on the Rationality of our Souls, by the Help of which we are capable of performing the most noble "atchievements" (Mandeville 1988a: 43). The lawgivers having gift of persuasion also offered reward for those who preferred the good of others rather than their own. In other words, self-denial or restraining natural inclinations was successfully rewarded by them. Lawgivers organized society through the instruction of honor as the highest good and shame as the worst of all evils. Hence, those who acted for public good were rewarded by honor and those who indulge and gratify appetites and desires rather than the good of others were punished by shame. Therefore moral virtues, imposed upon by "skillful politicians" were "political offspring which flattery begot upon pride" (Mandeville 1988a: 51). Yet man is "extraordinary selfish and headstrong as well as cunning animal" (1988a: 42) says Mandeville and not separable from his essence, entire conquest of passions and appetites is hard to achieve. Mandeville also affirms Bayle's thesis and says "[what] Mr. *Bayle* has endeavour'd to prove at large in his Reflexions on Comets: That Man is so unaccountable a Creature as to act most commonly against his Principle" (Mandeville 1988a: 167). Headstrong and cunning man acts against his principle and learns how to conceal and hide his passions subtly. If one is able to conceal his lust, pride and selfishness, he accomplishes to keep himself a distance from the feeling of shame. Therefore, under the word of modesty, there lies a habit of hypocrisy and perfect disguise of passions.

9.3 Adam Smith's Theory of Sympathy

In *The Theory of Moral Sentiments*, Adam Smith touches up the dark picture drawn by Mandeville in *The Fable* by manifesting the mankind's natural capacity of fellow-feeling for happiness and misery of the others. The first chapter of this treatise entitled 'Of Sympathy' indicates the primary concern of his moral theory.

He claims that sympathy is the guiding and principal sentiment and constitutes groundwork for our moral judgments. The very opening passage of *TMS* clearly gives the main idea of which Smith will defend throughout the book. He states that man is not solely motivated by self-love but fortune and happiness of others also motivate him (Smith 1982: 9). He defines sympathy as our capacity for fellow feeling with others. But, Smith carefully distinguishes sympathy from the feelings of pity and compassion. "Pity and compassion are words appropriated to signify our fellow-feeling with the sorrow of others. Sympathy, though its meaning was, perhaps, originally the same, may now, however, without much impropriety, be made use of to denote our fellow-feeling with any passion whatever" (Smith 1982: 10).

Earlier, La Rochefoucauld interprets pity and compassion as a manifestation of self-love. In *The Maxims*, he says that "[p]ity is often a feeling our own ills, prompted by the ills of other people. It is a clever way of anticipating the misfortunes that could possibly befall us: we help other people so that they will be obliged to help us when comparable circumstances arise; and the services we render them are, strictly speaking, good deeds that we do for ourselves in advance" (La Rochefoucauld 2007: 75). Mandeville, in *The Fable*, appropriates La Rochefoucauld's interpretation in the sense that he defines pity as "the most gentle and the least mischievous of all our passions" (Mandeville 1988a: 56). Yet, natural act performed by compassion or pity, which consists in sympathy for calamities and tragedies of others is not a sign of our fellow-feeling but "frailty of our Nature" and "the weakest minds have generally the greatest Share of it, for which Reason none are more Compassionate than Women and Children" (Ibid). Despite pity's resemblance to virtue, without "considerable mixture of it" it may lead evil (Ibid). Mandeville says that "[I]t has help'd to destroy the Honour of Virgins, and corrupted the Integrity of Judges; and whoever acts from it as a Principle, what good soever he may bring to the Society, has nothing to boast of but that he has indulged a Passion that has happened to be beneficial to the Publick (Ibid).

Smith signifies sympathy as a natural fellow-feeling and its innateness in human nature by giving a set of instances and indicates what he actually aims at. For instance, "when we see a stroke aimed and just ready to fall upon the leg or arm of another person, we naturally shrink and draw back our own leg or our own arm; and when it does fall, we feel it in some measure, and are hurt by it as well as the sufferer" (Smith 1982: 10). And he gives other examples:

> The mob, when they are gazing at a dancer on the slack rope, naturally writhe and twist and balance their own bodies, as they see him do, and as they feel that they themselves must do if in his situation. Persons of delicate fibres and a weak constitution of body complain that in looking on the sores and ulcers which are exposed by beggars in the streets, they are apt to feel an itching or uneasy sensation in the correspondent part of their own bodies. (Ibid.)

His instances and his portrayal of sympathy as a natural feeling indicate that he obviously constructs his theory against the principle that all human beings are solely motivated by selfishness. Furthermore, he defends that fellow-feeling with any passion or sympathy cannot be connected to a selfish principle and supports his theory by giving mechanism of sympathy in detail with some important extra components. What he says is that we are naturally concerned with the fortune of others by a mechanism of sympathy and by means of imagination, actual sympathy occurs when sentiments of the spectator and those of agent correspond or coincide. Even if, at first sight, sympathizing with others seems to be founded in self-love (putting oneself in other's situation), but it arises from an imaginary change of situation and person.

> When I sympathize with your sorrow or your indignation, it may be pretended, indeed, that my emotion is founded in self-love, because it arises from bringing your case home to myself, from putting myself in your situation, and thence conceiving what I should feel in the like circumstances. But though sympathy is very properly said to arise from an imaginary change of situations with the person principally concerned, yet this imaginary change is not supposed to happen to me in my own person and character, but in that of the person with whom I sympathize... It is not, therefore, in the least selfish. (Smith 1982: 317)[1]

In other words, Smith tries to correct misapprehension about the mechanism of sympathy by underlining the precondition of the "imaginary change of situations and selves". Marshall states that "sympathy, according to Smith's formulation, involves a loss of self, a transfer and metamorphosis" (Marshall 1984: 600). When spectators judge the propriety of an actor's reaction to a situation, they put themselves in his shoes by means of imagination and see whether under the same situation they would have the same sentiment and reaction. Since imagined sentiments of the spectator are generally in weaker degree than the sentiments of the actor, in order to produce a sympathetic concord between them, an adjustment is required. Actor succeeds this concord by "lowering his passion to the pitch, in which spectators are capable of going along with him" (Smith 1982: 22). Smith states that we do not only have natural disposition to judge others but we also judge ourselves. How do we learn to become spectators of our own sentiments and conduct? We learn how to judge ourselves from judging others by means of same sympathetic process. The precondition to form any judgment regarding our own sentiments and motives is to depart our natural station and behold them from a certain distance from ourselves. This can be achieved only by looking at them from the eye of other people (Smith 1982: 109–110). In the first stage, we imagine other people sympathizing with us and consider their judgments concerning us. Our desire for sympathetic concord with others who observe and assess our conduct leads a certain balance and regulation over our feelings and actions. We try to observe how other people see us and we try to moderate our behavior and lower the tone down. Thus our first judgments regarding ourselves are shaped in the light of approval and disapproval of others. Since it would be impossible to have a reflection of our own sentiments, character, conduct and even our appearance in a solitary place, others are essential to

[1] I cite Smith's examples regarding imagination procedure: "When I condole with you for the loss of your only son, in order to enter into your grief I do not consider what I, a person of such a character and profession, should suffer, if I had a son, and if that son was unfortunately to die: but I consider what I should suffer if I was really you, and I not only change circumstances with you, but I change person and characters. My grief, therefore, is entirely upon your account, and not in the least upon my own" (Smith 1982: 317).

objectify these. Only in society one can find the mirror which reflects propriety and impropriety of his passions. In society, Smith says man "will observe that mankind approve of some of them, and are disgusted by others. He will be elevated in the one case, and cast down in the other; his desires and aversions, his joys and sorrows, will now often become the causes of new desires and new aversions, new joys and new sorrows: they will now, therefore, interest him deeply, and often call upon his most attentive consideration" (Smith 1982: 110–111). As society provides with certain awareness in the sense that the mankind approves some of their sentiments and passions and disapproves some of others, it constitutes the first stage of forming our moral judgments regarding ourselves.

While we are judging our own conduct we achieve a standard through process of internalization of social responses and at the same time we learn how to be a spectator of own sentiments and conducts. Although social responses ensure a standard for propriety and impropriety of our behaviors, most of our fellows are limited in their knowledge or misinformed about our situation. Therefore, we imagine a fair and impartial spectator who would examine our own conducts from an objective standpoint. If an agent can judge himself from the point of this ideal spectator, he can achieve an objective criterion for self-assessment. By means of this impartial spectator the individual becomes his own judge. Smith states:

> I divide myself, as it were, into two persons; and that I, the examiner and judge, represent a different character from that other I, the person whose conduct is examined into and judged of. The first is the spectator, whose sentiments with regard to my own conduct I endeavour to enter into, by placing myself in his situation, and by considering how it would appear to me, when seen from that particular point of view. The second is the agent, the person whom I properly call myself, and of whose conduct, under the character of a spectator, I was endeavouring to form some opinion. The first is the judge; the second the person judged of. (Smith 1982: 113)

The inner judge or inner voice of man which is the internalized impartial spectator is "reason, principle, conscience, the inhabitant of the breast, the man, the great judge and arbiter of our conduct" (Smith 1982: 137). Formation of ideal moral judge within us not only enables us to get certain distance from partiality of others and ourselves but also it leads to restriction and management of our self-love. According to Smith, if one acts according to the principles of the impartial spectator he can get free from the passions which distract the social harmony. One of these passions which impartial spectator humbles is self-love. "We learn the real littleness of ourselves, and of whatever relates to ourselves, and the natural misrepresentations of self-love can be corrected only by the eye of this impartial spectator" (Smith 1982: 137). Smith likens the correction of misrepresentations of self-love to the correction of misrepresented proportions of the distant objects seen through a window. We transport ourselves to a different situation through our imagination in order to judge their real proportions. In the same way, we put ourselves in a certain distance in order to see the real significance of our small interests. Thus, impartial spectator saves us both from the misguidance of society and subjectivity that our passions cause without resulting in social disharmony. The figure of impartial spectator understood as conscience also answers the most basic criticism of Smith's moral theory exemplified by Sir Gilbert Elliot's question how impartial spectator's perspective can be justified to be different from conventional rules. In the second edition of *TMS* in his reply to Elliot, Smith clarifies the fact that "real magnanimity and conscious virtue can support itself under the disapprobation of all mankind" (Smith 1987: 49).[2] This means that even if conscience seems to be a social product there is possibility to assume it as independent from public opinion.

9.4 Mandeville's Licentious System

In the Part VII of *TMS*, Smith designates Mandeville's thesis as a "licentious system". Not alone does he criticize Mandeville's basic argument that all our actions spring from either selfish passion or vanity and desire for applause, but his rigorism regarding morals. For Smith, Mandeville assumes that whatever done from a sense of propriety is done from a love of praise and applause. He aims to reduce individuals to praise-seeking beings. Against Mandeville, Smith claims that the desire of doing what is honorable and noble has nothing to do with vanity. Love of well-grounded reputation and desire for what is really estimable cannot deserve the name vanity; rather these are the best passions of the human nature called the love of virtue and love of true glory. Every man is naturally inclined to desire not only what is approved and praised but also desire what is approvable and praiseworthy. We learn by experience that not all our feelings and actions are always approved and praised by everyone. We determine the real merits of our actions by distinguishing what is actually approved and what is worth to be approved. Being proper object of praise does not depend on public approval but a higher tribunal. Man's jurisdiction is founded altogether in the desire of praise and in the aversion of blame without impartial spectator. On the other hand, with it, jurisdiction of man is based on the desire of praiseworthiness and in the aversion of blameworthiness. Even if our action is not praised by anyone, we are capable of acting in a praiseworthy manner which deserves exact approval of impartial spectator. Smith condemns man of vanity who seeks praise even though he does not deserve it. "It is only the weakest and most superficial of mankind who can be much delighted with that praise which they themselves know to be altogether unmerited" (Smith 1982: 117). But a wise man, Smith says:

> ...feels little pleasure from praise where he knows to be praiseworthy, though he knows equally well that no praise is ever bestowed upon it. To obtain the approbation of mankind, where no approbation is due, can never be an object of any importance to him. To obtain that approbation where it is really due, may sometimes be an object of no great importance to him. But to be that thing which deserves approbation, must always be an object of the highest. (Ibid.)

[2]Letter 40 (10 October 1759).

Smith also rejects Mandeville's system of morals because it "seems to take away altogether the distinction between vice and virtue" (Smith 1982: 308). He is strongly opposed to moral rigorism of Mandeville by which he identifies every passion as being vicious. Smith criticizes him that his portrayal of virtue as complete self-denial is not a conquest but "no more than a concealed indulgence of our passions" (Smith 1982: 312). Although in *TMS* Smith clearly distances himself from Mandeville by opposing his characterization of man solely motivated by selfish instincts, who runs after praise instead of praiseworthiness, in *The Wealth of Nations (WN)* Smith's usage of self-interest as a basic motive of individuals in commercial societies makes someone think whether there is possible effect of Mandeville on Smith or not. Smith's two conflicting views in *TMS* and *WN* sow the seeds of a well-known problem called "The Adam Smith Problem" and he became the target of polemics at the end of nineteenth century. The main contradictory passages in *TMS* and *WN* which indicate two opposed views have been quoted over a century. Two famous passages below are sparkles of the debate:

> How selfish so ever man may be supposed, there are evidently some principles in his nature, which interest him in the fortune of others, and render their happiness necessary to him, though he derives nothing from it except the pleasure of seeing it. (Smith 1982: 9)
>
> Man has almost constant occasion for the help of his brethren, and it is in vain for him to expect it from their benevolence only. He will be more likely to prevail if he can interest their self-love in his favour, and show them that it is for their own advantage to do for him what he requires of them... It is not from the benevolence of the butcher, the brewer or the baker that we expect our dinner, but from their regard to their own interest. We address ourselves not to their humanity but to their self-love, and never talk to them of our own necessities but of their advantages. (Smith 1977: 30)

Macfie and Raphael, editors of Glasgow Edition of *TMS*, interpret that self-love and sympathy serve for different purposes, that is to say, the former motivates and the latter conducts and governs; therefore they are not comparable. They also add that "sympathy is the core of Smith's explanation of moral *judgment*. The motive to action [namely self-interest] is an entirely different matter" (Smith (Introduction) 1982: 21–22). Therefore, they do not see a contradiction between these two sentiments and they evaluate it as a pseudo problem (Smith (Introduction) 1982: 20). Some scholars think that although there are two different and conflicting sentiments in *TMS* and *WN*, there is a unity between them. Morrow affirms the "unity of spirit and aim in Adam Smith's treatment of separate divisions of moral philosophy (i.e. ethics, political economy and jurisprudence)" (Morrow 1927: 324). And he adds:

> It is true that in *Moral Sentiments* Adam Smith opposed the egoistic doctrine that man acts only from self-love, and exalts benevolence as the highest virtue. There are other inferior virtues recognized, such as prudence, frugality, industry, self-justice, but when so regulated they are conductive to the welfare of the general public as well as of the individual. The important consideration is that these self-interested activities must be regulated by justice. Very little is said in The Wealth of Nations about the principle of justice... but justice is of course always presupposed as necessary for the existence of nations at all, especially wealth nations... In short, unregulated self-interest is no more advocated in the *Wealth of Nations* than it is in the *Moral Sentiments*, whereas in the latter work the moral value of the inferior virtues, when properly regulated, is fully recognized. (Morrow 1927: 330–331)

On the other hand, many scholars who accept that there is a problem differ from each other by the way of their approach, justification and explanation. According to the first group, in 1764, during his residence in France, Smith is said to encounter French materialist philosophers like Helvetius and Holbach. Since, this date also coincides with period between publications of *TMS* (1759) and *WN* (1776) they argue that a change in Smith's thought regarding human nature is highly probable. Brentano holds that Smith's explicit rejection of selfish hypothesis in *TMS*, and then corroborating the same hypothesis in *WN*, indicates the influence of Helvetius whom he met in Paris. He states that "[I]n the "Investigations into the Wealth of Nations," on the contrary, he [Smith] holds entirely to the views of the book of Helvetius upon the nature of man, and regards selfishness as the only motive of human action. The consequences of this dogma of selfishness permeate almost all parts of his work" (Brentano 1891: 64). However, according to Nieli it can be seen from Smith's *Lectures on Jurisprudence* (1763) that his formulation of all his basic doctrines in *WN* and also his idea that economic relations are motivated by self-interest were contemplated before his journey to France (Nieli 1986: 614). For another group of scholars, Smith uses contradictory concepts or two different anthropological views in *TMS* and *WN* but they assert that these views already belong to separate fields (ethics and economics). Hence, the anthropological views in *TMS* hardly help us to understand self-interest principle in *WN*. In addition to all these approaches, the problem can be analyzed bilaterally: first, how Smith describes individuals and characteristics and sentiments of them, second how social and economic systems affect individuals. On the one hand, the characters of individual and human psychology are principal cannons of Smith's system of moral philosophy; on the other, as he argues in *WN*, the nature of man is affected by social, political and economic systems. Man's sentiments and their way of expression changes in different places, circumstances and times. In recent years, for many scholars "The Adam Smith Problem" appears to be no more problem anymore. It seems that character and motive of the economic agent in *WN* do not conflict Smith's moral theory in *TMS*. In order to understand the motive of his moral and economic agent, it is important to look his usage of self-love in both of his works. In *TMS* he uses Stoic definition of self-love that everyone "is first and principally recommended to his own care; and every man is certainly, in every respect, fitter and abler to care of himself than any other person" (Smith 1982: 219). Therefore, man's care for his health and his happiness is natural and comes before anything and anyone else. Even if everyone first prefers his happiness to others, Smith underlines that man does not act according to this principle.

> Though every man may, according to the proverb, be the whole world to himself, to the rest of mankind he is a most insignificant part of it. Though his own happiness may be of more importance to him than that of all the world besides, to every other person it is of no more consequence than that of any other man. Though it may be true, therefore, that every individual, in his own breast, naturally prefers himself to all mankind, yet he dares not look mankind in the face, and avow that he acts according to this principle. He feels that in this preference they can never go along with him, and that how natural soever it may be to him, it must always appear excessive and extravagant to them. (Smith 1982: 83)

Smith states that instincts of self-preservation are not selfish or self-interested. But if one seeks to satisfy his basic needs at the expense of others or if self-love is excessive then it becomes selfish and it has to be balanced. As mentioned earlier, the impartial spectator assists man to humble of his self-love "bring it down to something which other men can go along with" (Smith 1982: 83). On the other hand, in *WN* Smith uses self-love as our desire to better ourselves and our condition. He says that since man is affected by social and economic systems therefore, desire to better our condition turns into seeking material interests. But, what Smith offers is not seeking our interest through dishonesty, avarice and greed. In *TMS* Part VI "The Character of Virtue" which is entirely new chapter, included in its sixth edition and before Smith's death in 1790, he prescribes a bundle of virtues as a treatment for commercial societies and gives a formula for human perfection. These virtues are prudence, benevolence, justice and self-command which not only balance conduct of individual but also have regulating effect on commercial societies. Moreover, through the approbation of our conscience we can find not only the right path for proper moral judgments and acts but also we can compare the interests of ourselves and those of others. Hundert states that "Mandeville's society which has morally threatening quality as a masquerade has been tamed by indifferent and impartial spectator of Adam Smith" (Hundert 1994: 227).

Self-command is a cardinal virtue by means of which Smith stresses capability of self-government and autonomous character of the individual. Self-command as our ability to control and regulate our passions mentioned usually as selfish-affections and it is shared by all mankind, only with different degrees. It is the power of self-disciplining. According to Smith, self-command is not only the greatest virtue, but it is such a virtue that "from it all the other virtues seem to derive their principal lustre" (Smith 1982: 241). In that sense self-command can be understood as the necessary condition for having a virtuous life. Smith discuses different accounts of virtue in *TMS* and writes that according to some authors "the virtuous temper of mind does not consist in any one species of affections, but in the proper government and direction of all our affections, which may be either virtuous or vicious according to the objects which they pursue, and the degree of 'vehemence' with which they pursue them. According to these authors, therefore, virtue consists in propriety" (Smith 1982: 266). For Smith, too, virtue consists in propriety of actions and this propriety is decided on the ground that the reasons or incentives of actions are right ones. While impartial spectator enables the individual to see himself from a distance and discover the real incentives of his actions, the virtue of self-command enables him to moderate the passions whose violent feature directs him wrongly in his actions. Through self-command, individual gains more authority over the incentives of his actions. Furthermore, prudence, as another principal virtue, treats the vices caused by commercial vanity. Individual's care for health, fortune and reputation is object of this virtue. Prudence advises us bear our prosperity with mere moderation and it teaches us to avoid envy. Prudent man has characteristics of esteem of modesty, discretion and good conduct. He is also supported by the entire approbation of impartial spectator.

In *TMS*, Smith portrays a moral agent who is self-determined, self-governed and who has a certain self-legislative power over his desires and selfish inclinations. Firstly, he downplays the determinative roles of passions in formulation his notion of sympathy and in explaining the sympathetic process in moral evaluations. Then he claims that by appealing to the impartial spectator which is settled in human breast, moral agents are able to judge their own actions and set their own moral standards for their moral judgments. The impartial status of the ideal spectator does not only liberate agent from external constraints imposed by society but also from the internal constraints caused by selfish desires and inclinations. So it enables the agent to be autonomous in forming his moral principles and be governed by self-imposed rules. Both the qualities of the sympathy and the possibility that the agent forms an indifferent perspective and becomes his own judge introduce a sense of autonomy into moral sphere. Moreover, he emphasizes that when the individual has the power of self-command, he can control his passions and moderate them to the point that he can act from right reasons and behave properly to achieve the right things in his life. Smith's real intention is to develop a comprehensive system in which morality, economics and politics are clamped together. He does not advocate unrestrained or unregulated self-interest in *WN*. He points out that human nature is not entirely selfish and human beings are capable of restraining their selfish affections. Ideal man in a commercial society is prudent and self-disciplined and also capable of interacting and competing in the light of dictates of justice, law and morality. Unlike Mandeville, Smith intends to show that strong commitment to moral and ethical foundations does not slow down economic activity. He gives principal virtues which have regulatory effect not only on individual as well as well-being of a society. In this way, he reconciles sympathy and self-love by linking them up with moral experience of individual and his character in the commercial world.

References

Primary Sources

Mandeville, Bernard. 1988a. *The fable of the bees or private vices, public benefit*, vol. 1. [1732]. Indianapolis: Liberty Fund.

Mandeville, Bernard. 1988b. *The fable of the bees or private vices, public benefits*, vol. 2. [1732]. Indianapolis: Liberty Fund.

Smith, Adam. 1977. *An inquiry into the nature and causes of wealth of nations*. [1776], ed. Edwin Cannan. Chicago: University of Chicago Press.

Smith, Adam. 1982. *The theory of moral sentiments*. [1759], ed. D.D. Raphael and A.L. Macfie. Indianapolis: Liberty Fund.

Smith, Adam. 1987. *The correspondence of Adam Smith*, ed. E.C. Mossner and I.S. Ross. Indianapolis: Liberty Classics.

Secondary Sources

Brentano, Lujo. 1891. *The relation of labor to the law of today*. Trans. P. Sherman. New York: G.P. Putnam's Sons.

Hundert, Edward J. 1994. *The enlightenment's fable: Bernard Mandeville and the discovery of society*. Cambridge: Cambridge University Press.

La Rochefoucauld. 2007. *Collected maxims and other reflections*. Trans. E.H. Blackmore, A.M. Blackmore, and Francine Giguere. Oxford: Oxford University Press.

Marshall, David. 1984. Adam Smith and the theatricality of moral sentiments. *Critical Inquiry* 10(4): 592–613.

Morrow, Glenn R. 1927. Adam Smith: Moralist and philosopher. *Journal of Political Economy* 35(3): 321–342.

Nieli, Russell. 1986. Spheres of intimacy and the Adam Smith problem. *Journal of History of Ideas* 47(4): 611–624.

Chapter 10
Mandeville on Pride and Animal Nature

John J. Callanan

Abstract Mandeville's first publication – the thesis *Disputatio Philosophica de Brutorum Operationibus* (1689) – advocated the Cartesian position that both denied feeling and sensation, let alone thought, to non-human animals and stressed the inherent distinctiveness of the conscious sensory and inferential capacities of human agents. Yet his later writings subscribed to a directly opposed Enlightenment position. His translation of La Fontaine's *Fables* drew comparisons between humans and animals throughout, and by the time of the *Fable of the Bees*, Mandeville was clearly in the camp stressing the continuity of human and non-human animal nature, a tradition following Hobbes, Montaigne and La Rochefoucauld, and later to include Helvétius, de la Mettrie and Hume. The function of pride in Mandeville's ethics is examined in terms of this debate, framed by Bayle's famous 'Rorarius' entry in his *Dictionary*. With this background in place, Mandeville's claim regarding the psychological role of pride as the 'other Recompense . . . [of] the vain Satisfaction of making our Species appear more exalted and remote from that of other Animals' is then discussed. It is presented as a critique of Shaftesbury's discussion in the *Characteristics* relating to the norm of fulfilling one's human nature.

Keywords Mandeville • Animals • Bayle • Shaftesbury • Human nature

10.1 Introduction

It is well known that Mandeville's first piece of writing – the thesis *Disputatio Philosophica de Brutorum Operationibus* (1689) – endorsed Descartes's claim that non-human animals are incapable of higher consciousness, thought and reason.[1] It is equally well known that Mandeville's later writings seem premised upon a view of human beings as fundamentally closer in nature to non-human animals. Mandeville

[1]Mandeville (1689).

J.J. Callanan (✉)
Department of Philosophy, King's College London, London, UK
e-mail: john.callanan@kcl.ac.uk

E. Balsemão Pires, J. Braga (eds.), *Bernard de Mandeville's Tropology of Paradoxes*, Studies in History and Philosophy of Science 40, DOI 10.1007/978-3-319-19381-6_10

was not the first to suggest that the picture of human beings as higher than the animals plays a strategic role in philosophical and theological belief systems. This view would have been familiar to any reader of Montaigne (2003). I want to suggest though that the task of distinguishing oneself from non-human animals was a central theme in the *Fable of the Bees* (Mandeville 1924). I will also argue that the later development by Mandeville of the distinction between *self-love* and *self-liking* was important just because of the way in which it reinforced the former theme. These psychological mechanisms explain the prevalence of *pride* in our cognitive lives, and the latter is for Mandeville the source of our sense of self-importance compared to other animals.[2]

One might have questioned why pride alone should have such an influential role. A first thought is that Mandeville neglects the sense in which taking pride in one's characteristics might itself just be an instance of a more general capacity peculiar to human beings, which is that of self-conscious, critical and evaluative judgment upon their own mental states and character. It is this capacity, as Butler maintained, that ultimately *does* render us a distinct type of creature (Butler 1983). Once one allows such a distinct capacity of evaluative judgment is possible, it is not at all obvious that it functions primarily in accordance with the motivation of pride. There is no doubt that pride can on a particular occasion be the motivating factor that determines why we make the self-evaluations that we do. Nevertheless, this does not support the stronger thesis that the fundamental or central motivation for positive self-evaluations is itself that of pride. It also seems to disregard the obvious point that our prideful motivations can themselves become the object of our critical evaluations. We can make an evaluative appraisal of our own susceptibility to pride, and direct our behaviour in opposite ways as a result.

Mandeville's response, I'd suggest, is that this entire capacity to take an evaluative view upon one's desires is itself a fundamentally natural phenomenon that has its roots in the proto-evolutionary disposition of self-liking.[3] The primary origin of this response is the valuing of oneself and one's own interests. In this way, Mandeville seeks to re-naturalize that which seemed distinctly *non*-natural about human beings, and to reinforce his initial claims. In arguing for this claim, I'll first outline briefly the problem of the status of animals in Early Modern philosophy. Secondly, I'll consider Shaftesbury's notion of a 'higher self' as a possible target of Mandeville's attack. Thirdly, and finally, I'll outline what I take to be Mandeville's central objection.

[2]For a discussion of the importance of pride in Mandeville's theory see (Heath 1998).

[3]A similar claim is made in (Welchman 2007).

10.2 The Problem of Animals

There is a somewhat standard Early Modern narrative regarding animals that I would claim is of special relevance to this theme. I'm mentioning the following varied themes as I think they all play a role for understanding the context of Mandeville's critique in the *Fable*. It begins with Montaigne's opposition to the scholastic view of the human as possessing a peculiarly rational soul over animals' 'sensitive' souls. His reasoning went broadly along the same lines that Hume would adopt, drawing upon the observable analogies between human and animal behaviour. It is Montaigne who is the immediate target of Descartes's denial that animals have *any* higher representational capacities resembling that of humans, and that 'after the error of those who deny God ... there is none that leads weak minds further from the straight path of virtue than that of imagining that the souls of the beasts are of the same nature as ours ...'.[4]

Descartes's rhetorical grouping of atheism and immorality with the denial of a demarcation between humans and animals is notable, and it was arguably the theological implications of the Cartesian characterization of animals that interested Pierre Bayle, who in his *Dictionary* entry 'Rorarius' detailed, sometimes sardonically, the purposes that Descartes's position could be put to. One unpleasant angle concerned theodicy: infant pain and premature death could be explained as an evil that God allowed in the world on account of those infants' original sin. As such, the conceptual linking of possible pain to creatures with souls that are capable of sin is maintained. The possibility of animal pain thus presents a theological problem. One must either attribute souls to animals (and what's more, sin) in order to explain their apparent pain, or one must simply deny that the apparent pain behaviour they manifest *is* real pain behaviour. Descartes's endorsement of the latter option and his denial of pain to animals then comfortably fit a theological agenda.

As Bayle points out though, this move has the drawback of being entirely unbelievable. We simply do make true judgments based on observed behaviour when attributing such conscious capacities to other *human* beings, and as Montaigne, Hume and others point out, it is just this same kind of evidence that is at stake when observing non-human animal behaviour. The risks run in two directions. One can either just deny that the types of observed behaviour – person-recognition, inference, anticipation of events, communication, and so on – are evidence of a conscious soul, or one can accept that they are. If one accepts that they are good evidence, then the world is vastly more populated with souls than initially appeared to be the case. If one denies that they are good evidence, then the worry is that they are no longer good evidence for the existence of human souls either, and then the same reasoning could be adopted by a materialist who renders the world far less populated with souls than initially appeared to be the case.

[4]Discourse on Method, Part Five in Descartes (1985, 141).

The question of just what occurs in animal consciousness was discussed at length too, and here too we can find disturbingly *ad hoc* demarcations. For Locke for example, judgment is "the putting *Ideas* together, or separating them from one another in the mind".[5] For Locke, all judgment is the act of seeing when two ideas that we have acquired through sensation 'agree' or 'disagree'.[6] The most explicit connection between human perception and that of animals occurs in Book II, Chapter XI of the *Essay*. Having attributed some basic memory capacity to animals in the previous chapter, Locke now considers whether animals are capable of the slightly higher cognitive functions of the *comparison* of ideas, the *compounding* (i.e. the process of complex representation formation) and the *abstraction* of ideas (i.e. the formation of concepts from non-conceptual representational input. For Locke the function of comparison is the base cognitive capacity which allows for all the representations of relation. In summary, Locke maintains that '[b]rutes compare but imperfectly'; '[b]rutes compound but little' and '[b]rutes abstract not'.[7] It is the last of these, abstraction, that Locke focuses upon, since it is this activity that marks the distinctness of human being's higher cognitive capacity- it is the having of 'general ideas' that 'puts a perfect distinction betwixt man and brutes'.

Locke's theory of judgment though proved to be a different and highly influential theory amongst the hyper-empiricist tradition of French materialism. In his *Traité des Sensations*, Condillac would praise Locke for his empiricist account of the sensory origin of our ideas, but also criticizes him – Locke should have seen the next obvious step, that 'they [the faculties of the soul] could derive their origin from sensation itself'.[8] Condillac has no doubt that [j]udgment, reflection, desires, passions, etc. are only sensation itself which is transformed differently.[9] An even more radically sensationist tract came from Helvétius, whose 1758 *de l'Espirit* similarly argued in a reductive model of human beings' judgment to the capacity for sensation. Helvetius's *de L'Espirit*, De La Mettrie's *Machine Man* and d'Holbach's *System of Nature* were viewed by many, (for example by both Rousseau and Kant), as the over-exuberant nadir of the trend of opposing scholastic models of the self with a reductive model comparable to non-human animals.

For Kant for example, it is the human being's capacity for rational *judgment* that is key. As he puts it, 'reason raises him above the animals, and the more he acts according to it, the more moral and at the same time freer he becomes' (29: 900).[10] This latter idea, that through our reason we can become *more free* is part of an Augustinian tradition that is retained in the Early Modern period. Crucial to

[5]Locke (1975, IV.xiv.4).

[6]Of course it is a more complicated question as to what Locke really thought was involved in the act of judgment – for a discussion of some of the difficulties, see (Owen 1999).

[7]Locke (1975, II.xi.5 ff.).

[8]Condillac, *Traité des Sensations*, quoted in (O'Neal 1996, 16–17).

[9]Condillac, *Traité des Sensations*, quoted in ibid., 19.

[10]Kant (1997, 267, Ak. 29: 900).

this picture is Augustine's distinction between *libertas minor* and *libertas maior*.[11] The former indicates the power of free choice that is available to human fallen subjects capable of sin. The latter indicates the perfection of our power of free choice whereby the representation of the good is so evident to the subject's consciousness that it is constitutionally incapable of freely choosing otherwise. Peter Lombard gives a typical expression to the position in the claim that 'a choice [*arbitrium*] that is quite unable to sin will be the freer'.[12] The progression of human moral improvement involves the aspiration to transform the human *libertas minor* into the *libertas maior* of the angels, whereby 'after the confirmation of beatitude there is to be a free will in man by which he will not be able to sin'.[13] The theme was picked up in Leibniz's *Nouveaux Essais*, where Locke's representative claims – to Leibniz's representative's approval – that to 'be determined by reason to the best is to be the most free'[14] and moreover that 'those superior beings . . . who enjoy perfect happiness . . . are more steadily determined in their choice of good than we and yet we have no reason to think they are . . . less free, than we are'.[15] Kant's later distinction between the power of human choice and that of a pure 'holy will' clearly echoes that of the scholastic distinction.

There are two familiar traditions then with regard to the relation between human and non-human animals. On the one hand, there are more theologically inspired accounts whereby human beings carry something of the divine in them. On this account human beings have duties firstly to identify what aspects of their nature are the higher ones, and secondly to conform their conduct to the standard of that higher nature. The other tradition self-consciously attacks this position, and insists either on the falsity of the picture of the higher self, or of the folly of aspiring to conform one's behaviour to a picture of angelic perfection, or both. For example, Montaigne concludes the *Essais* with an admonition: whatever one's religious beliefs, the mimicking of some construed divine standard of moral perfection produces an entirely opposite effect than the one initially intended. When one has the ambition to behave as a higher being would, one is left with nothing of substance and in fact the result, Montaigne famously claims, is a distortion of our moral behaviour:

> They want to be besides themselves, want to escape from their humanity. That *is* madness; instead of changing their Form into an angel's, they change it into a beast's; they crash down instead of winding high. These humours soaring to transcendency terrify me as do great unapproachable heights.[16]

---

[11]See for example see (De Coorreptione et Gratia, 12:33 in Augustine 2010, 214) and (Enchiridion, Ch. XXVIII, para. 105, in Augustine 2006, 402).

[12]Lombard (1981 Book 2, Distinction 25, Ch. 4 463, quoted from Pink 2011, 548).

[13]*Ibid.* Cf. Aquinas's *Summa Theologiae*, Part I, Q. 62, Art. 8, and Anselm's *De libertate arbitriii*, I.

[14]Leibniz (1997, Bk. II, Ch. XXI, 198).

[15]*Ibid.* It is similarly claimed 'that God himself cannot choose what is not good; the freedom of the Almighty hinders not his being determined by what is best' (*ibid.*).

[16]Montaigne (2003, 1268 Bk. III, Ch. 13, "On Experience").

Montaigne's claim amounts to an ironic inversion of the theocentric paradigm: by having the correspondence of one's will with a divine standard as the proximate goal of moral improvement, one in fact undermines the very possibility of that improvement.[17]

More often then not though, human beings' autonomous capacity for rational evaluative *judgment* was viewed as distinctive of the higher self. Thus in Kant's famous claims in the introduction to the *Critique of Pure Reason*, that '[o]ur age is the genuine age of criticism, to which everything must submit'[18] and here Kant is clear that religion is no exception. Kant's 'tribunal of reason' metaphor echoes Bayle's claim that 'Reason, speaking to us by the Axioms of natural Light, or metaphysical Truths, is the supreme Tribunal, and final Judg without Appeal of whatever's propos'd to the human Mind' (Bayle 2005, 67 First Part, Chapter 1). Yet of course the same metaphor was appealed to in d'Holbach's *System of Nature* in 1770 where he taunts believers to 'cite the Divinity himself before the tribunal of reason'.[19]

10.3 Shaftesbury's Naturalism and the Higher Self

Shaftesbury's *Characteristics* in many ways hinges on this theme of the distinction between higher and lower animals. He focuses upon the idea of the capacity for evaluative judgment upon our desires as key to that demarcation. For Shaftesbury, the distinction is supposed to be one made *within nature* and yet is still in favour of there being a special higher place for human minds. In nature there is 'a system of all animals, an animal order or economy according to which the animal affairs are regulated and disposed' (Shaftesbury 1999, 169). What it is to be a human being in

[17]This theme reaches a conventional climax in Kant's Critical Philosophy, where as per usual, a middle position is put forward: the idea of such a perfect being (which Kant calls 'holy wills' is entirely coherent, and can serve as some kind of indeterminate aspirational target; however, Kant's restrictions on the scope of our knowledge entails that we cannot know anything about how that perfect being reasons or what courses of action might be pursued. As such, the demand to derive practical guidance from one's own rational resources is retained.

It might be noted that frequently something akin to the same complaint is leveled from one tradition to the other. This is that the practical reactions that are involved in each conception of proper human agency are in some sense automatic. The objection to the humanist tradition is that the purely animalistic conception of human beings reduces them to purely reactive agents, unfree creatures responding to sensory stimuli in increasingly complex, but nevertheless fully determined manners. The objection to the theological tradition is that it renders human beings automata in their unquestioning deference to theologically determined moral norms. Thus we find in Bishop Butler's sermons an attempt to circumvent this worry by appeal to a fundamental capacity that is distinctive human beings to take an evaluative view upon their evidence and to form their own judgment.

[18]Kant (1998, Axi).

[19]Holbach (1889, 312, Part II, Ch. 10, "Is Atheism Compatible with Morality?").

this animal order is to be a creature who can take a view on the various desires and impulses that it otherwise shares with animals. In fact, Shaftesbury and Mandeville appear to be in agreement with the thought that the desires and interests that we hold are themselves morally neutral, and that they only receive a moral value in virtue of the intentions that lie behind them:

> So that if a creature be generous, kind, constant, compassionate, yet if he cannot reflect on what he himself does or sees others do so as to take notice of what is worthy or honest and make that notice or conception of worth and honesty to be an object of his affection, he has not the character of being virtuous.[20]

Shaftesbury argues that the only way to realize 'divineness of a character' is with an inward turn to examine the motives behind one's judgments and that 'it is hard to imagine what honour can arise to the Deity from the praises of creatures who are unable to discern what is praiseworthy or excellent in their own kind'.[21] Here Shaftesbury links the theme of the self-evaluation of motivations with that of the aspiration already discussed, that of there being a duty to examine the higher aspects of one's own distinct species and to maximize those aspects in one's behaviour.

Shaftesbury even echoes Augustine's *libertas maior* tradition but unlike Augustine, Locke and Leibniz, Shaftesbury appears to think that such moral perfection is possible, and more so by virtue of cultivation and education:

> A man of thorough good breeding, whatever else he be, is incapable of doing a rude or brutal action. He never deliberates in this case or considers of the matter by prudential rules of self-interest and advantage. He acts from his nature, in a manner necessarily and without reflection, and, if he did not, it were impossible for him to answer his character or be found that truly wellbred man on every occasion. It is the same with the honest man. He cannot deliberate in the case of a plain villainy.[22]

These elements relate to an overall Stoic theme in Shaftesbury's thought, which is that happiness and virtue align when the individual is following the essential nature of one's own self. In the *Soliloquy*, he writes:

> [T]here is no expression more generally used in a way of compliment to great men and princes than that.... 'they have acted like themselves and suitably to their own genius and character'. The compliment, it must be owned, sounds well. No one suspects it. For what person is there who in his imagination joins not something worthy and deserving with his true and native self, as often as he is referred to it and made to consider 'who he is'?[23]

Shaftesbury compares human beings who have lost the understanding of who their 'true and native self' to animals with birth defects, those 'animals [who] appear unnatural and monstrous when they lose all their proper instincts....[and who] pervert those functions or capacities bestowed by nature'. When this happens to a human being, even the effect, Shaftesbury claims, can only be misery for the person:

---

[20]Shaftesbury (1999, 173).

[21]Shaftesbury (1999, 22).

[22]Shaftesbury (1999, 60).

[23]Shaftesbury (1999, 125).

> How wretched must it be, therefore, for man, of all other creatures, to lose that sense and feeling which is proper to him as a man and suitable to his character and genius.[24]

Someone who is realizing all his first-order desires but not by attending to his second-order evaluation of them is, Shaftesbury contends, as miserable as a human being can be. Conversely, we can attribute to him the Stoic thought that someone who denies themselves their first-order desires can nevertheless be content in a higher sense, just because that self-denial is a result of his following his second-order evaluation of what 'is proper to him as a man'.

10.4 Pride and Self-Liking

Of course, in order to live in harmony with one's true higher self, one must first identify one's true nature. This in turn presupposes that there is a higher self with which we can identify.[25] Many thinkers before Mandeville had the thought that human beings differ from other animals only in degree of rational capacity and not in kind. Mandeville however was among the first to argue that our desire to *think* of ourselves as higher than non-rational animals was itself the covert motivating factor behind a range of seemingly different behaviours. In the *Fable of the Bees*, the very idea of virtue is provided a genealogy that has its origins not in the state of nature, or in the very idea of civil society but rather is a concept that is formed purely for the functional role of demarcating human nature from that of other animals.

As is well known, in the *Enquiry Into the Origin of Moral Virtue*, Mandeville presented an account whereby clever politicians manipulated human beings' susceptibility to flattery for the purposes of creating behaviour that was more beneficial to those in power. The trick was to convince those subjects to willingly endorse the idea that 'it was more beneficial for every Body to conquer than indulge his Appetites'.[26] It is surely possible to force people to abstain from some desires in order to realize a collective good, but here Mandeville is considering a different project. This is the project of bringing people around so that they themselves endorse a contradictory notion of human self-fulfilment. The notion is contradictory to the degree that it requires convincing someone that it is in that individual's own interest to ignore the satisfaction of his own other interests.

This is a trickier proposition than that of forcing them to abstain from certain desires, since it in effect requires turning those individuals into the most enthusiastic practitioners in the blocking of their own interests. As Mandeville says 'it is

---

[24] Shaftesbury (1999, 215).

[25] If Mandeville is maintaining a sincere Augustinian position, then he might still identify with the denial of a higher self. We are fallen creatures after all. The idea that firstly we can on our own identify the higher self and secondly that we can then again on our own realize that higher self, is the hubris that Mandeville might be opposing.

[26] Mandeville (1988a, 1:42).

impossible by Force alone to make [the human being] tractable'. The goal in any case is not that of imposing a desire to resist another particular desire, but is rather more ambitious. The goal is to inculcate a desire to resist all desires. The desire that is inculcated must be flexible to the infinite varieties of desire that can be afforded us. In this way the goal is to create in human beings a disposition to be infinitely self-denying. The demand that human beings be made 'tractable' is a high demand, and so could only be done by appeal to some of the deepest features of their actual nature. The way humans are made tractable is through flattery, by pointing out that the best they could do was to be themselves and not to be a lower kind of creature than the kind that one is:

> Which being done, they laid before them how unbecoming it was the Dignity of such sublime Creatures to be solicitous about gratifying those Appetites, which they had in common with the Brutes, and at the same time unmindful of those higher Qualities that gave them the pre-eminence over all visible Beings.[27]

The clever politicians then 'extoll'd the Excellency of our Nature above other Animals', On Mandeville's account there is raised then the feature of the shame in the idea of acting as a different kind of creature than the one that one really is. If humans do act differently, they only maintain 'the Shape of Men, differ'd from Brutes in nothing but their outward Figure' (Mandeville 1988a, 1:44). The concept of virtue itself is then explained in the *Enquiry* as defined in terms of the human/animal demarcation:

> [T]hey give the Name of V I R T U E to every Performance, by which Man, contrary to the impulse of Nature, should endeavour the Benefit of others, or the Conquest of his own Passions out of a Rational Ambition of being good.[28]

Since animals are incapable of resisting their passions, and since human beings are so capable, it is put forward that it is not only a positive thing to resist the passions, but in fact the definitive characteristic of human beings. The clever politicians simply baptize behaviour that distinguishes humans from animals with a concept and thereby creates a notion of moral behaviour.

The advantage of this theory is that it presents an account whereby a new desire is created, the desire to resist one's desires in order to aid of becoming an authentic self. Since this latter desire is presented as the pre-eminently human one, it means that those who had the most boisterous self-belief in their own importance will now become the agents who are the most willing to deny themselves, since 'being human' has now been reconceived as a competition in self-denial. Therefore, 'the fiercest, most resolute, and best among them, [will] endure a thousand Inconveniences, and undergo as many Hardships, that they may have the pleasure of counting themselves Men . . . '.

Mandeville of course adapts this theory in the later edition of the *Fable* with his distinction between *self-love* and *self-liking*. The earlier account stressed the

[27] Mandeville (1988a, 1:42–43).

[28] Mandeville (1988a, 1:48–49).

vulnerability of human beings to their sense of pride. The later account emphasized the natural mechanism in virtue of which this vulnerability arises. Self-liking involves that 'that every one should have a real liking to its own Being, superior to what they have to any other' and that 'Nature has given them an Instinct, by which every Individual values itself above its real worth' (Mandeville 1988b, 2:130).

There are at least three important elements to this conception. Firstly, for Mandeville, self-liking is as natural to human beings as self-love. Self-liking is a biological evolutionary response that inspires a person with 'a transporting Eagerness to overcome the Obstacles that hinder him in his great Work of Self-Preservation'.[29] As such, it cannot be shaken off – it forms a bedrock disposition for human beings, one in accordance with which they co-ordinate and manage their other beliefs. It can no more be abandoned, Mandeville thinks, than the simple attitude just to do the things that please us can be abandoned from our consciousness. What's more, just like self-love, it is not exclusive to human beings. Mandeville takes pains to make the comparison here with non-human animals. He suggests that self-liking behaviour is ubiquitous among other animals and that 'many Creatures shew this Liking, when, for want of understanding them, we don't perceive it: When a Cat washes her Face, and a Dog licks himself clean, they adorn themselves as much as it is in their Power'.[30]

Secondly, it is a valuing activity just like the one identified by Shaftesbury – it is not merely the first-order interests in our consciousnesses but the second-order concern that we take towards those first-order interests. Thirdly, there is the fact that self-liking is essentially *non*-rational. It is a biological trait that provides a helpful role with regard to the demand for self-preservation. However, there is no obvious intrinsic value to one's own interest that makes *its* satisfaction more valuable than the satisfaction of another's. Yet we each naturally believe that it is so. Thus Mandeville holds that it is an entirely natural phenomenon to engage in an entirely non-rational evaluation or qualitative weighting of one set of desires against another.

The consequence of this picture is that human beings' are naturally well-positioned for manipulation. They are primed to accept a belief that will explain the priority and preeminence of their self-centered value. What's more, given the cognitive dissonance that is experienced upon one's failure to satisfy all of his desires, the subject has two options: either give up on the idea that one's own interests are in fact peculiarly important, or invest in a belief system that explains how the non-satisfaction of one set of interests can in fact be an instance of realizing a different and more valuable interest that the subject possesses. As such the subject is naturally disposed to engage in a re-evaluation of which desires ought to be satisfied and which desires ought not to be satisfied as part of one's overall account of the preservation of one's elite status as a human being. It is for an entirely *natural* reason that human beings are willing

[29]Mandeville (1988b, 2:176).

[30]Mandeville (1988b, 2:132).

to be happy with 'the vain Satisfaction of making our Species appear more exalted and remote from that of other animals than it really is'.[31]

The ironic theme – that human beings' need to deny their natural origins itself has a natural origin – is retained in the second volume of the *Fable*. To give just two examples: in the Fourth Dialogue, the origin of politeness is summarized as 'the Management of Self-liking set forth the Excellency of our Species beyond all other Animals'.[32] Similarly, when discussing the human tendency to express anger through scolding and insulting others in the Sixth Dialogue, Cleomenes claims that the effect of insulting is twofold. On the one hand, it makes the recipient of the insult feel degraded; on the other hand, it makes the insulter seem self-controlled, because they have chosen to express their anger by merely engaging in verbal insults and not through unlawful violence:

> Therefore where People call Names, without doing further Injury, it is a sign not only that they have wholesome Laws amongst them against open Force and Violence, but likewise that they obey and stand in awe of them; and a Man begins to be a tolerable Subject, and is nigh half civiliz'd, that in his Passion will take up and content himself with this paultry Equivalent; which never was done without great Self-denial at first: For otherwise the obvious, ready, and unstudy'd manner of venting and expressing Anger, which Nature teaches, is the same in human Creatures that it is in other Animals, and is done by fighting.[33]

Cleomenes goes on to say that since it is horses that kick and dogs that bite, there is a value in expressing anger verbally, which is that one distinguishes oneself from those animals.

10.5 Conclusion

On Mandeville's later account, human beings are already naturally in a position whereby they are willing to accept some belief system that can offer a coherent narrative that explains their importance to themselves. On the one hand, it must explain what the subject really wants to believe – namely, that one's own agency has a priority over that of others. On the other hand, it must explain why the non-satisfaction of one's own desires might have come to be thought of not as a real value in itself. What the human subject demands is a narrative that can justify *ex post facto* this default commitment to his own egotism while he himself undermines its own realization. As such Mandeville uses the Early Modern theme of the distinction of animals in a radical and imaginative way, as a crucial element in his own explanation of the source of the concept of virtue.

[31]Mandeville (1988a, 1:145).

[32]Mandeville (1988b, 2:175).

[33]Mandeville (1988b, 2:295).

References

Augustine. 2006. *Confessions and Enchiridion*, ed. Albert Cook Outler. Trans. Albert Cook Outler. Kentucky: Westminster John Knox Press.

Augustine. 2010. *On the Free Choice of the Will, On Grace and Free Choice, and Other Writings*, ed. Peter King. Trans. Peter King. Cambridge: Cambridge University Press.

Bayle, Pierre. 2005. *A philosophical commentary on these words of the Gospel, Luke 14.23: "Compel them to come in, that My house may be full"*. Indianapolis: Liberty Fund.

Butler, Joseph. 1983. *Five sermons, preached at the Rolls Chapel and a dissertation upon the nature of virtue*. Indianapolis: Hackett Publishing Co.

de Mandeville, Bernard. 1988a. *The fable of the bees*, vol. 1, ed. F.B. Kaye. Indianapolis: Liberty Press/Liberty Classics.

de Mandeville, Bernard. 1988b. *The fable of the bees*, vol. 2, ed. F.B. Kaye. Indianapolis: Liberty Press/Liberty Classics.

de Montaigne, Michel. 2003. *Essays*. Trans. M.A. Screech. London: Penguin Books.

Descartes, René. 1985. *The philosophical writings of Descartes volume I*, ed. John Cottingham, Dugald Murdoch, and Robert Stoothoff. Cambridge/London/New York [etc.]: Cambridge University Press.

Heath, Eugene. 1998. Mandeville's bewitching engine of praise. *History of Philosophy Quarterly* 15(2): 205–226.

Holbach, Paul Henri Thiry. 1889. *The System of Nature: Or, Laws of the Moral and Physical World.* Trans. H.D. Robinson. Boston; [Ann Arbor: J.P. Mendum [University Microfilms].

Kant, Immanuel. 1997. *Lectures on metaphysics*. Cambridge: Cambridge University Press.

Kant, Immanuel. 1998. *Critique of pure reason*, ed. Paul Guyer and Allen W. Wood. Cambridge/New York: Cambridge University Press.

Leibniz, Gottfried Wilhelm. 1997. *New essays on human understanding*, ed. Peter Remnant and Jonathan Bennett. New York: Cambridge University Press.

Locke, John. 1975. *An essay concerning human understanding*, ed. P.H. Nidditch. Oxford: Clarendon Press.

Lombard, Peter. 1981. *Sententiae in IV libris distinctae*. Grottaferrata: St. Bonaventure.

Mandeville, Bernard, and F.B. Kaye. 1924. *The fable of the bees: Or, private vices, publick benefits*. Oxford: The Clarendon Press.

Mandeville, Bernard, VOLDER, Burcherus de.,. 1689. *Resp. Disputatio ... de brutorum operationibus ... Præs. ... B. de Volder.* Leiden: Lugduni Batavorum.

O'Neal, John C. 1996. *The authority of experience: Sensationalist theory in the French enlightenment*. University Park: Pennsylvania State University Press.

Owen, David. 1999. *Hume's reason*. Oxford: Oxford University Press.

Shaftesbury, Ashley Cooper. 1999. *Characteristics of men, manners, opinions, times*. Cambridge: Cambridge University Press.

Welchman, Jennifer. 2007. Who rebutted Bernard Mandeville? *History of Philosophy Quarterly* 24(1): 57–74.

Chapter 11
«Remarks Upon that Wonderful Chapter»: The Controversy on Luxury Between Mandeville and Dennis

Matteo Revolti

Abstract In the debates triggered by the *Fable of the bees*, John Dennis was one of the earliest critics of Bernard Mandeville. In the essay *Vice and luxury* Dennis attacked Mandeville's text, paying much attention to the economic system elaborated by the Dutch author. Specifically, the English writer denied the beneficial effects generated by luxury in enriching society and increasing the wealth of the nation. According to Dennis, this perverse model was responsible of the corruption of English society and was related to slumps such as the South Sea Bubble. In this perspective, Dennis appealed to Machiavelli's civic humanism as the main bulwark against the *Fable*. In particular, he considered liberty and moral virtues as the main defense of civil society. From this point of view, the controversy between Dennis and Mandeville assumed a political meaning by stressing the clash between the values of the ancient constitution and the new economic model promoted by the *Fable*.

Keywords Luxury • Corruption • Machiavelli • Civic humanism • Public spirit • Foreign customs • Liberty • Mercantilism • Laissez-faire

11.1 Introduction

On 9 April, 1724 the *Daily Journal* advertised a book entitled *Vice and luxury public mischiefs: or, remarks on a book intituled, The fable of the bees*. The book's author was 66 year old John Dennis, an English critic, who in those years was engaged in a literary controversy with Alexander Pope (Hooker 1943: xxvi–xxx.). In the 124 pages of *Vice and luxury*, Dennis proposed to criticize some arguments of the *Fable*, focusing his attention on its "Remark L", where Mandeville defended the presence of luxury in England, and praised its beneficial effects for society. In particular, the Dutch physician was of the opinion that the presence of foreign luxury did not diminish the wealth of the British nation. From Dennis's point of view

M. Revolti (✉)
Department of History, Goethe Universität, Frankfurt am Main, Germany
e-mail: matteo.revolti@gmail.com

E. Balsemão Pires, J. Braga (eds.), *Bernard de Mandeville's Tropology of Paradoxes*, Studies in History and Philosophy of Science 40, DOI 10.1007/978-3-319-19381-6_11

Mandeville's remark was a symptom of England's corruption. Dennis ironically referred to Remark L as «that wonderful chapter» (1724: 51), mentioning that the dispute on luxury was at the origin of his essay against the *Fable*.

Within the British public debate caused by Mandeville's *Fable*, though, Dennis's critic *Vice and luxury* was not a great success.[1] Kaye, in his classical edition of the *Fable*, described Dennis as «an extreme rationalist author» (1924: 407–409) and accused him of misunderstanding Mandeville's text. Nevertheless, Dennis's contribution to the topic of luxury in the *Fable* is important for several reasons. First, the topos of luxury is strongly anchored within Dennis's literary production. Second, Dennis was the first critic of Mandeville interested in this topic. As Martin Stafford remarked (1997: 139), his work came out within months of the early attacks against the *Fable* by Law and Fiddes.[2] Unlike these authors, Dennis's criticism did not only concern itself with the moral aspects of the *Fable*, but also with the problem of luxury exposed in Mandeville's text. Moreover his book against the *Fable* symbolized the clash between the values of the ancient constitution and the new social model promoted by Mandeville.

In this contribution I intend to clarify the controversy between Dennis and Mandeville, analyzing first of all Dennis's literary and political background. As we shall see, those elements are important to understanding his polemic against the *Fable*. Later on I want to pay attention to his response to the *Fable* by explicating the arguments and the cultural tradition to which Dennis referred.

11.2 Sir Tremendous: Dennis Between Literature and Politics

Born in London into a saddler's family in 1657, Dennis attended Caius College in Cambridge and received his Master of Arts in 1683 at Trinity Hall.[3] After his degree he began to spend some time in the Will's Coffee-house. Situated at the crossing of Russell and Bow Streets, this coffee-house was the principal meeting point for many poets and literati. Here he met the famous dramatist John Dryden, who was to become his patron later on. Dennis was soon known in literary circles for his ferocious reviews against contemporary authors such as Blackmore and Steele.[4] Mandeville himself in his *Letter to Dion* described him as «a noted Critick, who seems to hate all Books that sell, and no other, has, in his Anger at that

---

[1]This is also evidenced by the absence of a second edition of *Vice and luxury*.

[2]As reported the *Monthly catalogue* for the year 1724, Law's *Remarks* were published in January, whereas Fiddes's *Treatise* in February.

[3]For an account on Dennis's life, see The life of Mr. John Dennis 1734; Paul 1911; Lenz 1913; Tupper 1938: 211–217; Hooker 1943: vii–cxliii.

[4]On the controversy between Dennis and Blackmore see Dennis 1696. On Dennis's relationship with Steele see Hooker 1943: xxxi–xxxiv.

Circumstance, pronounced against the *Fable of the Bees* in this Manner: it is a wretched Rhapsody; the Wit of it is low; the Humour of it contemptibly low, and the Language often barbarous» (1732: 46).

In his literary works Dennis showed himself to be a supporter of the protestant succession. At the death of William III he wrote an epitaph where the defunct sovereign was called the best and greatest of kings (Dennis 1702). According to Dennis the Glorious Revolution restored the values of the ancient constitution repressed by the Stuarts and established the vigour of Christianity in England once again. Politically Dennis was a supporter of the Whig party. This political involvement is especially apparent in his tragedy *Liberty asserted* presented in London in 1704, where he applauded the Whigs as «one who is for the Present Establishment, and the Protestant Succession [...]» (Dennis 1704: a1). In addition his political sympathy is proved by Dennis's friendship with some important members of Whig circles like Charles Montagu, William Cavendish and the Duke of Marlborough. Thanks to those friendships Dennis was engaged not only in literary works but also received official appointments. In 1701 the Duke of Marlborough procured him the place of one of the Queen's Waiters in the Custom-house. During that time Dennis published an essay entitled *A Proposal for putting a speedy end to the war* that concerned the conflict between England and France. In the essay he proposed boycotting French maritime commerce and strengthening the power of the English fleet in international waters (Dennis 1703). He advanced the idea that English merchants could have the monopoly of the market, financing the English fleet against their foreign competitors. Particularly in the *Proposal* Dennis presented a plan concerning the English independence from foreign customs and commerce.

During the years of the debate on the *Fable*, the old critic was losing his ancient prestige. His financial resources became insufficient: he was accused of bankrupt and he was forced to sell his title of Queen's waiter (Paul 1911: 58). In addition Dennis stood more and more in opposition to the literary establishment and entered a controversy with Alexander Pope. Pope and other members of the Scriblerian Club described him as a lunatic writer, calling Dennis «Sir Tremendous or the greatest critic of our Age» (Gay 1717: 18). Dennis responded by claiming that Pope and the Scriblerian Club represented the lowest level of English literature. Dennis's isolation from his literary environment was manifested in his works, where he made up analogies between the ancient Roman world to the present-day situation in England. Themes like the resistance of Coriolanus or the murder of Cesar were used by him to show his opposition against the corruption of British literature (Dennis 1720, 1722).

The shock set off by the *Fable* turned out to be a good occasion for Dennis to redeem his reputation. With *Vice and Luxury* he hoped to cleanse his literary position from the aspersions of the Scriblerian Club. That essay also offered him the opportunity to deal with some matters that interested him, as we shall see. Judging by the lukewarm reception by the British press, *Vice and luxury* did not appear as one of the most significant works of Dennis's production. The fortune of that essay is recorded by the Dennis's first biographer who defined it as «a religious controversy» (The Life of Mr. John Dennis 1734: 53). As late as 1911 a different judgement on Dennis's essay was offered by Harry Gilbert Paul. In his work entitled *John Dennis, his life and criticism*, Paul marked *Vice and luxury* as «his last and longest political tract» (1911: 62). This different evaluation of Dennis's work may be clarified through an analysis of its title, and especially by paying attention to the word "luxury" contained in it.

11.3 The Reception of Luxury in Great Britain and Dennis' Work

Historically, the term "luxury" did not have a monolithic meaning, but assumed several senses over the years. As many studies demonstrated, in England at the end of the seventeenth century that topic concerned not only the moral dimension: luxury was also tied to the problem of commerce on British soil (Sekora 1977; Berry 1994; Grugel-Pannier 1996; Berg and Clifford 1999; Reith 2003; Berg 2005). According to that approach luxury showed itself through the introduction of foreign products in England. Consequently, luxury appeared as the negative effect of commercial activities and its foreign character was the cause of its bad reputation among the inhabitants of the British Isles.

This argument was raised many debates. In the essay *Usury at six per cent* (1669) Thomas Manley criticized the proposal by Thomas Culpeper in the *Tract against the high rate of usurie*, where he had suggested the reduction of taxation from ten to six per cent. Manley replied to Culpeper arguing the reduction would provoke the vain consumption of foreign commodities imported by English money. In other words, the depletion of British currency reserves. This position against "foreign commodities" was emphasised particularly at the beginning of the war between France and England. The consumption of foreign products, especially from France, was perceived as a betrayal of English interests. This fear was launched in the year 1700 by *A Proposal for remeeding our excessive luxury*, where the anonymous author asserted that «the promoving of our Trade, and amending of our Coin, will not much increase our Stocks, untill we amend our Manners: for if Luxurie bring in more Goods than by our Native product and industrie we export» (A Proposal for remeeding our excessive luxury 1700: 1).

The idea of luxury as an exogenous element appeared in many literary works by Dennis. In 1705 he argued that luxury had a precise geographic counterpart in French customs and in the following year he even identified the music of Italian operas as a sign of luxury (Dennis 1706). Despite the literary attention devoted to luxury, the theme achieved a political meaning only in Dennis's *Essay upon publick spirits* published in 1711. In the 31 pages of that essay he denounced British submission to continental fashions as a form of prostitution. This submission was responsible for the death of public spirit, which is «the ardent Love of one's Country, affecting us with a zealous Concern for its Honour and Interest, and inspiring us with Resolution and Courage to promote its Service and Glory» (Dennis 1711: 2). According to Dennis the lack of this spirit was provoked by several factors, such as the importation of foreign customs and the impotence of the Church towards the dangerous influences of freethinkers. This state was emphasized by Dennis by his comparison with the kingdom of Henry VII: the frugality and sobriety of the first Tudor king was a counterexample to the contemporary death of public spirit (Dennis 1711: 8). Of course, the dispute on corruption and the public spirit was a common theme among Augustan moralists, satirists and political pamphleteers. As Goldsmith reminds us, the controversy combined two distinguishable strands of thought: on one side they referred to the values of the country ideology, on other side they saw a support against corruption and luxury in Puritanism (Goldsmith 1985: 3–4). The first element appeared in Dennis's *Essay*: we find the names of Lycurgus and Plato as champions of the republican side. Nevertheless, Dennis did not restrict himself to condemning contemporary customs. Remembering his experience at Customs-house, he also offered several proposals to contain the spread of foreign luxuries such as the introduction of sumptuary laws. In the *Preface* to his *Essay* Dennis argued that:

> If what is here is publish'd is favourably receiv'd, I shall endeavour to shew in a Second Part the mighty Mischiefs that the introduction of foreign Manners and foreign Luxury hath done to this Islands, and to rest of Europe; and the proper Methods that are to be us'd to restrain Luxury, and to retrieve Publick Spirit. (Dennis 1711: v)

11.4 Republican Virtues, Modern Vices: Dennis vs Mandeville

Thirteen years later, Dennis returned to the topic with *Vice and luxury*. We do not know whether this work was conceived by Dennis as the continuation of his previous essay. However, many of the *Essay*'s arguments also appeared in *Vice and luxury*. This hypothetical link can be proved partly from the structure of the text. In the appendix to the comment on the *Fable* Dennis also published four letters he had addressed to Sir Richard Blackmore and John Potter earlier. The letters were written some months before the scandal of the South Sea Bubble and they returned to the topic of the corruption of British spirit.

The choice of *Vice and luxury*'s addressee fell on the influential Earl of Pembroke, Thomas Herbert. The English politician had been the patron of authors like John Locke and had been elected as the president of the Royal Society in 1689. Pembroke's name in Dennis's work is significant for two reasons. First, Dennis himself dedicated his *Essay on the navy* to Pembroke in 1702. Second, Pembroke was chosen as addressee not only by Dennis but also by Fiddes, another critic of the *Fable*.[5] In his *Preface* Dennis wrote that his essay intended to defend the religion which was understood as a bulwark of the civil society and British liberty. The defense of religion was explained through a typical Whig syllogism. Liberty and society depend upon a contract between king and citizens; this contract is sealed by

[5]For the work against Mandeville see Fiddes 1724.

religion; therefore religion is the fundament of a civil society. *Ergo* infidelity is the principal threat to current liberty. That threat especially accrued from the issue of irreverent books, a phenomenon that had reached its peak with the *Fable*:

> But a Champion for Vice and Luxury, a serious, a cool, a deliberate Champion, that is a Creature intirely new, and has never been heard of before in any Nation, or any Age of the World. And to make it farther appear, how widely Infidelity, and how diffusively its Offspring Vice and Luxury have spread, the Work which this Champion has publish'd in their Defence, has found great Success, tho'a very wretched Rhapsody, weak, and false, and absurd in its Reasoning; aukward, and crabbed, and low in its Wit; in its Humour contemptibly low, and in its Language often barbarous. (Dennis 1724: xvi–xvii)

The preface continued with a defense of the charity schools. The defense of these institutions is conducted by Dennis by means of two arguments. First Dennis claims the formative character of charity schools, stating that the knowledge imparted by them is a necessary attribute to direct the actions of the young people to virtue. Dennis countered the criticism of their professional uselessness by accusing Mandeville of depriving the poor of the education, forcing them to be cheap labourers. Dennis concluded the preface by informing his readers that he will contest the assertions of the *Fable* with three remarks.

In the first remark Dennis offers a semantic analysis of the title of the *Fable*: he argues that private vices did not provoke public benefits, but only public mischief (1724: 2–6). For Dennis, public benefits are addressed to liberty, whereas vices and luxury are caused by public mischief, as shown by the recent scandal of the South Sea Bubble. In this range he quotes Sidney and Machiavelli as champions of virtues and authorities of the republican tradition. The names Sidney and Machiavelli in Dennis's essay were a reference to the tradition that Hans Baron called civic humanism. A confirmation of that adhesion is proved by the fact that Dennis also declared «I may be branded with the odious Name of Republican, and pass for an Assertor of Democratical Government» (1724: 8). Civic humanism linked the ideas of freedom and independence. This view has its fundament in Aristotelian conception of man as "a politikoon zoon" who improved his virtue through his political participation and in the practice of citizenship. In particular, this tradition elevated Rome as champion of ancient prudence: its fall was a consequence of the corruption and the feudalization by emperors and their mercenaries. The theme was central in Machiavelli's *Discorsi* and was transferred in England by Harrington in the seventeenth century. As Pocok demonstrated in his studies, the followers of Harrington, the so-called Neo-Harringtonians, saw the main source of British corruption in the growth of credit and commerce at the end of seventeen century. Dennis saw himself in this tradition, calling to his defense Sidney and Machiavelli against the corruption brought by the *Fable*. From Sidney, Dennis cited the *Discoures concerning government* where Sidney asserted the supremacy of popular government built on virtues and civic qualities against Filmer (Dennis 1724: 7). Conversely, the name of the Florentine secretary is linked to his *Discorsi sopra la primo decade di Tito Livio*: Dennis particularly alluded to the twelfth chapter of the first book where Machiavelli suggested how the government from corruption could be preserved.

Dennis concluded his first remark by observing that luxury was a danger for religion and the cause of the actual divisions in England. As he asserted:

> That the People of *Great Britain* were never so divided as they are at the present, appears to be manifest; for not only the whole Body of the Nation is divided into *Whig* and *Tory*, but *Whigs* are divided against *Whigs*, and *Tories* against *Tories* [...] The Clergy is not only divided into *Whig* and *Tory*, but the *Tory* Clergy is divided again into *Jurors* and *Nonjurors*; the latter of which abhor the former, and the former despise the latter. (Dennis 1724: 24)

The second remark is devoted to notes about the *Enquiry into the origin of moral virtues*. Against Mandeville's argument about the a-moral genesis of civil society, Dennis argued that religion and virtues are the bases of society. Mandeville declared in his *Enquiry* that virtues are the political offspring which flattery begot upon pride (1729: 22). Dennis replied that religion and virtues could be considered the pillars of every society as the Roman republic demonstrated. Especially, he argued, the existence of two different kinds of religion: natural and revealed religion. Dennis held the opinion that the first was stronger than the second, because the heathens followed laws, without being Christians. In agreement with Machiavelli, Dennis recognized that «Roman Virtue was the Effect of their Religion, and not of the Contrivance of wary Politicians» (1724: 48).

In the third note Dennis analyzed the famous remark *L* of the *Fable* dedicated to luxury: as the English author remembered, the praise of luxury was at the origin of his essay because «it has rais'd so much fresh Indignation in me, that I cannot help bestowing some particular Remarks upon that wonderful Chapter» (Dennis 1724: 51). In his remark Mandeville backed the beneficial effects of luxury with three reasons: first, that in one sense everything may be called luxury; and that in another there is no such thing. Secondly, that under a wise administration all people may swim in as much foreign luxury as their product can purchase, without being impoverished by it. Finally, its presence did not bring about negative effects in the army (Mandeville 1729: 73–85).

Dennis refuted the first thesis on the relative meaning of luxury, alluding indirectly to the Aristotelian doctrine of *metriotes*. According to him moderation was the best canon to identify what is excessive and what is not. As an example to prove his theory, he uses the image of a diet, as the mid-point between abstinence and gluttony (Dennis 1724: 54). Therefore he held the opinion that every man could distinguish vice from virtue as well temperance from luxury. But the most important of Dennis' considerations on luxury was the one that criticized a laissez-faire *ante litteram*, as some of Mandeville's critics have called it. Although there has been (and is still) a great debate on whether Mandeville was a mercantilist or an advocate of laissez-faire (Heckscher 1931; Viner 1953; Rosenberg 1963; Chalk 1966; Hayek 1966; Goldsmith 1977; Horne 1978: 51–57; Carrive 1994: 301–322), the criticism formulated by Dennis seemed to tend toward the second assumption.

A significant example of the model was expressed by Mandeville in mentioning the commerce between Turkey and England. In his remark Mandeville criticized the assumption that only domestic commerce could improve the economy of England. This mercantilist argument was denied by Mandeville through the example of trade between Turkey and England. If Turkey decreased her importations from England, the same Turkey would have less possibility to buy English products and would abandon its commercial partnership with England. Although Mandeville stated that the imports never exceed exports, he declared that no nation could be impoverished by the foreign luxury (Mandeville 1729: 74–79).

Dennis responded to Mandeville that if a man, with a £ 500 in income spends all his savings on foreign products like champagne, Tokaj or oil, he will become poor and the nation will receive no benefits (1724: 55). In addition, Dennis criticized Mandeville's argument that luxury was not related with corruption. In the *Fable* the Dutch physician maintained that corruption and robbery derived only from bad administration and not from luxury. Dennis asserted instead that the political administration was not guilty; for him the state had only a guarantor role and had to defend the properties of the citizens. In this perspective the state cannot intervene into private affairs, therefore the presence of corruption is due only to individual luxury (Dennis 1724: 56–59).

Finally Dennis dealt with the argument about the benefits from the circulation of luxury in the military sphere, showing the absolute incompatibility between luxury and army. That dissonance stood out with the example of Pompeius debacle in Farsalo due to the indolence of his army. For Dennis a free society could maintain its existence only through its warrior qualities: therefore he stressed the importance of the fighting values which marked the republican societies like Rome.

In conclusion *Vice and luxury* represents a key piece of the puzzle within the controversy on luxury raised by the *Fable*. Dennis was the first author to discuss Mandeville's considerations in the framework of economic system, reading Mandeville's text as an extended form of a perverse model which had spin-offs into social episodes like the South Sea Bubble.

Dennis identified himself with the values of the republican tradition, as a bulwark against the social and economic model proposed by Mandeville. His constant references to authors like Machiavelli and Sidney symbolized Dennis's defense of the republican tradition and his discomfort vis-à-vis the corruption of the society. Dennis's concern is constantly demonstrated through the defense of civil liberty: according to Dennis, England's liberty was threatened by her economic and cultural dependence on foreign customs and public debts. This stress on freedom and independence places Dennis within the circle of the Country Whigs, who fought against the financial revolution and the moneyed interest. In this perspective Dennis with *Vice and luxury* showed the contrast between the nostalgia for a virtuous and traditional society and the new economic model promoted by the *Fable*.

References

Primary Sources

A proposal for remeeding our excessive luxury. 1700.

Dennis, John. 1696. *Remarks on a book entituled prince Arthur, an heroick poem: With some general critical observations and several new remarks upon Virgil*. London: Samuel Heyreck/Richard Sare.

Dennis, John. 1702. *The monument: A poem sacred to the immortal memory of the best and greatest of kings, William the third. King of the Great Britain*. London: Daniel Brown/Andrew Bell.
Dennis, John. 1703. *A proposal for putting a speedy end to the war, by ruining the commerce of the French and Spaniards, and securing our own, without any additional expense to the nation*. London: Daniel Brown/Andrew Bell.
Dennis, John. 1704. *Liberty asserted. A tragedy. As it is acted at the new theatre in little Lincoln's-Inn-Fields*. London: George Strahan/Bernard Lintot.
Dennis, John. 1706. *An essay on the opera's after the Italian manner, which are about to be establish'd on the English stage: With some reflections on the damage which they may bring to the publick*. London: John Nutt.
Dennis, John. 1711. *An essay upon publick spirit; being a satyr in prose upon the manners and luxury of the times, the chief sources of our present parties and division*. London: Bernard Lintot.
Dennis, John. 1720. *Coriolanus, the invader of his country: Or, the fatal resentment. A tragedy; as it is acted at the Theatre-Royal in Drury-Lane*. London: John Peele.
Dennis, John. 1722. *Julius Caesar acquitted, and his murderers condemn'd. In a letter to a friend. Shewing, that it was not Caesar who destroy'd the Roman liberties, but the corruptions of the Romans themselves. Occasion'd by two letters in the London journal, the one of the 2d, the other of the 9th of December*. London: J. Mack-Euen.
Dennis, John. 1724. *Vice and luxury public mischiefs: Or, remarks on a book intituled, The fable of the bees; or private vices, publick benefits*. London: William Mears.
Fiddes, Richard. 1724. *A general treatise of morality, form'd upon the principles of natural reason only. With a preface in answer to two essays lately published in The fable of the bees. And some incidental remarks upon an inquiry concerning, virtue, by the right honourable Anthony Earl of Shaftesbury*. London: Samuel Billingsley.
Gay, John. 1717. *Three hours after marriage. A comedy, as it is acted at the theatre royal*. London: Bernard Lintot.
Mandeville, Bernard. 1729. *The fable of the bees*. London: Jacob Tonson.
Mandeville, Bernard. 1732. *A letter to Dion, occasion'd by his book call'd Alciphron, or the minute philosopher*. London: James Roberts.
The life of Mr. John Dennis, the renowned critik. In which are likewise some observations on most of the poets and criticks, his contemporaries. Not written by Mr. Curll. 1734. London: James Roberts.

Secondary Sources

Berg, Maxine. 2005. *Luxury and pleasure in eighteenth-century Britain*. Oxford: Oxford University Press.
Berg, Maxine, and Clifford Helen. 1999. *Consumers and luxury: Consumer culture in Europe 1650–1850*. Manchester: Manchester University Press.
Berry, Christopher J. 1994. *The idea of luxury: A conceptual and historical investigation*. Cambridge: Cambridge University Press.
Carrive, Paulette. 1994. *La pensèe politique anglaise: passions, pouvoirs et libertés de Hooker à Hume*. Paris: P.U.F.
Chalk, Alfred F. 1966. Mandeville's fable of the bees: A reappraisal. *Southern Economic Journal* 33: 1–16.
Goldsmith, Maurice M. 1977. Mandeville and the spirit of capitalism. *Journal of British Studies* 17: 63–81.
Goldsmith, Maurice M. 1985. *Private vices, public benefits: Bernard Mandeville's social and political thought*. Cambridge: Cambridge University Press.

Grugel-Pannier, Doris. 1996. *Luxus: eine begriffs- und ideengeschichtliche Untersuchung unter besondere Berücksichtigung von Bernard Mandeville*. Frankfurt am Main: Lang.

Hayek, Friedrich A. 1966. Dr. Bernard Mandeville. *Proceedings of the British Academy* 52: 125–141.

Heckscher, Eli F. 1931. *Merkantilismen: ett led i den ekonomiska politikens historia*. Stockholm: Norstedt. English edition: Heckscher, Eli Filip. 1934. *Mercantilism* (trans: Mendel Shapiro). London: George Allen & Unwin.

Hooker, Edward N. 1943. *The critical works of John Dennis*, vol. 2. Baltimore: Johns Hopkins Press.

Horne, Thomas A. 1978. *The social thought of Bernard Mandeville. Virtue and commerce in early eighteenth-century England*. London/New York: Macmillan/Columbia University Press.

Kaye, Frederick B. 1924. *The fable of the bees: Or private vices, publick benefits*, vol. 2. Oxford: Clarendon.

Lenz, Hermann. 1913. *John Dennis, sein Leben und seine Werke, ein Beitrag zur Geschichte der englischen Literatur im Zeitalter der Königin Anna*. Halle: Ehrhardt Karras.

Paul, Harry G. 1911. *John Dennis, his life and criticism*. New York: The Columbia University Press.

Reith, Reinhold. 2003. *Luxus und Konsum. Eine historische Annäherung*. Münster: Waxmann.

Rosenberg, Nathan. 1963. Mandeville and laissez-faire. *Journal of the History of Ideas* 24: 183–196.

Sekora, John. 1977. *Luxury: The concept in western thought, Eden to Smollett*. Baltimore: Johns Hopkins University Press.

Stafford, Martin. 1997. *Private vices, publick benefits? The contemporary reception of Bernard Mandeville*. Solihull: Ismeron.

Tupper, Fred S. 1938. Notes on the life of John Dennis. *English Literary History* 5: 211–217.

Viner, Jacob. 1953. Introduction. In *A letter to Dion*, Augustan reprint society publication 41, ed. Mandeville Bernard, 1–15. Berkeley/Los Angeles: University of California.

Chapter 12
Mandeville and the Therapeutics of Melancholic Passions

Cláudio Alexandre S. Carvalho

Abstract As modern variants of melancholy, Hypochondria and Hysteria are illnesses to which Bernard Mandeville devoted the major part of his medical writings. In framing and treating mental illnesses, Mandeville is sensitive to the increasing importance of new commodities, habits and social expectations. This will lead him to read those modern variants of melancholy as a palimpsest composed of symptoms and narratives whose main origin is the praise of individual's strive for pleasure and singularity. Detached from early eighteenth century academic and advising manuals, the dialogical model will grant access to a new grammar of physiological occurrences, complaints and treatments. Such grammar not only accepts but also requires patients' enrolment and active voice. The dialogical model adopted by Mandeville will also enable a questioning of medical expertise at the same time that increasingly demands from it the relief of social anxieties. Mandeville's approach to mental illnesses is centred on digestive problems, literally but also in the figurative sense, referring to the lack of discipline and the various consequences of the difficulty in processing and assimilating large amounts of goods and information.

Keywords Melancholic illnesses • Therapeutic theory and practice • Clinical roles • Social dimensions of mental illness • Semantics of complaint and relief

---

This article benefited from engaging discussions and suggestions made by various specialists in Mandeville's life and work, particularly Dr. Mauro Simonazzi, an expert in the cultural and scientific context underlying Mandeville's conception of melancholic passions. Thankful for such guidance and encouragement, I was lead to explore new aspects of Mandeville's view of medical practice deepening my initial views on the therapeutics of melancholy.

C.A.S. Carvalho (✉)
LIF – Language Interpretation and Philosophy, University of Coimbra, Coimbra, Portugal
e-mail: kraftcasc@gmail.com

E. Balsemão Pires, J. Braga (eds.), *Bernard de Mandeville's Tropology of Paradoxes*, Studies in History and Philosophy of Science 40, DOI 10.1007/978-3-319-19381-6_12

12.1 Introduction

Mandeville's *Treatise of the Hypochondriack and Hysterick Passions*, first edited in 1711, incorporates a rich understanding of the organic, psychic and social dimensions of those passions, or diseases as he will call them, and their treatments. An analysis of this text enables us to grasp (a) a reflexion upon Medicine's management of passions, here including a new model of clinical relation, (b) a survey on the strict relation between physical and mental disorders and, in connection with Mandeville's larger project, (c) the way some diseases have at their origin excesses of passion related to deep changes in society's, mainly in commerce and in the division of labour.

The value of Mandeville's medical writings has firstly been summarized by G. S. Rousseau (1975), that detailed various points where the *Treatise*, revising and reordering themes previously present in the academic theses Mandeville submitted at Rotterdam and Leiden, raises questions that still relevant in contemporary Medicine, especially in psychiatric expertise. The same reader of the *Treatise* emphasized its value as a sociological and literary document. An important question constantly reappearing in the text, and to which Mandeville is very sensitive, can be frame within cultural Psychiatry and consists in clarifying whether diseases can be reduced to objective signs of decay of bodily and mental functions or if each society has its own illnesses influenced either by specific aetiologies and social normativity.

Aside this particular questions, the *Treatise* cannot be considered a minor or merely a curious work among the profusion of similar texts in late seventeenth and eighteenth centuries, set apart from Mandeville's larger project. I follow E. Hundert's suggestion that this literary return to Physics is a necessary moment in Mandeville's project of a Science of Man, obscured by the polemic generated around *The Fable of the Bees*. Such project should understand society's formation and functioning proceeding through "anatomizing the invisible Part of Man" (Mandeville 1988a: 145),[1] going beyond moral assumptions discredited with the emergence of a commercial society. Medicine is a privileged instrument to observe Mankind, taking nature no longer as a source or a mean to improve morality but to evaluate it and liberate us from its "tyrannical demands" (Hundert 1994: 47). In this venture concerning illness and its management, Mandeville assumes himself as an "empyrick" who observes and critiques the multiple appearances of the "idol of Reason", itself a result of pride (Mandeville 1711: 53), correcting his previous conceptions along the way. At the same time, his "dramatization" of medical observation is accompanied by an acute evaluation of the current state of the art of medical theories and therapeutic practices. In the edition of 1730,[2]

[1] When quoting I decided to adapt the original text to modern English.

[2] Even though Mandeville affirms to have "made great alterations" to the first editions of the *Treatise*, we find only circumstantial differences along with dispersed addendums embedded in the dialogical sequence, mostly leaving unaltered previous arguments and theses (cf. Mandeville 1730: xxii). This only strengths the urgent need for a full critical edition this work deserves.

Mandeville still presents an analysis of the distortions of a practice due to pursuit of profit, glory or the imposition of one's own idols over others. Such critique is clearly beneficiary from his previous work on the *Fable*, and, both the illnesses considered as the complex web of therapeutic assistance, are seen as a by-products of society's evolution and flourishing.

12.2 Historical Background

Some biographical aspects (which I cannot present in due detail) such as his Dutch scientific and political backgrounds, his affinity with Pierre Bayle among others, and his report by the London College of Physicians,[3] will certainly enhance Mandeville's sceptical position towards not only "galenist" speculation and conjectures on anatomy and chemistry of physiology, but also to naïve mechanicist theories.[4] According to Mandeville those opposing views of science were both impelled by medical instruction and Guilds preoccupied solely in maintaining their knowledge and prestige undisputed. Such labour in constructing an image of virtue, was to be corrected so that modern Medicine could progress, assuming its social functions, released from religion's leash.

In accordance with the *Fable of the Bees* observations, Mandeville becomes increasingly aware of the mechanisms underlying social approval and "favour of the public". He not only considers the signs and values that society generally praises for a given social role, but also the way certain norms, sometimes dependent on the current prevailing trend, impose on the individual the need to display a set of outward behaviours (cf. Mandeville 1730: 176). Concerning medical practice, Mandeville further details his first sketch on the non-necessary coincidence between certified knowledge (and attributions) and efficiency in therapeutic practice. Indeed, Mandeville understands "quackery" as the reliance on various schemes of obscurity and superstition used to confuse and control the patients without dealing with the disease. One of those schemes is rhetorical and is mainly imputed to academic doctors and small conmen that intently try to explore patients' or patrons' money. The other

[3]In this text, we find a repercussion of the part Mandeville took in defending his fellow Joannes Groenwelt in 1703 (cf. Gordon 1931).

[4]In his *Treatise*, Science is compared to clothing fashion and the conflict between different theses or paradigms is considered to be fuelled not only by experience and evidence gathered but also envy (cf. Mandeville 1711: 114). Latter he will stress how reputation is frequently build upon spurious motives, chief among them the search for novelty: "[t]here is a vast pleasure in saying something that is not recorded to have been said before" (Mandeville 1730: 327). That search benefits from partnerships with the wonders arriving from the new world. Misomedon gives as example the bezoar stone (*Lapis Bezoar orientalis*), brought from the Indies into Europe by Dutch and Portuguese merchants, promising among others the cure for Epilepsy and Melancholy (cf. Cook 2007: 191–203).

consists in resorting on some of the chimeras of iatrochemistry, created by experimentalists like Paracelsus and Jan Baptist van Helmont, as effective means of cure.

Certain authors considered Mandeville's *Treatise* as being ahead of its time, not so much for its general method to consider and treat illnesses, which we can include in the approach of iatromechanism, but for its new understanding of clinical relation as part of an adequate treatment of melancholic "distempers". The form of dialogue, so distant from conventional medical treatises of late seventeen and early eighteen-century England, made easier the exposition of mental and spiritual symptoms of a disease, in a process of discovery made accessible to any lay reader (cf. Ingram 2011: 186–187),[5] but, as we will see further, it also served a therapeutic purpose. In the "Preface" he states that his intention is to surpass the initial barrier of impatience "one of the surest Symptoms of those who seek relief" (Mandeville 1711: vii).[6] One cannot help being surprised by the importance Mandeville puts into dialogue as a vital part of health's promotion, especially considering the physiological and even the materialist bias of his conception.

However, this work is not apart from its time, not in terms of terminology nor procedures, both consistent with the path towards Medicine's full differentiation. It provides a non-moral reading of organic processes, and an autonomy from religious dogmas.[7]

Since antiquity, Melancholy refers both to a constitutive temperament marked by the abundance of black bile, or to an acquired imbalance of humours that comes to alter one's original temperament. Such condition, innate or acquired, becomes aggravated in certain seasons, ages and in some living conditions or regimes, and with exception of a tradition relating it to genius (recaptured in the poetic and religious theme of enthusiasm), it refers to a withdrawal from the world, a deep and uncaused sadness, sometimes considered an essential detour to become closer with the divine, generally trough mortification and despair.

When Renaissance authors came to compile the long and disputed list of characters and symptoms associated with the illness they erected a complex web of physiological, moral and cosmological associations. Since then, melancholy has been problematized as being an imprecise condition susceptible of figment and also a way to cover or excuse moral weakness (cf. Schmidt 2007: 168–170). In his encyclopaedic work, R. Burton affirmed that "the Tower of Babel never yielded such confusion of tongues, as this Chaos of Melancholy doth variety of symptoms" (Burton 1883: 240).

---

[5]This directedness to lay readers was improved in the edition of 1730. In contrast with the previous editions, in the third enlarged edition, Mandeville adds footnotes translating Latin and Greek passages whose meaning was accessible only to a few learned doctors.

[6]At the same time, we must acknowledge that, despite its valuing of close dialogue and interview with the patients, the *Teatrise* presents several digressions close to the monological form of speech.

[7]According to the history of medical thought, particularly psychiatry, this aspects contribute to the creation of its own self-reference on distinguishing between illness – non illness (cf. Steinebrunner 1987: 368 ff).

From the various symbolic assemblages, dispersed on arts, satirical and medical literature, two symptoms will stand out in Hypochondria (also known as the Spleen) and Hysteria (also called Vapours), its variants. Those main symptoms are lethargic fear and sorrow, accompanied by various kinds of physical pain and soul's afflictions. Burton referred that this dual nature of melancholy required a "whole physician" capable to treat the body without neglecting the soul. Its various prophylaxis and palliatives focused (like in most of Elizabethan literature) on the taming of cruel affections through the power of reason, capable of instructing the will (Burton 1883: 329–331). Such care should at least complement the various galenic cures which aim to regulate humour's balance. According to medieval doctrines that still prevailed, the deregulation of bodily humours, was the occasion for the devil to tempt one into sin.[8]

For Philoperio (and his interlocutor) most of the *worries and suffering no longer derive from the signs of eternal damnation, so conspicuous in religious melancholy, but were expressed in complaints about everyday difficulties and pains that lead to a certain suspension of ordinary life*. Enthusiasm, once a sign of devotion, a means to represent a privileged connection with God, becomes a disorder caused by confused or exhausted fancy, closer to Frenzy than to Melancholy in its strict sense.

From the second part of the seventeenth century onwards we see a tendency to ground bewitchment in the physical level, sometimes with the survival of the idea that devils could interfere in humours equilibrium (Schmidt 2007: 131 ff).[9] The idea that melancholic forms of disease were a result of sin, especially falling into devil's temptation, was "upgraded" by some eighteenth century physicians that saw afflicted mind as being punished by material excesses, promoted by a generally affluent society.[10] Mandeville will obviously argue that the care of the body is dependent from the care of the soul, but will refuse to identify the latter with a religious or medical concept of virtue. Therapy is intended as a way to restore health and promote individual well-being. Hypochondria in specific was manifested in soul's 'labouring against herself', a passion that tends to force individual's withdrawal from normal social relations, and similar to modern concept of repression, dominating all other passions.

Post-Restoration society will emphasize the need to enforce moral perfection and promotes emerging charity schools and Societies for the Reformation of Manners. The underlying conviction of such movements, abundant in London, was the identity between reason -particularly obtained through high education- and the

[8]See for instance the subsection "A Digression of the nature of Spirits, bad Angels, or Devils, and how they cause Melancholy" in Burton, *The Anatomy of Melancholy* (1883: 115 ff).

[9]This idea is submitted to inspection in the *Treatise of the Hypochondriack and Hysterick Passions* (cf. Mandeville 1711: 105).

[10]This vision will underlie George Cheyne's *The English Malady*. In her work devoted to *Obesity and depression in the Enlightenment: the life and times of George Cheyne*, Anita Guerrini acknowledges the influence of Mandeville's theory of passions, not only medical but also "sociological", on Cheyne's analysis.

good character and works. According to this general framework of *recta ratio*, living in sin is being mentally and spiritually ill. The satire towards these ideals as founders and promoters of moralist hypocrisy is not entirely absent from the *Teatrise.*[11] Critique, sometimes ironical others comical or plain, remains within the attempt to circumscribe good, doubtful and bad medical practice. And here we must underline the term "practice", for Mandeville says, through Philopirio's words,[12] that since antiquity the greatest contributions to the different branches of Medicine (Anatomy, Physiology and Therapeutics) derived from continuous and rigorous observation of individual manifestations of disease. Observation of those signs should be complemented with dietetic prescriptions and pharmacopeia that experience proved good in promoting nature's force, always taking into account patient's idiosyncrasies, constitution/temper and biography. Along the *Treatise*, Sydenham is considered one of the greatest representatives of such procedure and his resistance to give explanations for phenomena that, given the limitations of human science, remain inexplicable, is very much praised. He understands the products of reason as supplement to experience and practice. This attitude is contrasted with those that characterize the "prolific brain" always prepared to find answers and models able to provide explanations to every symptom of any given illness (cf. Mandeville 1711: 107–108). The frequent resort on metaphor and speculations is considered highly dangerous by Philoperio, since figurative descriptions of a disease can lead to delays in the pursuit of adequate cures and to damaging prescriptions (cf. Mandeville 1711: 86–90; 1730: 226–231). So, even if Philoperio reviews some theories praising them for their ingenuity, he ironically remarks: "What a pity it is they won't cure sick people" (Mandeville 1711: 86).

Mandeville will add in the last edition of the *Treatise* that, in spite of his admiration for algebra and geometry: "to the practice of Physic, I mean the cure of diseases, there is no part of Mathematics that can be a greater help, or give more light in the mysteries of it, than it can in those of revealed Religion" (Mandeville 1730: 173). Its study is praised in the formation of doctors, for without it the detailed truth of human constitution and mechanism cannot be accessed and the absence of certainty [*Wiskonst*] would leave diagnoses bound to mere guessing. However, given the variety of organisms and diseases, the principles of Newton philosophy couldn't find no firm soil to ground a systematic therapeutic for any disease (see Mandeville 1730: 180–201). Here we find the touchstone of Mandeville's view of medicinal

[11]Indeed, has shown by Ph. Hilton in *Bitter Honey. Recuperating the medical and scientific context of Bernard Mandeville*, a book Sir Malcolm Jack had the kindness to recommend me, the legacy of satire is transformed and, as comedy in general, is considered to be a resource to refresh animal spirits and prevent madness.

[12]The *Treatise* is unequivocal in identifying Mandeville with Philoperio, and in the last edition of the work the biographical points are supplemented by his self-description: "a Foreigner and a Physician, who, after he had finished his studies and his Degree beyond sea, was come to *London* to learn the language; in which having happened to take great delight, and in the meantime found the Country and the manners of it agreeable to his humour, he has now been many years, and is like to end his days in *England*" (Mandeville 1730: xiii).

practice, identified with the ability to cure, it must not be so vain as to presuppose a regularity of nature's manifestations (Mandeville 1711: 140). Claiming its classic legacy, Medicine is conceived as an Art based on recurrent observation, learning of the mysterious ways nature manifests its vigour in human body and mind, depending on the variety of one's constitution, temper and passions.[13]

But paradoxically, Mandeville seems convinced that free market can discipline medical services, regulating Doctor's strive for reputation (pride's fulfilment) and income.[14] In this sense, in the third dialogue, Philoperio has no reluctance in admitting to his patient that he places himself within a fierce marked to obtain personal gain.[15] Virtue is no more a predicate of a good Physician then it is of a Politician (cf. Cook 1999).

12.3 A New Physiological Model

Mandeville was aware that until the present day no one had given a convincing account of the interaction between mind and body. However, he sustained that medical observation and experience, despite such mystery, had proved effective in relieving patients. This was particularly so in the various kinds of melancholy, which covered a wide range of mental illnesses, that, in the pronounced transition to commercial and polite society, would "discover" new worries and highlight old ones.

A new understanding and grammar of illness will support the treatment of melancholy. With the gradual overthrow of the humoreal aetiology of melancholy that took place along seventeenth century, and the appearance of mechanicist models of passions, research on mental illness will be directed toward the nervous system and the communication between the mental and the "organick" involved in passion. But Cartesian or mathematical models of the human body were incapable to grasp the relation between the body and the soul, their reciprocity and participation.

[13]In Mandeville's view, nowhere has such practice been more reliable than among the Hippocratic School. In the edition of 1730, Philoperio discredits the narratives of Melchisédech Thévenot that referred Chinese physicians as exemplary of the dedication and patience a doctor must have in reading the natural and non-natural signs of the body. Such narratives had its origins in missionary incursions and resulted from the wonder of Jesuits witnessing prodigies and ritual customs. He maintains that medical information gathered from inspection of urine and feeling of the pulse were proficient only in the eminent practitioners of antiquity (cf. Mandeville 1730: 78–79).

[14]As remarked by Anthony Francis McKee (1991: 6), this reference to selfish passions as a motive to pursue medical studies can already be found on Mandeville's *De Medicina Oratio Scholastica*, written when he was only 15 years old.

[15]"It is indeed no exaggeration to say that Mandeville, writing almost 300 years ago, has prophetically anticipated the medical problems facing such a nation as the U.S.A., for example, in which present-day discussions about socialized medicine evolve around the physician's conception of himself as an economic creature, a social creature, an ethical and, of course, a professional creature diagnosing and curing illness" (Rousseau 1975: 20).

Mandeville repeatedly states that the problem with mechanical explanations is their tendency to suppose a regular functioning of the body, overlooking "organick" nature's dependency on minute and volatile elements, specialized in different movements and functioning. This leads their representatives to bold generalizations on tempers and passions, based on the scarce knowledge provided by anatomy. Philopirio sustains that an adequate level of analysis of the body and its passions functioning requires a certain "sentiment of things, words cannot express" (Mandeville 1711: 140). He calls into attention the knowledge, provided by the microscope, that matter is infinitely multiform. Focusing on the stomach, that in his view replaces the spleen, 'usual suspect' of melancholic troubles, Mandeville holds that "stomachic Ferment" is not always equal and, like animal spirits, is provided with different functions (Mandeville 1711: 127).

De anima brutorum, represents a major advance, since in this work Th. Willis proposes, from the standpoint of experimental neuroanatomy, that like other animals, Man has a bodily soul, vital (distributed through the blood) and sensitive (rooted in the Brain), in connection with the material world of motions, being capable to accumulate and order various images "collected" by the senses, and then judge them through his rational soul.[16] This teaching, directly influenced by Gassendi's recovery of the sensory pleasure (*voluptas*) "from historical obscurity and moral opprobrium" (Hundert 1994: 45), legitimates the research on the circuits of "organick" affections and instincts. Misomedon, the patient of Mandeville's *Treatise*, resumes by saying that the human body is affected/excited by "disorderly motions not subjected to the rational soul" (Mandeville 1711: 133).[17]

This new emphasis on the nervous system and its circuit "restricted" to the communication between sensory organs and the brain is at the source of a crucial transition in iatromechanics. It is also one of the origins of British tradition of medical and philosophical empiricism that, at least since Locke, a student of Willis, conceptualizes sensations and perceptions as legitimate sources of conscience and knowledge. According to such perspective, individual experience and the impressions stored in memory along one's life are the ground of individual identity, not some ideal essence. In a similar manner, Mandeville, through Philoperio's mouth, insistently rejects the idea that we can prove the existence of an indivisible, unchangeable and immortal substance that safeguards our very essence and enables an afterlife (cf. Mandeville 1711: 129). According to the physician of the *Treatise*, the conception of something as a soul may well result from our pride as a species, our inability to cope with the idea of our own decay.

Both his medical formation and his deep political engagements, will play a decisive role in Mandeville's views on the treatment of distempers. Leiden was

[16]Philopirio states that the question of knowing if the Soul is "seated in Some part or diffused through all the Brain, the Blood or the whole Body, is likewise not easy to determine" (Mandeville 1711: 121).

[17]Mandeville is certainly aware of the disputes between the new medical theories and their accusation of impiety by Theologians and the Cambridge Platonists (see Henry 1989: 98–102).

at the forefront of new theories and research and his professors were not only scholars on the gallenic-aristotelic framework and Cartesians, but also authors with spinozist sympathies, giving attention to bodies' constitution, sensitivity and affection directed to heal fallen humanity.[18] From that period to the edition of the *Teatrise*, Mandeville revises his mechanicist view, abandons ontological dualism,[19] and remains attentive to the progresses made in understanding digestion and animal spirits' production and circulation, as he had done in his medical thesis, *De chylosi vitiate* (defended in 1691).

Mandeville's materialist tendency is confirmed by the idea that digestion is the main responsible for Hypochondria and Hysteria, for on it depends the production of nutritious chyle for animal spirits specialization. Like in ancient physiology this conception is based on an equilibrium that will be disturbed. It stresses that digestion and concoction depend upon the *Menstruum* (nervous ends surrounding the stomach), and are made difficult when the animal spirits, running in the blood stream are too thin or too crude. According to this new paradigm, the equilibrium is affected, and with it all the nervous circulation between the brain and the body, in two main cases: (1) through consumption of raw, noxious or indigestible aliments and (2) when the most volatile animal spirits are consumed or wasted by their participation in higher functions of the mind (reasoning or imagining) or in sexual activities. In this last case, Mandeville seems to follow the tradition of "love melancholy" and its two main variants, kinds of illnesses that can be considered as historical antecedents of hysteria and hypochondria: "furor uterinus" and "satyriasis" (see Ferrand 1623: 75 ff). These relate to peak illnesses that have their origin in excessive venery occasioning the production of vapours in the groin and abdominal zones, arousing compulsive sexual desire.

Some thoughts involving the "volatile economy of the Brain" can affect the Stomach, for its "airy velicitous agents (...) officiating between the soul and grosser Spirits of the Senses have always access to her invisible self" (Mandeville 1711: 131). Through the mechanisms of sympathy to discourse or by surprising news (joyful or unwelcome) the balance of the spirits is affected and the nerves surrounding the stomach will make digestion more difficult. In the latter edition of the *Treatise* Mandeville continues to emphasize the importance the dynamic relation between brain and stomach adding that "the Stomach is the conscience of the Body" (Mandeville 1730: 326).

The treatment of melancholic diseases centres on exam and integrates guidance and discipline of the soul according to reason or faith, is reconsidered in order

[18]Composed in Rotterdam, Mandeville's *De Medicina Oratio* from 1685, already professed this design of medicine to restore humanity to its pre-fall state. For a comprehensive resume on the formation of Leiden's particular scientific environment see *Matters of Exchange* of Harold Cook (2007: 226–266).

[19]Such influence dates back to his thesis *Disputatio Philosophica de Brutorum Operationibus* presented in 1689 where he defends animal automatism as a valid model of comprehension (cf. Mandeville 1689/1991: 379–381).

to explore another path available on the physical side of disease.[20] But even if Mandeville recognizes the importance of thought on the ensuing distempers, for a Physician's promotion of cure: "the only guide must be Nature, preventing infinite Wandering" (Mandeville 1711: 137). This requires, throughout the treatment and convalescence, the ability to discern between the efforts of one's own nature (for self-preservation) and the destruction operated by the distemper. Such detour follows the principle that Philoperio frequently mentions, that emotions and thinking cannot take place without a material substract. This means that even if he fully acknowledges first person experience as informative on the particularity of the distemper, determining if it is due to one's constitution or to some kind of damaging experience or habit, Mandeville privileges individual physiology as the ground for operative of change.

Mandeville goes as far as possible in dissecting passion without imposing moral judgements on its description. One of the most striking examples can be found in the "Remark (R.)" of *The Fable of the Bees* where, in defining courage he resorts on the intensity of blood circulation, its composition in terms of spirits and its distribution through bodily parts. However, even if we can describe the passions of shame or pride within a physiological discourse that points their contrastive signs, Mandeville is aware that the causes of passions are not entirely physical nor their bodily repercussions indifferent to their cognitive processing. It is truth that in certain constitutions "the orderly or disorderly mixture of the Fluid in our body" (Mandeville 1988a: 211), seem to fix the boundaries within which appetites, feelings and behaviour will be framed. But, in the varieties of melancholic disease, the mental representation of passions, especially the way they are affected by socially established desires, plays an importance role in aetiology of physiological decay or dysfunction.

12.4 The Discovery of Symptoms and Diagnose

In the three dialogues forming the *Treatise*, the therapeutic relation, deeply personal and inserted in a medical service centred on the Household, passes through the alternation between articulation of complaints and punctuation (and reordering) of such complaints (by the physician). In a third moment, we see explicit recommendations as well as prescriptions to cure or, at least, make chronic distempers more bearable. This process can also be found in John Purcell's appeals to his reader's capacity to express their troubles with clarity, in a language increasingly autonomous towards moral issues and centred on the "organick", even inviting patients to judge about the correctness of their physicians prescriptions (Purcell 1707; see Schmidt 2007: 162). This new tendency is clear in Misomedon (but also in his wife Polytheca) that after

[20]"Putting right reason aside might be allowed to physicians, who were after all concerned with treating bodies rather than guiding souls" (Cook 2007: 392).

serious deceptions with medical practitioners (both Physicians and Apothecaries[21]) begins years of self-instruction in the "bookish" knowledge of medical theories alongside readings of classical *Humanioris Literaturae*. In fact, this patient's life is divided into two halves – his early life in which profligacy is his main occupation and his later years in which his only occupation is consulting and studying of his own illness. Misomedon will articulate his complains no longer in the medical and religious grammar of the afflicted consciousness that strives for divine consolation to temperate its passions, but in classical and modern medical expertise, expressed in a refined language. In his journey we see the great importance hold by apothegmatic formulas, useful in periods of transition or difficulties in the path towards cure or care of oneself. Immoderate reading of conflating theories leads to waste the finest bodily spirits,[22] but also to a sense of scattered self, that Misomedon will surpass into a coherent self-image through the guidance of Philoperio.

In the dialogues, (especially in the third), the idea of relief and joy produced by talking about one's pains and suffering is clearly stated by both Philoperio and Misomedon.[23] This relates with the fact that both Hypochondria and Hysteria had gradually turned into fashion diseases, for man and women of refined taste. They are based on a strong idea of individuality, particularly of individual discourse, where *one comes to terms with real physical, mental and spiritual problems, but at the same time tends to favour a certain infatuation of the self.* Here, we find Mandeville's stress on pride's skewed expressions. Both hysterics and hypochondriacs take their condition as worse than others say them to be and don't like it being discredit (cf. Mandeville 1711: 265), displaying a kind of exacerbated sensitivity to their physical and psychological "boundaries".[24] Mandeville is not indifferent to this fashion, and remarks the frequency in which the Hypochondriacs take satyr as a fulfilling practice. In this sense, satire can be read as a social function -questioning, critiquing and mocking certain human actions and state of affairs in order to acknowledge their full dimension and eventually change them- supported in individuals with a particular constitution.[25]

[21]Specially these last ones are accused by Misomedon of dealing into a trade "where knaves have a great latitude" (Mandeville 1711: 218).

[22]This idea is reinforced in the last edition with a citation from the *Philosophical Translations* of 1673 (cf. Mandeville 1730: 218–219).

[23]Misomedon says "if your medicines do me no good, I am sure your company will" (Mandeville 1711: 41). Also Polytheca praises an apothecary named Pharmenio that despite considering her to be incurable "has the Patience to weight my complaints, or at least the good manners to hear them, and seldom fails of giving me ease (. . .)" (Mandeville 1711: 200).

[24]Hyper-sensitivity to all the interfaces and occurrences of his organic state, and frequent fear associated, are distinctive traces of hypochondria preserved in today's colloquial meanings associated with this condition.

[25]"It is a general Observation that the beloved theme of all Hypondriact is Satyr; which I know is worth nothing unless it bites" (Mandeville 1711: 235). I must thank the kindheartedness of Professor Frank Palmeri that called my attention to the seminal work of Mary Claire Randolph on the semantic evolution interconnecting satire and medical language from Renaissance onwards.

Following his father Michel de Mandeville, Bernard devoted to nervous and gastric specialities, kinds of chronic disease, since unlike acute diseases, that require long and boring standing on the bedside, they are wholesome. On the other hand, despite their great demands of constant observation, this never cease to excite wit and curiosity (furthermore they propitiate gratifying chattering).[26] Aside this coincidence between personal and professional predilections, a substantial part of the *Treatise* deals with the issue of professionalization and accreditation of medical professionals, and Philoperio highlights the necessity that each Physician specializes in a restricted subject and become expert in a distemper (cf. Mandeville 1711: 41).

An important aspect central in the dialogues and concurring with Mandeville's wider conception of the passions, is that hysteric and hypochondriac passions must not simply be suppressed. Dialogue serves to explore their symptoms, some conspicuous some difficult to dissect, but always coming to discourse in the mode of complain.[27] And Philoperio stresses the importance of such encounter. Against the common practice at the time to prescribe without even seeing the patient,[28] he defends that each disease manifests itself differently according to its progress, the patient's complexion, passions and biography. *The physician must gather information inviting the patient to narrate his history with the adequate detail.* He advises Misomedon to look into his past, inciting him to find the "causes of your distemper" (Mandeville 1711: 142).

Here we see another important aspect in Mandeville conception. Diagnosis is already a part of the therapeutic process enabling a work through the passions, and not simply their repression. In fact, the *Treatise*, especially the second dialogue, can be read as a "dramatization" of the theses earlier exposed in his *Disputatio Philosophica de Brutorum Operationibus*. Adapting those theses to dialogue enables both their convincing presentation, since they orbit around a particular case, and

Whereas in ancient times satire is paralleled with the power of Blacksmiths, in modern times it acquires a new sense becoming closer to the incisions of the barber surgeon (cf. Randolph 1941: 125–157).

[26]I cite the expressive account of Philoperio that we can take as being close to the convictions of Mandeville himself: "I could never go through a multiplicitude of Business. Everybody ought to consult his own temper and abilities in all undertakings. I hate a crowd, and I hate to be in a hurry. Besides, I am naturally slow, and could no more attend a dozen patients in a day, and think of them as I should do, than I could fly. I must own to you likewise, that I am a little selfish, and cannot help minding my own enjoyments, and my own diversion, and in short, my own good, as well as the good of others. I can, and do heartily admire at those public-spirited people that can slave at an employment from early in the morning till late at night, and sacrifice every inch of themselves to their callings; but I could never have had the power to imitate them. Not that I love to be idle; but want to be employed to my own liking (. . .)" (Mandeville 1730: 351–352).

[27]There is some agreement even today, at least in Psychoanalysis, that some kind of mental illnesses like hysteria and neurosis, don't need compulsory measures, and the patient is the first to require the doctor's attention.

[28]In the first dialogue Philoperio states that "[w]hat I am against is, the speculative part of physic (. . .) that teaches Men to cure all manner of distempers in their closets, without even seeing a Patient" (Mandeville 1711: 52).

a careful examination of their effectiveness (cf. McKee 1991: 201–202, 218–221). Perhaps more importantly, the *Treatise* presents a corrective path towards a healthier consumption not only of food, but of goods and sexual joys.[29] When confronted with his past errors, some of them seemingly innocuous given their social acceptance and frequency, Misomedon notes the "comical ways we have of digging our own graves" (Mandeville 1711: 143). Another important aspect is that the language of error and repentance we find here is no longer based on religious premises, in spite of the fact that medical conceptions still informed by religious views on sexual pleasure and gluttony as in vice in general. At the same time, compassion tends to clearly surpass any other feeling towards the diseased, and when faced with Polytheca's excuses for the troubles she might have caused her husband, Misomedon states that he would never complain about such unfortune, not even in a case where illness would fall upon his domestic servants (Mandeville 1711: 209).

12.5 Treating One's Passions

Treating melancholy (and its various modes) requires, like other mental or nervous illnesses (like mania or frenzy), attention to the interrelation between physical states (and symptoms) and mental states. *But unlike other mental illnesses, such distemper has emotional manifestations -fear and sorrow-, expressions that must be taken as primary symptoms and not simply the result of circumstantial movement of bodily spirits*. This also means that their treatment, even if enhanced by pharmacopeia, diet, exercises and diversions, *requires an effective work on one's own emotions*. This change requires an action that cannot be delegated in anyone else but the patient. Such resolution will lead to modify those passions whose dangerous effects are occasionally counterbalanced by their productivity.

This means that nervous distempers are never completely alien to one's identity, and their removal, like in the attempt to strictly conform "self-love" to "self-linking", could mean the disappearance of the individual/subject itself. So it is understandable that even if praising Hippocratic School, Philoperio is very cautious towards some excesses in variants of extreme therapies, for they may go against the drive for self-preservation (Mandeville 1711: 242–243). Philoperio tends to favour non-intrusive therapies that respect one's habits and pleasures.

These issues call into attention the social dimension of disease, because some symptoms resulting from excessive pride and/or demands can be disguised under some eminent social roles and positions that made them acceptable and/or immune

[29]The constant idea of corrective or at least palliative measures, sometimes marked by austerity, is accompanied by feelings of regret towards wasted time due to ignorance, negligence or search of corporeal joys. In the last edition of the second Dialogue, we see Misomedon adopt Horace's interjection "Quae mens est hodie cur eadem non puero fuit?" (Mandeville 1730: 211).

to social or judicial persecution (cf. Mandeville 1988b: 129–130). The perfect example of such case takes place when a tyrant comes to assume power[30] . . .

When it comes to determine the cause of both hypochondria and hysteria, the listening and ordering of their patients discourse enables Philoperio to identify damaging ways of living. Such ways are frequently the product of habits related to social condition and/or profession, one could say, to a particular taming of individual passions. Hypochondriacs need a credible listener that can certify their individuality, and if like in Misomedon, they search for new doctors is not simply in order to end their suffering but to find a listener that eases their pains, so: "Novo medico gaudent omnes Hypochondiaci: sed quod remedium credis flagitare symptom morbi est" (Mandeville 1711: 280).

The case of Misomedon, despite included in a typified and common disease, is, like the case of any other patient, unique. The first manifestations of his disease were gastric, he was stricken with heartburn, but *he remained calm as long as his appetite and sleep remained unaffected.* With the persistence of symptoms he will search for practitioners that will only aggravate his condition at the same time as they exploit him until he is in a helpless condition. He gives a detailed account of the errors that lead the distemper to tyrannize his body and invade his soul, until he became a "hypochondriacus confirmatus". It is considered, in accordance with medieval medical Treatises on melancholy circulating, at least since William of Auvergne, that in cases of prolonged and violent excesses like the one's Misomedon endured, the affected individual is at the verge of insanity. But the origins of his disease weren't just dietetic. He had indulged in frequent sexual intercourse (*rex uxoria*), which damaged the qualities of his spirits. Besides, he made intense use of his brain during his 5 or 6 years of hard study to ease his suffering. Like in Ficino's account on the third chapter of the first book of *De Vita Triplici*, melancholy is the "disease of the Learned",[31] whose efforts (intense use of the brain at improper hours) waste their volatile animal spirits leaving their blood too crude to enable a good digestion and the making of nutritious chyle (Mandeville 1711: 95, 148 ff).[32] Greater intensity in intellectual exercises is proportional to greater waste of spirits, and in accordance with the tradition of "melancholia adusta" results from the combustion of humours – blood, yellow bile, black bile itself and phlegm- in the brain, that waste animal spirits and turned them cruder and sour, affecting any constitution not originally melancholic nor fragile, such as it happens with individuals "complexio sanguine". Symptom of such occupations are back pain and eye strain leading frequently to the

[30]This is a theme that goes back to the *Republic* (572c ff.), where Plato relates a particular upbringing with the acquisition of a vicious character whose extreme frenzy can lead do madness.

[31]Mandeville uses the High-Dutch designation of the disease: "Der Gelahrten Krankheydt" (Mandeville 1711: 94).

[32]Some, like Michael Ettmüller, related such case to "stopping and squeezing the Belly against the Books", which would hinder the circulation of the humours (Mandeville 1711: 149).

use of spectacles.[33] Explaining this conception, Philoperio notes that brain activity never ceases, not even during sleep, but there the images do not come directly from the senses, and are lighter so to speak, enabling a "wandering at leisure" (Mandeville 1711: 160–161).

Some "classes of Men", mostly the Labour/productive people and the generality of "Blockheads" are immune to such troubles due to their constant exercising and light thinking.[34] Mandeville is aware that the Low Countries, not famed for their manners, have no similar epidemic has the one beginning to spread among the English Gentry.

For Misomedon are prescribed traditional cures for sedentarism and excessive leisure. Rest and profiting from Country's air are recommended along with horse ride, light readings, moderate diversion and some kinds of sporting activity.[35] In his case, diet is decisive and Philoperio even advises Stock-Fish (cod dried in open air), the diet of Dutch sailors, which is capable of wonders and "fills the body with volatile as well as balsamic juices" (Mandeville 1711: 246),[36] much more bearable to his stomach than the refined foreign foods he delights and got used to. In this case Misomedon questions Philoperio if he isn't following a common tendency among Physicians already referred by Baglivi, in chapter of *De praxi medica* with the suggestive title "Falsa Medicorum idola", the prescription of diets and remedies that correspond to their own likings. Once again, Philoperio remarks that the art of Medicine proceeds from the careful analysis of a particular case history in order to select suitable measures to cure or ease one's illness. He recognizes that "[n]o *Regimen* of Life can be contrived that suits everybody" (Mandeville 1730: 321), which means that it must take into account, as much as possible, one's sensitivity to a specific practice, remedy or diet, sometimes being able to accommodate individual tastes. So the main reference for simple advices or for a detailed regime isn't doctor's *gusto*, but the adjustment between one's constitution on the one side and experience gathered by observation of previous cases on the other.

[33]Here, Mandeville makes a veiled reference to Spinoza, whose occupation was precisely to grind lenses, remarking how frequent it is the use of spectacles among the Spanish (Mandeville 1711: 159).

[34]On the same theme Philoperio ironically says in the second dialogue: "so there's another advantage of fools that Erasmus never dreamt of, and the muddiest and most fuliginous blood is only to be looked for in Men of brightest parts (...)" (Mandeville 1711: 100). However, Mandeville seems aware that, given the diffusion new habits of consumption and the generalization of some expectancies, the subsistence of something as a segregation of diseases according to one's rank tended to be mitigated (see Foucault 2003: 28–30).

[35]Mandeville's views on exercise are influenced by the innovative theory proposed by Francis Fuller (1705) which explored not only the general benefits of exercise but also their therapeutic benefits on some distempers.

[36]Misomedon cannot hide his disappointment for such a simple and unpretentious prescription, exclaiming: "Pro Thesauro Carbones" (Mandeville 1730: 316).

As to the cases of Polytheca and their daughter, both suffering from different forms of hysteria, Philoperio's inquiry will lead him to emphasize damaging diets that eventually resulted in cachectic and hysteric states, with profuse imagination, sometimes accompanied by Fits.[37]

In the case of their daughter, the lack of *amplexus virils* leads to capricious manifestations of the distemper. But the main reason must be the Menstrual Flux and Uterus, which by submitting the animal spirits to multiple injuries, origins "lack of constancy and resolution and infirmity of the mind" (Mandeville 1711: 172). Along with the recognition of the peculiar troubles of adolescence, considered as a differentiated phase in one's personal development, Philoperio is resuming the theme of women's greater sensitivity but less constancy. Due to their constitution women are at mercy of unbearable emotions coming from their senses and disturbing their peace of mind and ability to reasoning. Given the girl's youth, Philoperio proposes cutting medication, delineates exercises (especially horse riding) and a proper diet. Philoperio underlines the importance of exercise as benefiting appetite and digestion, it "removes obstructions, invigorates the blood, and strengthens the whole body" (Mandeville 1711: 135). However, traditional treatments are considered in the case this regime proves ineffective in restoring her health: baths, fomentations and even bleedings (cf. Mandeville 1711: 242).

In the case of Polytheca, in spite of refusing the classical jargon of the "Vapours", used by men to refer "Extravagant and Imperious women, when they are denied or thwarted in their unreasonable desires" (Mandeville 1711: 199), Philoperio considers her condition to be based on excessive fancy, and recommends leaving nefarious medicines and perhaps lighten her spirits with some wine. This recovery of the sacredness of wine will lead to considering its damaging effects to the frequent user. After that witty description of the addiction's process, from pleasurable consumption to compulsion, Philoperio holds that she may be saved by gentler methods than purging, bleeding, sweating, and commits himself to prepare her some medicines. A curious case happens when after Polytheca leaves the scene, Misomedon complains about his wife's stubbornness accompanying the instability of her moods, so similar to what is today typified as bipolar disorders, Philoperio assumes the role of a marital counsellor and invites Misomedon to acknowledge the same tendency he is imputing in his wife simply for her honest defence of the merits of apothecaries.

The way Philoperio combines Pharmacy, exercise and diet is conformed to Daniel Le Clerc's *Treatise of medical history*, recently translated and edited in England. He praises the easiest (here he refers to Arnold of Villanova), cheapest and famous remedies, discrediting hermetical pretenders: "vain promises of lying chemists" that robbed us from traditionally known remedies (Mandeville 1711: 254).

[37]Philoperio defends that desired "trash" food can be digested for it will activate the production of stomachic ferment, but if frequent such diet has catastrophic consequences often observed in "green sick girls" (Mandeville 1711: 167–168).

Along the dialogues, Philoperio plays with his patients' passions appealing to, or awakening, their pride through flattery or deference, so that they follow his advices and prescriptions. Here we must be cautious, avoiding projection of contemporary assumptions into a practice that can be read in various ways. If Mandeville text denounces the way academic physicians take advantage of anguished people by recurring to empty rhetoric, at some points in the dialogue, Philoperio himself has to exercise his expertise confined to the boundaries of polite manners. At the same time he is publicizing himself as a practitioner.[38]

Anyways, the analogy between Politicians cunning way to move passions and Physicians guidance and prescriptions would be misleading. A Physician has received "public trust", but his compromise is not simply to serve social utility.[39] This can explain why it is expressed in negative terms: "[h]is prescriptions by assisting some ought never prejudice others" (Mandeville 1711: 243).[40] Such amelioration may contribute to improve social relations, like marital or filial bounds, but is different in kind and extent from the Politician's dextrous management of passions.[41]

Mandeville's great innovation in these three dialogues is not to be located on the treatments he suggests, after discrediting some miraculous beliefs, but in his, sometimes implicit, social understanding and framing of distempers. Mandeville's medical work must be inserted in a larger "science of man" able to observe and understand the passions (and their organic substract) within a society with a particular mode of economic organization and moral normativity. He explores mental illness through the curious interplay between the physical, the psychic and the way some personal and professional roles make one more prone to a specific distemper. His mouthpiece Philoperio does not rely exclusively in the body as source of distemper, but each complain and every disease has bodily states or processes as the ultimate seat, even those related with distorted or excited fancy seeking relief.

---

[38]In the preface to the first edition he invites his readers to contact him through his bookseller, (cf. Mandeville 1711: xiv).

[39]However, there is an interesting analogy between the government of the physical body and the political body, especially through the metaphors of digestion, the balance between accumulation and spending, and the way new exchanges with the outside (other people or other countries) will foster more refined desires (see de Marchi 2001: 67–92).

[40]In the enlarged edition, Mandeville adds to Misomedon discourse the need to survey and adopt measures capable to exemplary repress some practices: "I hate cheats of all sorts; and in things of public concern, I think, a man ought to be hanged, who for Lucre endeavours to render that [simple recipe] mysterious, which in itself is plain, or may easily be made so" (Mandeville 1730: 350).

[41]"The short-sighted vulgar, in the chain of causes seldom see further than one link; but those who enlarge their view, and will give themselves the leisure of gazing on the prospect of concatenated events, may, in a hundred places see good spring up and pullulate from evil, as naturally as chickens do from eggs" (Mandeville 1988a: 91).

12.6 No Flourishing Without Some Thorns

In his last writing on the melancholic diseases Mandeville follows some of the ideas present along the *Fable*, namely the contrast between an unchangeable human nature and the evolution of societies towards flourishing in commerce of consumption goods. And he is very clear on this point: "Human nature is the same in our Age that has been in all others under the same circumstances. All lovers of their Country, and even the best of Men, have always wished and prayed for Wealth and Power, with the increase of Knowledge to the Nations they belonged to; and they have no sooner enjoyed what they wished for, but they have always grumbled and showed themselves impatient to bear those Evils which ever were and ever will be the consequences of those Blessings in all large and flourishing societies" (Mandeville 1730: 332). As Anthony McKee expressed it in his magnificent thesis "Mandeville's patients, revelling in the profits of the rising British Empire, are suffering from various forms of indigestion, having consumed too many texts, too many consumer goods and too many exotic foods that are new to the British diet" (1991: 203). No wonder we find a remarkable change in the classification of diseases but also on their understanding and treatment.

The socio-historical account given by Foucault remains valid: "[i]n the Middle Ages, at a time of war and famine, the sick were subject to fear and exhaustion (apoplexy, hectic fever); but in the sixteenth and seventeenth centuries, a period of relaxation of the feeling for one's country and of the obligations that such a feeling involves, egotism returned, and lust and gluttony became more wide-spread (venereal diseases, congestion of the viscera and of the blood); in the eighteenth century, the search for pleasure was carried over into the imagination: one went to the theatre, read novels, and grew excited in vain conversations; one stayed up at night and slept during the day (hysteria, hypochondria, nervous diseases)" (Foucault 2003: 27–38).

Analysing the generalized association of depression with English culture, Foucault referred the way health and happiness were generally considered to follow from nature's equilibrium. Deviation from the happy mean could no longer be derived solely from one's complexion/constitution and the exposition to "milieu" temperatures, but also from idleness associated with a new wealthy class. In the work of Johann Spurzheim, the mentor of phrenology in Britain, we find another interesting aspect of melancholy's aetiology and its frequency in England, namely the unparalleled freedom of conscience, religion and commerce that one could find in this nation. According to that renowned doctor, contrary to societies that repress unorthodox opinions and beliefs, leaving belief to one's consciousness could lead to an endless and extenuating search for one's truth. One the other hand, a mercantile society leaves room to some private interests and passions such as egotism and envy that go well beyond those considered legitimate within the strict boundaries of "natural freedom" (Foucault 2006: 366–367).

Mandeville's work can indeed be read as one of the strongest attempts to observe not only the rising of a new culture of sensibility and politeness, but also the problems affecting the social order of an affluent society still holding worldviews based in the conciliation of the different social spheres with an all-encompassing morality. In this project, the care of the self's body and mind is transposed into modern terms and will remain attentive to numerous aspects demanding constant reflexion upon the values that must regulate the relation to oneself, and between individual and society. Mandeville's medical conceptions and therapeutic approach will be a key contribution to understand the paradox consisting in the fact that the more refined morality still has an unmoral source.

References

Primary Sources

Burton, Robert. 1883. *The anatomy of melancholy. What it is, with all the kindes, causes, symptomes, prognostickes, and severall cures of it*. Philadelphia: E. Claxton & Company.

Cheyne, George. 1733. *The English malady: Or, a treatise of nervous diseases of all kinds, as spleen, vapours, lowness of spirits, hypochondriacal, and hysterical distempers*. London: G. Strahan and J. Leake.

de Mandeville, Bernardus. 1689. *De Brutorum Operationibus: Disputatio Philosophica*. Leyden. English edition: *Animal Functions*. In 1991 *An Anatomy of Power: The Early Works of Bernard Mandeville*, ed. and trans. Anthony Francis McKee, 384–410, Glasgow: University of Glasgow.

de Mandeville, Bernard. 1711. *A treatise of the hypochondriack and hysterick passions vulgarly call'd the hypo in men and vapours in women ... in three dialogues*. London: Dryden Leach and W. Taylor (Facsimiles from Gale ECCO Print Editions).

de Marchi, Neil. 2001. Exposure to strangers and superfluities: Mandeville's regimen for great wealth and foreign treasure. In *Physicians and political economy: Six studies of the work of doctor economists*, ed. Peter D. Groenewegen, 67–92. London: Routledge.

Ferrand, Jacques. 1623. *De la maladie d'amour ou melancholie erotique. Discours curieux qui enseigne à cognoistre l'essence, les causes, les signes, & les remedes de ce mal fantastique*. Paris: Denis Moreau.

Fuller, Francis. 1705. *Medicina Gymnastica: Or, a treatise concerning the power of exercise, with respect to the animal economy*. ed. Robert Knaplock, London.

Gordon, Ward H. 1931. An unnoted poem by Mandeville. *Review of English Studies* 7: 73–76.

Le Clerc, Daniel. 1699. *The history of physic*. Robert Knaplock London: Dr. Drake and Dr. Baden.

Mandeville, Bernardi à. 1685. *De Medicina Oratio Scholastica*. Rotterdam. English edition: *Oration on Medicine*. In 1991 *An Anatomy of Power: The Early Works of Bernard Mandeville*, ed. and trans. Anthony Francis McKee, 332–360, Glasgow: University of Glasgow.

Mandeville, Bernard. 1726. *The true meaning of the fable of the bees*. London: William and John Innys.

Mandeville, Bernard. 1730. A treatise of the hypochondriack and hysterick diseases. In *Three dialogues*, 3rd ed. London: J. Tonson (Facsimiles from Gale ECCO Print Editions).

Mandeville, Bernard. 1988a. *The fable of the bees or private vices, publick benefits*, vol. 1. [1732]. Indianapolis: Liberty Fund.

Mandeville, Bernard. 1988b. *The fable of the bees or private vices, publick benefits*, vol. 2. [1732]. Indianapolis: Liberty Fund.

Purcell, John. 1707. *A treatise of vapours, or, hysterick fits*, 2nd ed. London: for Edward Place.

Secondary Sources

Burtt, Shelley. 1992. *Virtue transformed: Political argument in England, 1688–1740*. Cambridge: Cambridge University Press.

Cook, Harold J. 1989. Physicians and the new philosophy: Henry Stubbe and the virtuosi-physicians. In *The medical revolution of the seventeenth century*, ed. Roger French and Andrew Wear, 246–271. Cambridge: Cambridge University Press.

Cook, Harold J. 1999. Bernard Mandeville and the therapy of the 'clever politician'. *Journal of the History of Ideas* 61: 101–124.

Cook, Harold J. 2007. *Matters of exchange. Commerce, medicine, and science in the Dutch golden age*. New Haven/London: Yale University Press.

Den Uyl, D.J. 1987. Passion, state and progress: Spinoza and Mandeville on the nature of human association. *Journal of the History of Philosophy* 25: 369–395.

Foucault, Michel. 2003. *The Birth of the Clinic: An Archaeology of Medical Perception*. Trans. A.M. Sheridan. London: Routledge.

Foucault, Michel. 2006. *History of Madness*. Trans. J. Murphy and J. Khalfa. London: Routledge.

Guerrini, Anita. 2000. *Obesity and depression in the enlightenment. The life and times of George Cheyne*. Norman: University of Oklahoma Press.

Henry, John. 1989. The matter of souls: Medical theory and theology in seventeenth-century England. In *The medical revolution of the seventeenth century*, ed. Roger French and Andrew Wear, 87–113. Cambridge: Cambridge University Press.

Hilton, Phillip. 2010. *Bitter honey. Recuperating the medical and scientific context of Bernard Mandeville*. Bern: Peter Lang.

Hundert, Edward J. 1994. *The enlightenment's fable: Bernard Mandeville and the discovery of society*. Cambridge: Cambridge University Press.

Ingram, Allan. 2011. Deciphering difference: A study in medical literacy. In *Melancholy experience in literature of the long eighteenth century: Before depression, 1660–1800*, ed. Allan Ingram, Stuart Sim, Clark Lawlor, Richard Terry, John Baker, and Leigh Wetherall-Dickson, 186–187. Basingstoke: Palgrave Macmillan.

Israel, Johnathan I. 2006. *Enlightenment contested. Philosophy, modernity, and the emancipation of man 1670–1752*. New York: Oxford University Press.

Klever, Wim. Bernard Mandeville and his Spinozistic Appraisal of Vices. *Foglio spinoziano* 20. http://www.fogliospinoziano.it/articolispinoza/Mandeville.PDF. Accessed 10 Oct 2013.

McKee, Anthony Francis. 1991. *An anatomy of power: The early works of Bernard Mandeville*. Glasgow: University of Glasgow.

Porter, Roy. 1985. *Patients and practitioners Lay perceptions of medicine in pre-industrial society*. Cambridge: Cambridge University Press.

Randolph, Mary Claire. 1941. The medical concept in English renaissance satiric theory: Its possible relationships and implications. *Studies in Philology* 38: 125–157.

Rousseau, George S. 1975. Mandeville and Europe: Medicine and philosophy. In *Mandeville studies. New explorations in the art and thought of Dr. Bernard Mandeville*, ed. Irwin Primer, 11–21. The Hague: Nijhoff.

Schmidt, Jeremy. 2007. *Melancholy and the care of the soul: Religion, moral philosophy and madness in early modern England*. Aldershot: Ashgate.

Schoenberg, P.J. 1976. A dialogue with Mandeville. *British Journal of Psychiatry* 129: 120–124.

Steinebrunner, Bernd. 1987. *Die Entzauberung der Krankheit. Vom Theos zum Anthropos – Über die alteuropäische Genesis moderner Medizin nach der Systemtheorie Niklas Luhmanns.*. Frankfurt a.M.: Peter Lang.

Wild, Wayne. 2006. *Medicine-by-post. The changing voice of illness in eighteenth-century British consultation letters and literature*. Amsterdam/New York: Rodopi Press.

Chapter 13
The Exchange Between Mandeville and Berkeley

Mikko Tolonen

Abstract George Berkeley directed an *ad hominem* attack on Bernard Mandeville in his *Alchiphron*. Although rarely analysed in secondary literature, this and the following exchange, was an important occasion in history of philosophy that contemporaries probably followed closely. The idea of this paper is to offer an analysis of Mandeville's subsequent answer to Berkeley's accusations offering an interpretation that situates this in the context of Mandeville's intellectual development. The relevance that this paper claims to have is that it shows in practice what Mandeville's intellectual development meant in eighteenth-century debates on political economy and how this relates to an equally important question about nature of moral knowledge. The paper will also take into consideration John Hervey as an outside commentator on the polemic between Mandeville and Berkeley.

Keywords History of political thought • Political economy • Book history • Atheism • Enlightenment

George Berkeley seems to have been an angry man. As we know, his *Alchiphron* (1732) includes an attack on Bernard Mandeville and his *Fable of the bees* (among other lashes that he delivered elsewhere).[1] Mandeville's answer to Berkeley, *Letter to Dion* was published that same year. It has received relatively little analytical attention in scholarship, but it was the last of Mandeville's known published texts and an important document because he actually answers his critics in print, something that he did not do at the height of the controversy that *Fable of the bees* caused in 1723–1725. This paper will be divided into two: first part is a historical reconstruction of the path to *Letter to Dion* in order to highlight a change

[1]Nature of Berkeley's attack on Mandeville as directed against "a strawman" is noted in Berman (1993: 2–3). Also John Stuart Mill and Leslie Stephen criticised Berkeley for missing his mark on Mandeville (Berman 1993: 13).

M. Tolonen (✉)
Department of Modern Languages, University of Helsinki, Helsinki, Finland
e-mail: mikko.tolonen@helsinki.fi

E. Balsemão Pires, J. Braga (eds.), *Bernard de Mandeville's Tropology of Paradoxes*, Studies in History and Philosophy of Science 40, DOI 10.1007/978-3-319-19381-6_13

in Mandeville's thought relevant to our reading of the piece.[2] Second part is an analysis of the text itself in relation to the objections put forward especially in the second dialogue of *Alciphron*, but not limited to it. The relevance that this paper claims to have is that it shows in practice what Mandeville's intellectual development meant in eighteenth-century debates on political economy and how this relates to an equally important question about nature of moral knowledge. The paper will also take into consideration John Hervey as an outside commentator on the polemic between Mandeville and Berkeley.

13.1 Path to Dion

Mandeville was a consciously provocative author starting from his first publications in Britain. *Pamphleteers* in 1703, for example, is a satirical, topical and direct work. Nevertheless, when the reactions to *The Fable of the Bees* started pouring in 20 years later, they must have been a shock to the author, even when, for example, with the "charity school" essay Mandeville was practically begging for trouble by suggesting that the free education of the poor might turn them into a more crafty kind of pick-pockets rather than benefiting the public. After the second edition of *The Fable* was published, it took less than 3 months for the book to end up in front of the grand jury.

The presentment of the Grand-Jury of 1723 includes accusations of 'diabolical attempts against religion'. It is possible that the political nature of the book is one reason why *The Fable* received such a hostile welcome (Speck 1978).[3] However, it is undeniable that the "charity school" essay (combined to some of the more controversial passages of the book) was unusually provocative regarding current affairs that concerned many. Another serious accusation of the presentment was that *Fable of the bees* had 'a direct tendency to propagate infidelity, and consequently to the corruption of all morals', which effectively means the debauching of the nation. These are also the main accusation that Berkeley directs at Mandeville in his *Alciphron*.

After the presentment in 1723, Mandeville's own approach rapidly changed from provocation to a careful defence of his reputation. A vindication was published in the *London Journal* (Kaye 1924: xi). It was also included in the later editions of *The Fable*. In *Letter to Dion*, Mandeville says that he himself published the vindication (Mandeville 1732: 7). According to Mandeville, if Berkeley had only read it, he had not written against him at all, or at least shown how this vindication was insufficient. Clearly, the vindication meant much to Mandeville. Mandeville started to worry about his reputation and what that *The Fable* might have been doing to it. Berkeley

[2]This part relies on the account that I have put forward in more detail in Tolonen (2013).

[3]Speck's suggestion is that much of the controversy regarded the fact that the makeup of the jury was Tory at heart.

did not take notice of it and Mandeville repeated several times in *Letter to Dion* that Berkeley had not read the *Fable*, but trusted false reports about it.

Another contingent matter that we should take into consideration is that it is likely that in 1724–1725 Mandeville sold the copyright of the work to a famous publisher called Jacob Tonson jr. and did not have much say on further editions or possible changes. The third edition, which is the first "Tonson edition" is in effect the last edition of the work. The rest are reprints with some typographical changes that do not affect the copytext. Publisher Tonson's vision about bookselling was rather cynical. For him, good controversy always meant good business. It is interesting, in this context, that Tonson is the publisher of *Alciphron* and he also owned the copyright of that work.[4] This is an important factor and something that to my knowledge has not been taken into consideration in earlier scholarship. As a publisher of both *Fable* and *Alciphron*, Tonson would naturally benefit from controversy and, to use a cliché, "from any publicity". These kinds of issues are philosophically relevant because they influence the matter more than is often realised.

About different ways of defending himself, Mandeville mentions that he 'once thought' of compiling 'a list of the adversaries that have appeared in print'. The reason why this plan was given up, according to Mandeville, was that antagonists were too many and the points they were making too few. The reason given by Mandeville for not providing any further answers in print was that simply reading 'some part or other, either of the Vindication or the book it self' should prove the raised accusations wrong. During this time, Mandeville had probably also compiled a manuscript defending himself that was finished by 1726. It is surprising how little notice this has received in Mandeville scholarship. 'I have wrote', Mandeville says in 1728, 'and had by me near two years, a Defence of the *The Fable of the Bees*', in which I have stated and endeavour'd to solve all the objections that might reasonably be made against it, as to the doctrine contain'd in it, and the detriment it might be of to others' (Mandeville 1729: ii).[5] Hence, one simple explanation why making alterations to the first part of *The Fable* was not necessary for Mandeville was that he started writing a separate defence, which however was never published as such.[6]

One reason of talking about Mandeville canon is that even when The *Letter to Dion* was designed as a piece engaging Berkeley, in fact it was not completely

[4]Indicated in the Longman sale records.

[5]Kaye argues that *Remarks upon two late presentments of the Grand-Jury of the county of Middlesex* would be the defence of *The Fable* that Mandeville is discussing in the preface (Kaye 1921: 457–458). I find this unlikely. Particularly because Mandeville's main point (in 1728) is that it is a work that has not been published and the *Remarks* was already published in 1724.

[6]Kaye has proved that Mandeville was not the author of *True meaning of the Fable of the bees* of 1726 and certainly this is not the defence of *The Fable* that Mandeville refers to. This work has been mistakenly attributed to Mandeville (Kaye 1921: 463–464). The anonymous author of *True meaning of the Fable of the bees* defends a view that all moral distinctions are made by politicians tricking men to act against their passions. He also tries to reduce all the passions to self-love (*True meaning* 1726: 10, 71). These are the kind of claims that Mandeville wanted to take distance from by writing *Part II*.

written for that occasion. *The Fable* is quoted extensively to prove that Berkeley's criticism was misplaced. But more interestingly, we can infer that Mandeville also used his unpublished Defence of the *Fable* of 1726 in his answer to Berkeley. This was nothing new for Mandeville, for example, he did a similar thing with some collected poems: he had a habit of publishing all sorts of miscellaneous pieces that he had composed earlier. To understand the composition of the *Letter to Dion* is important to realize that the dialogue between Berkeley and Mandeville was more of a hybrid of different discussions and at the same time, in Mandeville's part, it was a culmination of finally printing his answer to his many critics embodied now in George Berkeley.

13.2 1732 and *Part II* of 1729

What Berkeley did in 1732 was to re-intensify the tradition of denouncing *the Fable*. We must see Mandeville's answer to *Alciphron* in the context of this long line of accusations. But what makes it interesting is that instead of defending his *Fable* properly in public before Berkeley's *Alciphron*, Mandeville had already published a new work, *Part II* of the *Fable* that changed things considerably.

The reason why *Part II* is important for the story of Mandeville's answer to Berkeley is this: Here we have a new work where Mandeville who had been looking at the criticism poured onto him reflects on this criticism and puts forward what he clearly considered to be a new theory of the formation of civil society. In this work the question of passions, sociability and morals is more complicated than previously. And in this sense, what Mandeville took as representation of himself in the third dialogue of *Alciphron*, denouncing all moral beauty in the form of a follower of *The Fable* named Lysicles, this must have been irritating to him. What is also significant about the title of the *Fable* is that while the memorable part (and what everyone including Berkeley criticized) was the subtitle, "private vices, public benefits", it does not play a central role in *Part II*.

But to turn back to *Alciphron*, the point is that Berkeley was not reading some of these Mandeville's later works when writing it. It might be that in Rhode Island where he composed the work, he did not have access to *Part II* and most definitely not to Mandeville's *Origin of honour* that was published in 1732. This asymmetry between the development of Mandeville's thinking and what Berkeley was focusing on is important. It enables us to understand how Mandeville approaches Berkeley's accusations. First of all, Mandeville had given a new version of the theory of civil society that effectively disregarded the earlier infamous slogan of private vices and public benefits. Also, Mandeville had changed his opinion about the possibility of natural virtues in the case of natural affection. Hence, it is important that after modifying his view, Mandeville takes up again the task of defending the paradox of "private vices, public benefits", this time against Berkeley. Of course, since Mandeville's perspective was now somewhat different, at some points he is not in fact defending what he set forward in the original *Fable*, but instead what he advanced in *Part II*. Also, his attitude seems to be much more relaxed now in the *Letter to Dion*, because he does not have to struggle with the problem of reducing everything to self-love because he has already accepted some other-regarding affections as natural. So, what I want to say is that in one sense the Mandeville who answers Berkeley is not virtually the author of the *Fable of the bees*, but in this sense the author of *Part II*.

13.3 Dialogue Between Berkeley and Mandeville

In *Alciphron* Berkeley famously defines 'modern Free-thinkers' to be 'the very same with those *Cicero* called Minute Philosophers' who 'diminish all the most valuable things', 'act the reverse of all other wise and thinking men'; 'aim to erase the principles of all that is great and good from the mind of man, to unhinge all order of civil life' and 'to undermine the foundations of morality' (Berkeley 1732: 47–48). Mandeville refuses to recognize himself as such minute philosopher. In *Alciphron* these free-thinkers are supposed to be, or they profess to be, free thinking lovers of truth. Examination, however, proves that they are everything but that. They are a sect, much comparable to superstitious religious outfit. Berkeley paints them as atheists, particularly Lysicles who was at the time seen (also by other people than Mandeville) to represent the author of the *Fable of the bees*.

In *Letter to Dion*, Mandeville makes serious accusations about Berkeley's way of writing. His view about the characters of Alciphron and Lysicles is that 'they are fellows without feeling or manners' and no gentleman would ever act in the way they do. In Mandeville's words, Berkeley portrays them to 'always begin with swaggering and boasting of what they'll prove; and in every argument they pretend to maintain, they are laid upon their backs, and constantly beaten to pieces, till they have not a word more to say; and when this has been repeated above half a score times, they still retain the same arrogance' and 'immediately after every defeat, they are making fresh challenges, seemingly with as much unconcern and confidence of success, as if nothing has pass'd before, or they remember'd nothing of what had happen'd' (Mandeville 1732: 52–53). For a sceptic like Mandeville, to whom the question of identity and self were not immovable objects, uniform and unviolated character mattered instead a great deal.

Berkeley's *Alciphron* received, what might seem to us, a surprisingly hostile welcome in the eighteenth century. This regarded especially the way he had treated both Shaftesbury and Mandeville. Hervey, in his *Some remarks concerning the minute philosopher*, for example, writes that Berkeley was 'monstrously and manifestly partial' (Hervey 1732: 6). He particularly singles out that the handling of the *Fable of the bees* and its author has been 'injudicious and unfair' saying that Berkeley's book has done more harm to Christianity than the *Fable* (Hervey 1732: 44, 8).

Also for the contemporaries it was not only the question of how philosophy was presented that mattered. Equally important was how the characters that represented the authors were drawn. In the case of Shaftesbury, the treatment of Cratylus (singled out to be the character of Shaftesbury) was brought up not only by Hervey, but also by Mandeville in *Letter to Dion*. In fact, Mandeville defends Shaftesbury against Berkeley and says that he has been mocked without reason (Berkeley 1732: 190). Mandeville writes, 'I am fully persuaded' that Shaftesbury 'was in the wrong' in many things, 'but this does not blind my understanding so far, as not to see, that he is a very fine author, and a much better writer than my self, or you either', he tells Berkeley. Mandeville says he cannot understand the 'indignity and contempt, which you treat *Cratylus* with' (Mandeville 1732: 47–48).

Rhetorically Mandeville's strategy in the *Letter to Dion* is to assume that Berkeley, like so many others, had not read the *Fable of the bees*, but was leaning on false reports about it. Mandeville writes his answer to Berkeley in a very cordial manner. Mandeville is sure that if Berkeley had actually read his work, 'he would not have suffer'd such lawless Libertines as *Alciphron* and *Lysicles*, to have shelter'd themselves under my wings; but he would have demonstrated to them, that my Principles differ'd from theirs, as Sunshine does from Darkness' (Mandeville 1732: 3). Mandeville also draws a parallel to the grand jury that had also trusted the judgment of others (Mandeville 1732: 8). The reason why he is writing his answer is explicitly that he hopes that Berkeley revises his views in a future edition of *Alciphron*.

What Berkeley puts Alciphron and Lysicles to defend in the name of Freethinking, Euphranor (or others) by open-dialect will either ridicule or put to questionable light. Berkeley's reading of the *Fable* is that it aims to show that 'vice circulates money and promotes industry, which causeth a people to flourish'. Hence, any vice like drunkenness is supposed to produce this effect, because it causes 'an extravagant consumption which is most beneficial to the Manufacturers, their encouragement consisting in a quick demand and high price' (Berkeley 1732: 79). Thus, the counter-argument concentrates on the idea that drunkard is not necessarily as beneficial to the brewer as one might think and most definitely this is not as beneficial for the public as some claim. What Berkeley wants to say is that 'money spent innocently' circulates just 'as well as that spent upon vice' (Berkeley 1732: 81).[7] Vice here is to be understood in a wide sense of fashionable, luxurious living. Hence, what in Berkeley's opinion follows is that agriculture employs men just as well as money spent on fashion and luxury. What Berkeley is after is something called real happiness of the state. Riches are just means to produce it (Berkeley 1732: 94). Berkeley argues that men's primary pleasure is not sensual, but rational higher kind of pleasure. For this reason as well, Mandeville is wrong. What Berkeley wants to say is that man's 'true interest is combined with his Duty' (Berkeley 1732: 184).

We need to understand that the moral, social and economical questions under discussion are not hypothetical, but concrete ones. We are really talking about how to prove different hypothesis about flourishing in a particular country. This, of course, is true to Berkeley in the question of Ireland. What is also relevant is that the dialogue between Berkeley and Mandeville conforms to the same form that was

[7] Also, it is important that the number of inhabitants is underlined as an indicator of the flourishing of people.

structured around the grand jury presentment: questions of religion and debauching the nation. At the same time the question of what Mandeville actually means by vice becomes crucial.

13.4 Religion

Regarding Mandeville's answer, I will first treat the question how Mandeville defends himself in the *Letter to Dion* against the accusations that the *Fable of the bees* is contrary to religion or promotes atheism. Mandeville says on numerous occasions that the *Fable* was not written against religion, but it just exposes the hypocrisy of many Christians, taking notice of the scarcity of true self-denial among them (Mandeville 1732: 19, 24–25, 63). We need to remember that also Hervey backed up Mandeville on this particular question.

Now, we need not assume that Mandeville was an atheist. Or the point is actually that his personal religious views, one way or the other, do not have an effect on his moral and political thought. On several occasions he discusses what he terms true Christianity. This, some believe, is due to a shade of Armenianism that Mandeville derived from his Dutch background. In any case, Mandeville systematically defends a view that 'true Christianity' is a 'private concern of every individual'. Multitudes may join in outward, public performance of religion, but the actual religion is always an individual and private matter (Mandeville 1732: 40–41). Mandeville is continuously underlining that the "private vices, public benefits" argument concerns this world. Christian thinking must always consider the afterlife and this is a different matter than the question of the flourishing of the state. Mandeville utters that a man may go directly to hell because of his avarice and his wealth builds hospitals as soon as he is dead (Mandeville 1732: 39). The position Mandeville defends regarding religion comes also quite close to Jansenists and Pierre Nicole in particular. The common point is that worldly flourishing and afterlife have little to do with each other. We may see how Mandeville's and Berkeley's views clash regarding the question of virtuousness of an atheistic colony, which is quite contrary to Berkeley's own Bermuda project that again underlines that we are discussing concrete questions.

In *Alciphron*, Berkeley dismisses the Baylean question whether an atheistic society could be virtuous that also Mandeville had discussed (Berkeley 1732: 135, 148).[8] When Berkeley touches upon the question of setting up a colony of atheists, his suggestion is to go ahead and try. Without religious duty, you will not survive. What is important is that once again we are talking about practical measures. Mandeville answers: 'you think the multitudes among Christians to have better morals, than they were possess'd of among the antient heathens. The vices of men have always been so inseparable from great nations, that it is difficult to determine any

[8]"Question whether to have religion or not examined".

thing with certainty about the matter. But I am of opinion, that the morals of a people in general, I mean the virtues and vices of a whole nation, are not so much influenced by the religion that is profess'd among them, as they are by the laws of the country, the administration of justice, the politicks of the rulers, and the circumstances of the people' (Mandeville 1732: 55). And since it is the circumstances of the people that the private vices, public benefits argument mainly concerns, we may perhaps understand in what way this part of Mandeville's answer is justified.

The reason why Mandeville thinks that the difference between the virtuousness of heathens and Christians is more accidental is that it varies considerably why people actually follow certain rules. Some act according to laws and customs because of religious tenets, no doubt. Also, some because they value what others think about them. Some from whatever fleeting motives they might have. This is the point how customs function. What Mandeville is doing is deteriorating the argument that we may strictly define virtue (unless we use the concept of self-denial) and claim that proper motivation for certain actions is what matters the most. Mandeville does not find religion as an overriding motive making men sociable (Mandeville 1732: 60). It is one motive among many for people to act in a certain way, but it has little consequence to the flourishing of the people. 'In Great-Britain there are thousands that abstain from unlawful pleasures, who would not be so cautious, if they were not deterr'd from them by the expence, the fear of diseases, and that of losing their reputation' (Mandeville 1732: 56). Regarding the question of atheistic colony, Mandeville challenges Berkeley to show: 'If the laws and government, the administration of justice, and the care of the magistrates were the same, and the circumstances of the people were likewise the same, I should be glad to hear a reason, why there should be more or less incontinence in *England*, if we were heathens, than there is, now we are Christians' (Mandeville 1732: 55).

What Berkeley wants to argue (that interest and Christian duty should always go together and actually cause the best consequences) is in line with the reformation of manners movement that Mandeville had been opposing for some time. These same arguments simply would not do for Mandeville. Mandeville's objection regarding religion and the view that he thinks Berkeley defends is that if we are talking about duty as a Christian, these two things and how they relate to the prosperity of the state have nothing to do with each other. Vanity makes man profit his tailor, not his interest in helping him or the state. 'Religion is one thing, and trade another', Mandeville writes. 'He that gives most trouble to thousands of his neighbours, and invents the most operose manufactures is, right or wrong, the greatest friend to the society' (Mandeville 1732: 68).

What the difference between Mandeville and Berkeley boils down to is important. In one sense Berkeley represents the lot of thinkers that find a conception of virtue (and self) uniform belonging to the same continuum as the Cartesian moralists as well as Shaftesbury, Hutcheson and many others. Whereas the way towards a new formulation of personal identity, and analysis of what is agreeable or useful to self and others, is opened up by Mandeville (this can well be called Humean analysis of self and moral psychology). Important part of this new paradigm that challenges the public benevolence and real happiness of the state hypothesis is the uncertainty of moral knowledge. Berkeley is in the end couched to Christian conception of man and virtue, whereas Mandeville's contribution is to liberate Hume and others of this strain. Equally relevant in this scheme is not to argue that religion should be explicitly denounced because it is not what it claims to be. But what is important is to treat religion as one custom (and set of motives) among others.

In the *Letter to Dion*, Mandeville is quite explicit that the *Fable* has not been written 'for the encouragement of vice, and to debauch the nation', like Berkeley claims (Mandeville 1732: 1). Mandeville's suggestion is that even his discussion of whoring is in fact an innocent work because he has been careful not to say anything that might be hurtful to weak minds. Mandeville says that his texts are philosophical and they do not popularly address the people (Mandeville 1732: 14–15). This also makes it understandable how he sees himself justified to claim that he is not promoting the vices of the day, but ridiculing them (Mandeville 1732: 32–34). Therefore, Berkeley misrepresented him when he puts his idea as 'the more mischief men did, the more they acted for the publick welfare' because 'without vices, no great nation can be rich and flourishing' (Mandeville 1732: 54). In Mandeville's own perspective, he has always laid 'down as a first principle, that in all societies, great or small, it is the duty of every member of it to be good, that virtue ought to be encouraged, vice discountenanc'd, the laws obey'd, and the transgressors punish'd' (Mandeville 1732: 4). I believe this to be a sincere comment. What the *Letter to Dion* is supposed to do is to change things so that it is impossible for Berkeley 'to remain ignorant any longer of the innocence of my intentions, and the injustice that has been done me' (Mandeville 1732: 11). Mandeville's goal in writing the *Letter to Dion* was to establish a dialogue between him and Berkeley, not merely to strike back.

13.5 Private Vices Paradox

It is important to realize that in his attack on Mandeville in *Alciphron*, regarding consumption and the question of circulation of money, Berkeley is actually leaning on arguments about "innocent consumption" excluding luxury out of this realm. Mandeville, of course, is strongly opposed to this idea trying to explain to Berkeley why he is wrong and why the question is categorical about formulating public life according to natural human passions and not of degrees of innocent and not-so-innocent consumption.

Regarding the question of promoting vice, Berkeley keeps bringing up his accusation that the key demonstration of Mandeville's debauching principle is that he promotes drunkenness in the name of public benefits (Berkeley 1732: 69, 79). He insults Mandeville directly trying to show how he has cornered himself with these ideas (Berkeley 1732: 84–85). As I have repeatedly said, what we need to understand is that in the *Letter to Dion* there is a change in the presentation of the paradigm "private vices, public benefits". It is removed quite far from the most extreme cases of the *Fable* (robbers being useful to locksmiths etc.). In *Letter to Dion*, Mandeville says that 'What I call vices are the fashionable ways of living'

(Mandeville 1732: 31). Given the changing context, this should not be overlooked. From the provocations of the *Fable*, we have come almost to the level of promoting a national lottery. Mandeville also explains that the reason why he used the notorious paradox in the first place was to raise attention. If one reads the book, Mandeville says, he will find that it does not mean encouragement of vice (Mandeville 1732: 38). Hervey calls this a lame reading of the hypothesis, something that even he would be willing to submit to (Hervey 1732: 43–44). In this sense, what Mandeville says about the private vices, public benefits is somewhat dissimilar from the original *Fable*. He is now giving a new reading of the paradox. Mandeville says that he has never advanced anything comparable to an idea: drink and be rich. Yet, in his direct criticism of Berkeley, he does not give up the idea that some vices are necessary for a state to be flourishing.

Not only has the more radical edge disappeared from Mandeville's use of the paradox, he also advances different arguments defending its use. Most general use of the private vices and public benefits is the explanation that men are driven for example by pride or fickleness (which would be considered a vice by moralists) when they for example wear golden brocades. Their motivation to wear certain fashionable items is not to encourage trade or concern for the public in general, this follows as unintended consequence (Mandeville 1732: 20). An idea, as we know, made famous later by others.

This leads Mandeville to talk about 'positive evil' in more general terms, a theme first pursued in depth in 'A Search into the nature of society' in 1723. Mandeville writes that 'natural as well as moral Evil' can sometimes contribute to 'worldly Greatness' and furthermore 'a certain Proportion of them is so necessary to all Nations, that it is not to be conceiv'd, how any Society could subsist upon Earth, exempt from all Evil, both natural and moral' (Mandeville 1732: 21). In economic realm, Mandeville's example is that plenty of crops is not necessary a good thing for the state revenue. Plenty in one's country and want in others of course is (Mandeville 1732: 50). About positive evil, Mandeville's favourite example is that eighteenth-century London is dirty. The idea is that different professions necessarily make the streets as they are. The only way to make them less dirty would be to give up some of the opulence (or so it seemed). Mandeville's argument is that 'Dirty streets are a necessary evil inseparable from the felicity of London' (Mandeville 1732: 16–17).

Now, the point that I want to make is that while this economic extension of positive evil and "private vices and public benefits" is quite familiar and we understand what Mandeville means when he argues against Berkeley that 'To wish for the encrease of trade and navigation, and the decrease of luxury at the same time, is a contradiction', yet there is something more to the argument (Mandeville 1732: 49). The point is that we need to adopt a way of looking at the world where we are able to question ideas about certainty and embrace the opaque nature of moral knowledge. This will then open up a path leading to an idea of a system that incorporates all citizens equitably within its boundaries, instead of treating poverty as a personal shortcoming, for example. Compared to this new way of perceiving the moral and political realm, Bayle's earlier ideas about toleration, for example, seem almost conservative.

Although great emphasis is put on worldly greatness, Mandeville's argument concerns politics in a much more extensive way and he discusses this in his *Letter to Dion*. As I already pointed out, Mandeville's problem with Berkeley was the same that he had with reformation of manners movement, Shaftesbury, Hutcheson and all the other moralists that had preceded him. Although different in many ways, what they were doing was precisely what Berkeley was doing, combining man's interest with his alleged duty. This includes an idea of uniformity of self and I don't think we are very far off, if we claim that this resembles Cartesian and Christian conception of personal identity.[9] Moralists always started by defining the conception of a virtuous man and proper motivation and thought that this introspection was also the way to a flourishing society. For Mandeville this was not the way to approach large societies and as a moral principle it would restrict humanity to a certain, arbitrary mode without giving actual way to human flourishing that does not confine itself to one custom, but gives humans the opportunity to develop their own ways of living. *Alciphron* was particularly abusive of Mandeville because it accused him of renouncing all moral beauty when in fact what Mandeville was after was to formulate a new way of perceiving it.

There is a clear political dimension to the problem of uncertainty of moral knowledge. In his *Letter to Dion*, Mandeville was particularly worried about the role of education. He did not wish to deny that man could be taught to become virtuous, pious and good in some sense, what he wanted to say was that we should not care so much about particular motives, but concentrate on the different ways that people act in a beneficial manner. This emphasis on the external is what protects the way people cultivate their own moral judgment and identity. In political realm man simply should not be given an opportunity to knavery (Mandeville 1732: 33). 'It is the business of all law-givers to watch over the publick welfare, and, in order to procure that, to submit to any inconveniency, any evil, to prevent a much greater, if it is impossible to avoid that greater evil at a cheaper rate'. Thus, Mandeville gives a concrete example how the idea of "private vices, public benefits" functions also in a strictly political realm. He discusses a law that says, 'if a felon, before he is convicted himself, will impeach two or more of his accomplices, or any other malefactors, so that they are convicted of a capital crime, he shall be pardon'd and dsmiss'd with a reward in money' (Mandeville 1732: 42–43). According to Mandeville, 'this shews the usefulness of such a law, and at the same time the wisdom of the politician, by whose skilful management the private vices of the worst of men are made to turn to a publick benefit' (Mandeville 1732: 45). The point is that ever-changing legal system is always based on a pragmatic foundation. This is one example of the practical concern of the legislators and how political attempts to define what is a good man, and how he should act, will not do when we discuss the welfare of a state.

[9]Note the general contrast that Jonathan Lamb makes between Cartesian and Humean understanding of passions and identity (Lamb 2009).

One of Mandeville's positive contributions is to question the agent-oriented way of defining what is right and wrong and to extend the idea of private vices and public benefits to the political realm: to show how the separation of moral (and personal) and public (and political) realms is relevant for the existence of both. What we are in this manner preserving is an opportunity for everyone to become something based upon the principles of their own judgment and not on some predetermined attributes. Mandeville writes, 'we stand in need of the plagues and monsters I named, to have all the variety of labour perform'd, which the skill of men is capable of inventing, in order to procure an honest livelihood to the vast multitudes of working poor, that are required to make a large society' (Mandeville 1732: 67). What is said here (from the moral perspective) is nothing that we would consider controversial, the claim is that it is our passions that define who we are and we should be able to cherish this. One inevitable consequence is that cultivation of passions will also advance the economy, create new forms of labour and provide the ones in need with a livelihood. External attributes that enable this process are the pragmatic content and execution of laws.

13.6 Conclusion

I believe that the dialogue between Mandeville and Berkeley made a difference in Berkeley's thinking as well. One consequence is that the focus starts to switch to the idea of the labour of the poor and their condition. Mandeville took further the mercantilist idea that everyone needs to be employed and industrious. Could we not then conclude that the next questions to be asked are: 'Whether there ever was, or will be, an industrious nation poor, or an idle rich?' and 'Whether a people can be called poor, where the common sort are well fed, clothed, and lodged?' These are Mandevillean questions that Berkeley is asking in his *Querist*. The reason why I think this dialogue between Berkeley and Mandeville is important is that this is also the way we may see that the conception of justice eventually started to expand. The concentration on the poor and the question whether we can or should define strictly what moral motives people should have eventually leads to a question of different rights that also poor people have. Only when we give up the ideals of perfectibility and uniform self we start appreciating the ones in the margins. But moral judgment does not accomplish much before it is protected by concrete laws. Developing this line of thinking is where Mandeville did a better job than Berkeley. The flaws in his approach that prevented him from being more successful in this undertaking lay elsewhere.

References

Berkeley, George. 1732. *Alciphron in the works of George Berkeley, Bishop of Cloyne*, 9 vols, vol. 3, 1948–1957, ed. A.A. Luce and T.E. Jessop. London: Nelson.

Berman, David. 1993. *George Berkeley Alciphron in focus*. London: Routledge.
Hervey, John. 1732. *Some remarks concerning the minute philosopher*, ed. J. Roberts. London.
Kaye, F.B. 1921. The writings of Bernard Mandeville. *The Journal of English and Germanic Philology* 20: 416–467.
Kaye, F.B. 1924. Introduction. In *The Fable of the bees*. Oxford: Oxford University Press.
Lamb, Jonathan. 2009. *Evolution of sympathy in the long eighteenth century*. London: Pickering & Chatto.
Mandeville, Bernard. 1703. *Pamphleteers*, ed. R. Wellington. London.
Mandeville, Bernard. 1729. *Fable of the bees. Part II*, ed. J. Tonson. London.
Mandeville, Bernard. 1732. *Letter to Dion*, ed. J. Roberts. London.
Anon, Remarks. 1729. *Remarks upon two late presentments of the Grand-Jury of the county of Middlesex*, ed. A. Moore. London.
Speck, W.A. 1978. Bernard Mandeville and the Middlesex Grand Jury. *Eighteenth Century Studies* 11: 362–374.
Anon, The presentment. 1728. *The presentment of the Grand-Jury for the county of Middlesex, to his Majesty's court of King's-bench, at Westminster, against infidels and sodomites, and impious books*. London.
Tolonen, Mikko. 2013. *Mandeville and Hume: Anatomists of civil society*. Oxford: Voltaire Foundation.
Anon, True meaning. 1726. *True meaning of the Fable of the bees*. London: William and John Innys.

Chapter 14
Mandeville and the Markets: An Economic Assessment

Rogério Arthmar

Abstract This article analyses the debate that occurred in England in the second decade of the eighteenth century regarding the possibility of a shortage of demand for commodities in general. It identifies the points of contact of such exchanges with a controversy that, 100 years later, would lead to emergence of the so-called Law of Markets. Initially, it presents the opposite stances respecting parsimony and luxury consumption in the context of the mercantilist thought. Bernard Mandeville's views about the effects of vice and virtue on the stability of purchasing power are examined afterwards. Following this, the criticisms of Mandeville's ideas by George Bluet, Francis Hutcheson and George Berkeley are considered. Finally, we evaluate the extent to which the debate anticipated some crucial propositions put forward by the classical school of economics in the early nineteenth century.

Keywords Vices • Virtue • Demand • Savings • Investment

14.1 Introduction

The famous debate about the possibility of commercial crises that took place in England in the 1820s, known as the general glut controversy, would enter the history of economics annals as one of the highest points of nineteenth century political economy. The dispute originated with the concern of some writers, notably the Reverend Thomas R. Malthus, that a rapid accumulation of capital could create a universal surplus of goods with no corresponding demand. The phenomenon would be rooted in an insufficiency of consumer spending, driven by an excessive expansion in savings, which was deemed essential to an increase in investment. It was feared that the additional supply of goods flowing from the newly created capital stock would be unsellable at the going prices, owing to the previous fall in consumption outlays. Based on this diagnosis, Malthus advocated, among other things, the functionality of luxury expenditures by landowners as an alternative

R. Arthmar (✉)
Department of Economics, Federal University of Espírito Santo, Vitória, Brazil
e-mail: rogerio.arthmar@ufes.br

E. Balsemão Pires, J. Braga (eds.), *Bernard de Mandeville's Tropology of Paradoxes*, Studies in History and Philosophy of Science 40, DOI 10.1007/978-3-319-19381-6_14

source of demand to counter-balance the slackening in sales and thus avoid a general glut (Malthus 2004 [1820], 301–31). Oblivious to such concerns, David Ricardo, Jean-Baptiste Say and James Mill, among others, argued that economic crises of major proportions were impossible because the purchasing power generated by the production process would be automatically spent on consumer goods or on investment in productive labourers and fixed capital. They agreed, therefore, that no artificial stimulus to demand was necessary beyond the one generated by the capital accumulation process in itself, a proposition that would be designated as the Law of Markets or, simply, Say's Law (Ricardo 2004 [1817], chap. XXI; Say 1850 [1803], B. I, chap. XV; James Mill 1808, chap. VI; for modern views of the subject, see Kates 1997; Baumol 1977; Sowell 1972, 79–145).

Going back further in time, we find that the controversy over the possibility of a lack of aggregate demand had distant antecedents in England. In 1705, Bernard Mandeville published his poem, *The Grumbling Hive: or Knaves Turn'd Honest*, where he praised vice and pomp as sources of the prosperity of a beehive, a picturesque allegory of the society of his day. This work, despite its originality, would go unnoticed in literary circles. Some years later, however, in 1714, a new edition of the poem was issued, with a comment by Mandeville on the origin of moral virtue and extensive explanatory notes on the meaning of the poem. Now called *The Fable of the Bees, or Private Vices, Publick Benefits* (hereafter, *The Fable* or *FB*), the book would again fail to capture widespread attention. However, in 1723, with a revised text, supplemented by two polemical essays dealing, respectively, with charity schools and the nature of society, the volume aroused such a wave of indignation among the critics that the unexpected publicity, ironically, lead to five further editions of the work in less than a decade, a record at the time (Kaye 1924, ix–x; Stafford 1997, xii).[1] Among the long list of sermons, pamphlets and other treatises that appeared after the second edition of *The Fable* and directed against the content of the book – the majority, of moral or religious inspiration[2] –, some sought to directly refute the economic doctrines developed in the book. In this latter group are the works by the lawyer George Bluet, the philosopher Francis Hutcheson and the bishop George Berkeley. As we shall see, such attacks on Mandeville and his reaction to them produced a fruitful collision of ideas that, in many important respects, would foretell the controversy over the markets' operation a 100 years later.

[1]In the eighteenth century, the original volume of *The Fable* had editions in 1724, 1725, 1728 and 1729. In the latter year, Mandeville launched the second part of the book, containing a foreword and six dialogues, published in two separate editions in 1730 and 1733. The two volumes were released together in the years 1733, 1755, 1772 and 1795, and translations into French and German appeared in 1740 and in 1761, respectively (Kaye 1924, xxxii–vii).

[2]A compilation of all written material contesting Mandeville's ideas published throughout the eighteenth century is found in Stafford (1997).

14.2 The General Context of the Debate

The first decades of the eighteenth century, when Mandeville's main works were published in England, fall within the phase of transition from mercantilism to liberalism. More precisely, it was during that evolutionary stage of economic thought that the latent contradiction between the belief in free trade, understood as the primary source of social wealth, and the regulation of commercial activities, implemented to strengthen the power of the state, came to the fore in full force (Grampp 1962, 61–89; Heckscher 1943, 757–64). The conflict between these two strands of thought would be manifest, like other issues, in the incompatibility between the distinct positions on the economic role of sumptuary expenses, widely recorded in the literature of the period. To better figure out the matrix of this divergence, it must be recalled here that the mercantilist concept of wealth encompassed the accumulation of objects with the capacity to store value and having great durability, which made the precious metals the most suitable vehicle for preserving one's possessions. The practice of frugality was seen as essential to the enrichment of society, a recommendation that, moreover, was often shrouded in precepts of an ethical or religious nature (Heckscher 1943, 637–40; Viner 1937, 26–32).

The main objections to sumptuous consumption arose from two sources: the first one, from the alleged weakening of the productive potential of individuals, in making them lazy, careless and unprepared; and, second, from the fact that luxury spending almost always went toward foreign exotic items, meaning thus an increase in imports and the loss of precious metals through national boundaries. Obviously, in such conditions, the country's trade surplus, so valuable in the eyes of mercantilist authors, would be sacrificed (Johnson 1960, 289–97). By way of illustration, Josiah Child, in enumerating the causes of Holland's prosperity in the seventeenth century, praised its "parsimonious and thrifty living", considered by him as something absolutely "extraordinary" (1668). Thomas Mun, more caustic, energetically reproved the excessive consumption of the Britons for preventing the Kingdom from achieving a higher level of exports:

> The summ of all is this, that the general leprosie of our piping, potting, feasting, fashions, and mis-spending of our time in idleness and pleasure (contrary to the Law of God, and the use of other nations) hath made us effeminate in our bodies, weak in our knowledge, poor in our Treasure, declined in our valour, unfortunate in our enterprises, and contemned by our enemies. (Mun 1664, chap. 19)

Such preaching in favour of virtuous behaviour, however, faced stern opposition long before Mandeville. On the European continent, especially in France, the first decades of the seventeenth century saw the development of the Jansenist theological movement, which professed the irretrievable downfall of man after the original sin, as well as man's ultimate inability to achieve redemption through a life of renunciation. Behind every meritorious action, as these religious authorities proclaimed, selfishness and self-interest were hidden. The Jansenist tradition would meet with a favourable reception, sometime later, in the works of French moralist philosophers such as François de La Rochefoucauld, Jean de La Fontaine and Pierre

Bayle, who portrayed the individual as a being fully controlled by passions and insensitive to reason. In the writings of these authors, all expressions of solidarity are reduced to simple disguises of the most ingrained feelings of self-love and pride. Insistent calls to resist temptation, they asserted, were incompatible with man's natural propensities because the harshness of existence could only be relieved in worldly pleasures (Horne 1978, 19–31; Kaye 1924, xcvi–cv). Rochefoucauld, for instance, condensed this line of thought in the following maxim: "Vices enter into the composition of virtues as poison into that of medicines. Prudence collects and blends the two and renders them useful against the ills of life" (Rochefoucauld 1871 [1665], 23).

Scepticism about the vaunted advantages of virtue would not be confined to France. Simultaneously, in England, Thomas Hobbes proclaimed the primacy of appetites in determining human behaviour, making desires and strong passions the fundamental sparks of resolute action. At the same time, Hobbes despised temperance because it acted as a disincentive to individual effort. "Frugality, though in poor men a virtue, maketh a man unapt to achieve such actions as require the strength of many men at once", wrote Hobbes, to which he added, "for it weakeneth their endeavour, which is to be nourished and kept in vigour by reward" (Hobbes 1999 [1651], 88). At the end of the seventeenth century, however, the defence of luxury spending by certain British authors assumed a less psychological and a more markedly economic tone. Interested in the customs revenues to finance the British royalty – the only Crown income free from rigid parliamentary control – Tory thinkers such as Dudley North and Nicholas Barbon challenged the notion of the intrinsic value of precious metals while explicitly advocating the importance of a strong demand for superfluous articles in promoting businesses and employment (Ashley 1987, 335–71). Barbon, in one section of his opuscule, noted incisively: "The use of trade is to make, and provide things necessary or useful for the support, defence, ease, pleasure, and pomp of life". In addition, he even observed: "It is not necessity that causeth the consumption, Nature may be satisfied with little; but it is the wants of the mind, fashion, and desire of novelties, and things scarce, that causeth trade" (Barbon 1934 [1690], 21, 35). The following year, North also made known his unconditional agreement with this provocative conception of the economic process: "The main spur to trade, or rather to industry and ingenuity, is the exorbitant appetites of men, which they will take pains to gratifie ... for did men content themselves with bare necessaries, we should have a poor world" (North 1691, 27).

14.3 Avarice and Luxury in Mandeville

As Mandeville's speculations on virtue and vice are scattered throughout his writings, interspersed with moral and philosophical ramblings about man and society, the exact content of his effective contribution to the evolution of economic thought has been subject to conflicting readings. He tends to be classified as either a proponent of individualism or as a typical mercantilist preacher.[3] For our purposes, however, it is sufficient to note that these disparate views stem from the fact that Mandeville's work is based on two central assumptions that lend support to both interpretations, namely: (i) individuals are driven by a complex of passions, always seeking their self-interest; and (ii) the State, through a proper regulation of such passions, especially pride and shame, induces people to perform acts consistent with the public interest (Rosenberg 1963). Indeed, Mandeville is categorical in describing man as an "extraordinary selfish and headstrong, as well as cunning animal", adding yet that it would be "impossible by force alone to make him tractable". In order to tame such impulses, he explains, lawmakers would have discovered in adulation "the most powerful argument that could be used to human creatures" (*FB* I 1992, 41–43). Bearing in mind this unique perspective on the relationship between the individual and society, we will examine how Mandeville constructed his spirited defence of luxury consumption.

In repeatedly addressing the issue of sumptuary expenses, Mandeville was keen to stress its vital importance as a source of employment and livelihood for various professional groups. In one of many passages in *The Fable* in which this subject is dealt with, he considers the situation of a hypothetical merchant involved in the export of cereals and fabrics and in the acquisition of wines and liquors from abroad. Such imports, clarifies Mandeville, in addition to providing the resources with which foreign countries finance their demand for domestic manufactured goods, would not only favour maritime commerce but also increase the national customs revenue. However, the need for these purchases abroad would ultimately be rooted in the licentiousness and drunkenness of the population. Despite the widespread recrimination of such vices, they would have the felicitous property of setting in motion an entire network of honourable people involved in the circulation and sale of alcoholic beverages within the country. At this point, it is useful to transcribe the words in which Mandeville put forward his case on this particular matter:

> It may be said, that virtue is made friends with vice, when industrious good people, who maintain their families and bring up their children handsomely, pay taxes, and are several ways useful members of the Society, get a livelihood by something that chiefly depends on, or is very much influenc'd by the vices of others, without being themselves guilty of, or accessory to them, any otherwise than by way of trade, as a druggist may be to poisoning, or a sword-cutler to blood-shed. ... The same may be said not only of Card and Dice-makers,

[3]Kaye (1924, xciv–ciii) considers the uncompromising defense of luxury spending, free trade and *laissez-faire* as Mandeville's main economic legacy. Hayek (1948), in his turn, credits Mandeville as the authentic precursor of individualism. More recently, Jack (1976) and Dumont (1975) situate the originality of *The Fable* in its ability to identify the contradiction between the moral recommendations of society on the one hand, and the unrestricted pursuit of material wealth on the other. Landreth (1975), following Heckscher (1943, 566–67, 734–35 *passim*), defines Mandeville as an unconditional mercantilist, interested, first of all, in ensuring the growth of production by the subordination of the individual to the State. The inherent difficulty in such classifications is well illustrated by Chalk, who, in a first article (Chalk 1951), shares with Kaye and Hayek the view of Mandeville as an individualist thinker, a judgment reworked later (Chalk 1966) in favor of the recognition of the mercantilist positions of the Dutch author.

> that are the immediate ministers to a legion of vices; but of mercers, upholsterers, tailors, and many others, that would be starv'd in half a year's time, if Pride and Luxury were at once to be banished the nation. (FB I 1992, 85 – emphasis in the original)

If this were everything Mandeville had to say about luxury spending, there would be little to add with regard to his contribution to the progress of economic ideas. Certainly, his views were not new in the literature, given that many authors had previously emphasised the connection between the facilities for marketing products and work opportunities. When translated into practice, the stress on employment creation led, in most cases, to the view that raw materials, instead of being exported in their crude state, must be manufactured within the country to obtain their maximum values as finished products. Imports, in turn, should be produced internally whenever possible. Thus, through the enlargement of the trade surplus, as the mercantilists writers insisted on, the increasing availability of precious metals would stimulate the demand for domestic production and, *a fortiori*, the crafts of the country's labouring classes. On the other hand, it must also be remembered that the recurring concerns about the plight of workers were unrelated to their personal wellbeing. Since labour was viewed just as one element of production, it was generally recommended that wages should be fixed at the subsistence level in order to keep low the costs of the exporting activities and force the workingman to constant exertion, depriving him of any incentive to idleness (Furniss 1957, 39–74; Gregory 1921).

The ability of vice to provide the livelihood for a large segment of the population, often taken as Mandeville's central message is, strictly speaking, nothing more than an aspect of his economic analysis. Indeed, his defence of luxury spending is more finely elaborated – although unfortunately, seldom noted in the literature – covering the crucial problem of the conditions underlying the stability of aggregate demand. Let us examine this issue more carefully. Any increase in frugality, Mandeville observes, would immediately occasion a reduction in demand for the services of tradesmen and manufacturers. The negative impact of an increased willingness to save, however, would not be confined to the unemployment of a restricted category of workers since it would spread itself to many related activities. Thrift, once adopted as the standard of conduct by the society, would lead, among other things, to a decline in purchases of new houses and, consequently, "three parts in four of *masons, carpenters, bricklayers,* etc. would want employment; and the building trade being once destroyed, what would become of l*imning, carving,* and other arts that are ministring to luxury" (*FB* I 1992, 223 – emphasis in the original). A nation of honest and frugal people, Mandeville tells the reader, would be indeed a poor society, lacking productive employment for its inhabitants. The parsimony of the Dutch, he also observes, much praised at the time as the main cause of the opulence of their country, merely reflected the need to fund the gigantic tidal containment works in their territories.

Given that savings promoted only economic backwardness, luxury spending was indispensable to put back into active circulation the funds subtracted from the markets by the tireless efforts of the thrifty, for the only role of money, as Mandeville understood it, was to pay the wages of the working people. He clearly realised, therefore, that if the acquisition of sumptuary items were confined to the limits of each individual's income, it would fail to neutralise the negative consequences of parsimony on employment opportunities. This is the ultimate reason that leads him to state the advantages derived from the theft of a miser, when his economies are thrown back into the monetary circuit, or even the squandering promoted by a prodigal heir, who quickly dissipates his benefactor's fortune (*FB* I 1992, 87, 104).

It should be emphasised here that, in *The Fable*, the economic utility of luxury relates to its power as an antidote to greed and avarice or, in more modern language, to its capacity to preserve the society's purchasing power This crucial role Mandeville assigns to luxury – thus anticipating the argument that would serve as a counterpoint to the Law of Markets in the next century – implies that sumptuary expenses, to have real significance, should extrapolate the current income of individuals to avoid a crisis of underconsumption. Ostentation or, for that matter, any other reproachable category of spending *per se*, does not possess any intrinsic quality that, in purely economic terms, makes it preferable to other types of expenditure. To be more precise, it becomes meaningful only when it counterbalances the opposing vice, that is, avarice.

The most representative example of this paradoxical symbiosis is provided by Mandeville in his story of the young spendthrift Florio who, anxious to emulate the standard of living of his wealthy friends, goes to the miser Cornaro, who alone is willing to lend him money – though at a high cost – to fund the profligacy of this improvident debtor: "Where would *Cornaro* ever have got such a prodigious Interest, if it was not for such a fool as *Florio*, who will give so great a price for money to fling it away?", Mandeville asks, immediately adding: "And how would *Florio* get it to spend, if he had not lit of such a greedy usurer as *Cornaro*" (*FB* I 1992, 102). In a clear tribute to Rochefoucauld, Mandeville's reflections on prodigality contain the following illuminating fragments:

> Avarice, notwithstanding it is the occasion of so many evils, is yet very necessary to the Society, to glean and gather what has been dropt and scatter'd by the contrary Vice. Was it not for Avarice, Spendthrifts would soon want materials; and *if none would lay up and get faster than they spend, very few could spend faster than they get.* ... I look upon Avarice and Prodigality in the Society as I do upon two contrary poisons in Physick, of which it is certain that the noxious qualities being by mutual mischief corrected in both, they may assist each other, and often make a good medicine between them. (*FB* I 1992, 101, 106 – emphasis added)

All the strength, but also the weakness, of Mandeville's economic rhetoric lies in this very point. The force of his argument comes from his awareness, albeit in a rudimentary way, of the depressive effects that avarice or, in a broader sense, savings, exerts on the economic system. The direct impact of frugality appears in the deficit of sales opportunities for domestic production, with poverty and unemployment of the workforce being the inevitable consequences. Thus, for a complex society formed by individuals with different inclinations and characters, Mandeville properly identifies the crux of the problem when he proclaims aloud the necessity, for the proper functioning of the economic organism, that every penny saved be replaced in the active circulation by an equivalent deficit spending somewhere else in the system. That said, however, it should be observed as well that Mandeville's solution fails exactly in assuming that only the consumption of luxuries or of vicious things could perform this function, as if this kind of expenditure were the only one capable of inducing an individual to spend beyond his regular income. It is on these contradictory ideas regarding luxury expenditure that Mandeville's critics, to whom we now turn our attention, would focus their criticisms.

14.4 Bluet and the Roots of Classical Orthodoxy

Appearing in 1725, *An Enquiry Whether a General Practice of Virtue Tends to the Wealth or Poverty, Benefit or Disadvantage of a People?*, authored by a young lawyer named George Bluet, was one of the more extensive and meticulous contemporary critiques of Mandeville. In the preface, the author describes the characterisation of virtue in *The Fable* as "monstrous", proposing to fully refute the supposed utility of villainy and vice. Each country's wealth, according to Bluet's definition, comprises the soil and everything that can be extracted from it, either in the form of staple goods, conveniences or ornaments, along with the gains obtained from trade. However, because the most basic needs are approximately equal for all people, the actual wealth of a society is determined by its greater or lesser access to the comforts of life. Differences between countries in the provision of land and natural resources, on the other hand, induce them to exchange with each other their respective production surpluses. In view of this, Bluet considers that only the activities directed to improve the soil, to enhance the value of its produce or to promote trade in general could create wealth. He concludes then that all remaining professions constitute a burden on the community, invariably leading to waste: "By the help of these hands [employed in agriculture, manufacture and commerce] then, the society will be as rich as it can be, and no sort of labour that does not contribute to one of those purposes can add at all to their wealth" (Bluet 1997, 242).

The disappearance of the occupations associated with vice as a result of the diffusion of virtue and honesty among the inhabitants of a country would not entail, according to Bluet, any weakening in the economic power of society. The provisions for the sustenance of the individuals who became unemployed by a transformation of this nature would continue to exist, but with one key difference: under the new conditions, the same workers could still be fed after their allocation to activities linked directly to the production of wealth. There would never cease to be opportunities for the better use of the labour displaced by a deficiency of demand for locks, padlocks or walls when assaults on property ceased altogether or, going a step further, when all other types of vice disappeared. The reason was that no country would lack areas that could be better cultivated, while the possibilities for the enhancement of goods in general would be virtually unlimited. "In the first place, there never was yet that country in the world where every part of the soil was so completely improved as not to be capable of much further improvement",

Bluet notes, complementing that "the skill and labour that might be employed in the improvements of commodities or in adding to the elegancies of life is almost infinite" (Bluet 1997, 244).

The universal practice of honesty and virtue would not only fail to generate poverty, as held by Mandeville, but, on the contrary, it would allow for a significant gain in social wealth due to the use of sterile workers in genuinely productive functions. To illustrate his point, Bluet uses the example of a sheep breeder forced to build fences and employ shepherds to protect his flock against wolves. Bluet then asks: if these predators were suddenly eliminated, would it not be a blessing for the breeder? Yes, he answers, for the laid-off workers could then be contracted to the development of other parts of the property, generating greater profits for their employer. Similarly, criminals and looters could be considered the wolves of society, and all decent men would certainly be grateful if such outlaws ceased to exist. In short, there could never be an absence of demand for the services of the workers made redundant after the spread of virtue across the society. Unlike what Mandeville prophesied, investment opportunities would not face any limits, and wealth could actually be enlarged in the same rhythm as probity finds shelter in the hearts of men:

> There could be no want of employment then, supposing this great change [to virtue] to be ever so sudden, and that a miracle intervened to effect it at once. But this is setting the present question in a very improper light. When this is applied to practice, ... the change must be supposed to be *gradual*; and then it will appear still plainer that there would necessarily arise a succession of new trades, or a greater number of the present trades that contribute to the ornamental parts of life, in proportion as the trades in providing against roguery grew useless and wore off. (Bluet 1997, 244)

The most notable aspect of Bluet's reasoning lies in his ability to anticipate, even without mastering the concept of capital, the central propositions that in the future would be the cornerstones of the classical theory of accumulation, as formulated by Adam Smith, David Ricardo and John Stuart Mill (Corry 1962, 14–25). First, he emphatically argues that increases in social wealth depend on the existence of prior savings held by individuals. The more is consumed in a given period, the less is available to be used in activities related to subsequent expansion of the national supply of commodities. All categories of consumption not directly linked to production, therefore, would reduce the growth potential of the economy, an idea that foreshadows the famous classical distinction between "productive" and "unproductive" labour. With regard to this particular point, Bluet expressly states: "In this sense of the word, the luxury of one age affects the future ones as certainly and necessarily as the extravagance of a *father* makes the *son* poor, or as the excess of expense 1 year produces a greater scarcity *the next*" (Bluet 1997, 266 – emphasis in the original).

Second, in the same chapter, we find the equally crucial proposition that production can never be constrained by a lack of demand, since the eradication of luxury spending would immediately be counter-balanced by a concomitant increase in the expenses to improve the conditions of supply in the economy as a whole. This conjecture should be understood, in the context of the time, as a primitive exposition of the process of capital accumulation, especially in the form in which it would later come to life in the writings of classical authors, that is, through the unconditional acceptance of the postulate that all savings are invested. When criticising Mandeville's thesis on the usefulness of vice and misfortune in sustaining employment, Bluet refers to the situation of the Dutch to support his reasoning, in a kind of rhetoric that would later provide the basis for Say's Law:

> Suppose yet further that such hands [employed in the repairing of dykes] were deprived of their present employment ever so suddenly, that Providence should in one night's time raise barriers against the sea, that were to last as long as the world in itself, in all probability, these wise rulers would not consider such a miracle as a national misfortune, but would find out ways enough to employ those hands who now work in their dikes, especially when, as the author [Mandeville] tells you, in some of their provinces there is *abundance of ground lying waste* for want of employment. (Bluet 1997, 246–47 – emphasis in the original)

Let us now examine how others argued against Mandeville in addressing this issue, i.e., the role of luxury spending in the optimal functioning of the economy.

14.5 Hutcheson and the Ascendancy of Morals over the Economy

Francis Hutcheson, a key figure of the Scottish Enlightenment[4] and one of the most influential thinker in the formation of Adam Smith's economic doctrines, would be the most relentless opponent of Mandeville. In a series of three letters to the *Dublin Journal*, published in February 1726, Hutcheson expresses for the first time his displeasure with the suggestion that luxury spending benefits the public, a concern that would resurface in Hutcheson's writings throughout his life. Despite his determination to refute Mandeville's theses, the economic reasoning he used to achieve this goal turned out to be inferior to that presented earlier by Bluet, as we shall see below.

Hutcheson begins his offensive against the alleged public utility of vice by questioning the very concept of luxury employed in *The Fable*. Individual happiness, he explains, in addition to requiring the fulfilment of basic material needs (appetites) and personal desires (the affections or passions), involves sincere concern for the collective welfare. Because the fulfilment of all aspirations is an impractical task, men would do better if they could soberly enjoy the objects of their satisfaction, consuming them to the extent of their contribution to a healthy earthly existence. The supreme good to which a person could aspire would consist of the esteem and affection for one's fellow human beings, along with reverence for God and trust in His purposes. But the love of family, country and friends, although the primary focus of concern for each individual, should not prevent or make condemnable the appreciation of the beauty contained in architectural masterpieces, music or poetry. Since mankind abandoned the caves, and agriculture enabled people to

[4]Regarding the Scottish Enlightenment, see Verburg (1991, 38–56) and Bryson (1945, 1–29).

turn to manufacturing and mechanical crafts, the continual progress in the arts and refinements of life ought to be viewed indeed as a symptom of wealth and public happiness. "If vice be the opposite to virtue", Hutcheson ponders, "we may easily conclude that the utmost improvement of arts, manufactures or trade is so far from being necessarily vicious, that it must rather argue good and virtuous dispositions" (Hutcheson 1997, 394).

However, if most of what Mandeville consecrated as vice might be interpreted as "virtuous dispositions", what can be said then about luxury spending? In his second letter, Hutcheson offers an answer by stressing the relativity of the concepts of intemperance, pomp and pride, as none of these can be weighed, measured or added. In the case of such human propensities, he remarks, it is necessary to take due account of the proportion between means and circumstances. Food, clothing or architecture of any style or sophistication should never be described as an absolute evil. The designation of a particular type of consumer good as luxurious or extravagant is subject to both the physical constitution and the richness of each person. Ostentation, properly understood, only encompasses consumption that is excessive, i.e., beyond what is consistent with one's own wellbeing or in an amount superior to the individual's income. Consequently, according to Hutcheson, common sense was to set the limits of worldly pleasures and to confine the expenses involved therein to every citizen's current earnings:

> If then in each constitution, station or degree of wealth a man of good sense may know how far he may go in eating and drinking or any other expenses, without impairing his health or fortune, or hindering any offices of religion or humanity, he has found the bounds of temperance, frugality and moderation for himself; and any other who keeps the same proportion is equally temperate, though he eats and drinks or spends more than the other. (Hutcheson 1997, 396)

After removing from vices almost all derogatory connotations attributed to them by Mandeville, Hutcheson acknowledges that the suppression of luxury, even when limited to what might be considered excessive spending, would nevertheless mean a decline in demand. And this gap would need to be filled by a different type of expenditure to support manufacturing and encourage trade. Hutcheson's solution to this critical economic problem, presented at the point where his analysis reaches its apex, is, to say the least, disheartening because he fails to transcend the realm of moral philosophy. From the point of view of the individual, Hutcheson suggests that a life of restraint and prudence would be longer and therefore bring about greater demand over time than a disorderly and generally shorter life. His confidence in behaviour of such kind, however, is not quite strong, as he also argues that, by restraining personal impulses and redirecting personal expenditures to better meet the needs of their families, households could achieve greater comfort and protection. In the case of wealthy gentlemen, to whom such a recommendation would be senseless, they should lend their savings free of charges to their less fortunate friends, enabling the latter to enjoy the pleasures of modern society that otherwise would be denied to them. "Unless therefore all mankind are fully provided not only with all necessaries, but all innocent conveniences and pleasures of life it is still possible, without any vice, by honest care of families, relations and any worth persons in distress to make the greatest consumption" (Hutcheson 1997, 399).

Three decades later, in his *System of Moral Philosophy* (1755), Hutcheson uses similar reasoning when investigating the likely consequences of a reduction in the consumption of luxury goods:

> And what if men grew generally more frugal and abstemious in such things? More of these finer goods could be sent abroad: or if they could not, industry and wealth might be equally promoted by the greater consumption of goods less chargeable: as he who saves by abating of his own expensive splendour could by generous offices to his friends, and by some wise methods of charity to the poor, enable others to live so much better, and make greater consumption than was made formerly by the luxury of one. (Hutcheson 1755, 320)

It is clear, therefore, that Hutcheson addresses the problem of aggregate demand through a strictly moral appeal, which can be summed up in one plain formula: decreased appetites, increased affections. Or, put differently: less personal spending, more expenditures with fellow human beings. Although some authors (Thweatt 1979; Taylor 1965, 106) claim that we find here a primitive formulation of Say's Law, this interpretation is somewhat exaggerated. The reason is straightforward. Hutcheson's analysis rests almost entirely on an improbable ethical prescription: that every individual able to save should do so because such an attitude is advantageous not to himself but mainly to other people. In summary, by resting the fate of the economic system on the grandeur of the human soul, Hutcheson shows his being unaware of the crucial missing link in the whole argument, so deftly presented by Bluet, namely: that parsimony is useful to society in serving, first of all, the interests of those who carry it out. In more objective language, savings are advantageous by the circumstance that it provides an economic – and not just a spiritual – gain to frugal people, embodied in the profit earned through the productive investment of the laid away income. Under such conditions, as the classical economists would further argue, a country was to reach the greatest possible growth of its national dividend and of employment opportunities within the national economy.

14.6 Bishop Berkeley and the Minute Philosophy

In 1732, the Irish empiricist philosopher and bishop of the Church of England, George Berkeley, published his *Alciphron or the Minute Philosopher*. In the second of the book's seven dialogues, he launches an onslaught against what he believes to be the true ideas of Mandeville, whom he deprecates as a free-thinker or a minute philosopher (in the bishop's definition, one who worships the material world and denies the eternity of life). Berkeley's economic critique is included here not for its originality, which left much to be desired, but because it is the only one to have provoked an explicit reply from Mandeville, to be assessed in the next section. According to the bishop's very peculiar reasoning, the free thinkers' message, of which *The Fable* was allegedly the most elaborate version, promoted atheism, incited the more abject vices and, above all, preached against both the civilian government and the respect for property. The unbridled pursuit of pleasure, when competing with virtue in society, Berkeley claims, leads to discord and to the ruin of the whole community.

If the prescriptions of such a philosophy prevailed, men would occupy themselves exclusively with satisfying their most pressing desires, becoming lazy and causing in this way the progressive degeneration of the race (Berkeley 1997 [1732], 542–46). It is idle to observe here that such alarmist concerns not only distort Mandeville's thought but also elide the central question raised in *The Fable* regarding the role of vice in stabilising purchasing power. When Berkeley decides to tackle the sensitive issue of luxury expenditure, he merely repeats the arguments previously elaborated by Hutcheson, without adding anything of his own. At a certain stage of the dialogue between the protagonists of his book, Euphranor, one interlocutor of Lysicles – advocate of the minute philosophy – contests the alleged usefulness of vice, objecting that dissolute men would experience a shorter life than righteous ones, whose health would give them a long existence of moderate consumption for themselves and their families. The conversation soon turns to an exchange of what type of spending would bring forth the greatest wealth for society. The relevant passage occurs when Lysicles declares he has irrefutably demonstrated the importance of vice in the consumption of manufactures:

> *Euphranor*: You seem to me to have proved nothing, unless you can make it out that it is impossible to spend a fortune innocently. I should think the public weal of a nation consists in the number and good condition of its inhabitants; have you any thing to object to this?
> *Lysicles*: I think not.
> *Euphranor*: To this end which would most conduce, the employing men in open air and manly exercise, or in sedentary business within doors?
> *Lysicles*: The former I suppose.
> *Euphranor*: Should it not seem therefore, that building, gardening, and agriculture would employ men more usefully to the public, than if tailors, barbers, perfumers, distillers, and such arts were multiplied?
> *Lysicles:* All this I grant; but it makes against you. For what moves men to build and plant but vanity, and what is vanity but vice? *Euphranor*: But if a man should do those things for his convenience or pleasure, and in proportion to his fortune, without a foolish ostentation or over-rating them beyond their due value, they would not then be the effect of vice; and how do you know but this may be the case? (Berkeley 1997, 541)

The extremely rigorous definition of vice adopted by Mandeville, embracing all that exceeds the individuals' basic needs, would eventually be the most criticised aspect of his doctrine. So, two flanks of attack on his ideas were quickly developed. The first sought to refute the sweeping classification proposed by Mandeville, condemning the exaggeration of listing under the head of vices all that, in his own words, "man should commit to gratify any of his appetites" (*FB* I 1992, 48). The second flank, also explored by the authors reviewed here, aimed to prove that if sumptuary expenditures were eliminated altogether, an equivalent amount of spending of a more laudable character could be generated to meet other needs. Mandeville's response to such criticisms is discussed in the next section.

14.7 Mandeville's Reaction

As regards the first line of attack above indicated, Mandeville pointed out the hypocrisy of those who, while raging against vice, have not hesitated to enjoy the blessings of life without the slightest remorse, caring only to affect a disguised indifference to earthly temptations. In his answer to Bishop Berkeley, *A Letter to Dion* (1732), Mandeville replies to his opponents with the following comment: "No pleasure is denied them", he writes, "forsooth, that is used with moderation, and in cloaths, houses, furnitures, equipages and attendance, they may live in perfect conformity with the most vain and luxurious of the fashionable people; only with this difference, that their hearts must not be attached to these things" (Mandeville 1997, 588).

Regarding the second kind of criticism, Mandeville had already anticipated, in *The Fable*, an answer to the moralistic assault conducted by Hutcheson and Berkeley when he reasserted the expediency of vices to ensure the economic dynamism of wealthy nations. The naive uproar for the abandonment of luxury in favour of a virtuous behaviour, he said, would be irreconcilable with a world ruled by the logic of commerce. Those who defended such transformation had in mind only an abstract man who, in reality, never existed. In the introduction to his essay *An Inquiry into the Origins of Moral Virtue*, Mandeville clarifies his differences with the moralists of the time: "One of the greatest reasons why so few people understand themselves, is, that most writers are always teaching men what they should be, and hardly ever trouble their heads with telling them what they really are" (*FB* I 1992, 39). The call to frugality, he insists, would be suitable only for small, poor communities accustomed to a peaceful and tedious life. In a powerful and teeming nation, by contrast, parsimony could easily be excessive, generating unemployment amid a large and work-starved population. "Prodigality", as one reads in *The Fable*, "has a thousand inventions to keep people from sitting still, that frugality would never think of" (*FB* I 1992, 105).

It remains to examine Mandeville's reaction to the idea that ostentation, after all, would encompass only the citizens' deficit spending. The analogy of the financial situation of a country with that of a household was too attractive to be ignored in a discussion of such scope. Bluet and Hutcheson tirelessly warned against the risk of economic ruin that would befall those who dared to live beyond their means. Consequently, if everyone followed Mandeville's prescriptions, it would be impossible to conceive the welfare of society amid the poverty of its members. As Bluet joked about this possibility, if the collective well-being is formed by adding the happiness of each individual in society, the adoption of Mandeville's recommendations implies that "an Army may be well cloathed, though every single man in every regiment were forced to go naked" (Bluet 1997 [1725], 252). However, this reasoning totally misses the target, as Mandeville never proposed the indiscriminate practice of vice. Instead, he repeated time and again that the driving force of the whole economic engine comes from the combination of luxury and avarice, which generate a compensatory succession of gains and losses that relentlessly reproduces itself, in an unceasing process indispensable to the nation's economic strength.

14.8 Concluding Remarks

The great merit of Mandeville's economic doctrine lies in his commitment to bring to the centre of the discussion the issues related to the preservation of purchasing power in an affluent society. The long mercantilist tradition associating the growth of national wealth with savings called forth the question of who absorbs the corresponding non-consumed production. A positive trade balance appeared to be the most common alternative, although it suffered from the defect that – as many authors observed – no country could sustain an imbalance in its external transactions for an extended period of time. During a period when the message of free trade started to make strides, there was a pressing need to specify the conditions under which the productive potential of society would be fully mobilised. The fact that the discussion revolved around the problem of how to employ the existing population does not mean, as noted, that this concern was linked to some view of the working population as a source of consumption. There remained the mercantilist perspective that saw the available labour force as merely a component of the production process, one whose remuneration should be kept at the lowest possible value to stimulate the industriousness of employees and keep the costs of the exported articles low.

Mandeville's solution to the provision of employment, i.e., luxury expenditure, did not add much to what had already been said before. More importantly, with respect to economic theory, was his intuition about the importance of deficit spending as an indispensable counterpart of parsimony. Vice, while inducing expenditure, was not relevant by involving a malignant action *per se* but, instead, because it assumed the form of effectual demand, that is, of a buying decision backed by money. This, however, as Mandeville's critics easily realized, could be said about every type of spending, whether necessary or superfluous, virtuous or condemnable. The arguments regarding the moral substratum of expenditure, as raised by Hutcheson and Berkeley, only scratched the surface of the issue. The controversy, as it took shape at the time, required a more solid response, framed in strictly economic terms and capable of reconciling private and public interests. The task of guiding the discussion to its proper context was performed by the young Bluet, who helped pave the way for the future advent of classical economics. In his conception – which anticipated indeed the Law of Markets – demand would dispense with any stimuli other than those arising from the wealth expansion process in itself. By appealing to a notion close to Say's Law, Bluet was able to achieve a reconciliation between virtue, embodied in savings, and its economic reward, measured by profit. Private enrichment, through the productive investment of saved resources, could now be seen as synonymous with collective welfare.

References

Ashley, William J. 1987. The tory origin of free trade policy. *The Quarterly Journal of Economics* 11(4): 335–371.

Barbon, Nicholas. 1934 [1690]. *A discourse of trade*. Baltimore: Johns Hopkins Press.

Baumol, William. 1977. Say's (at least) eight laws, or what Say and James Mill may really have meant. *Economica, New Series* 44(174): 145–161.

Berkeley, George. 1997 [1732]. Alciphron. Dialogue II. In *Private vices, publick benefits?*, ed. Stafford, 528–571.

Bluet, George. 1997 [1725]. An inquiry whether a general practice of virtue tends to the wealth or poverty, benefit or disadvantage of people. In *Private vices, publick benefits?*, ed. Stafford, 228–382.

Bryson, Gladys. 1945. *Man and society. The Scottish inquiry of the eighteenth century*. Princeton: Princeton University Press.

Chalk, Alfred. 1966. Mandeville's fable of the bees: A reappraisal. *The Southern Economic Journal* 33(1): 1–16.

Chalk, Alfred F. 1970 [1951]. Natural law and the rise of economic individualism in England. In *Readings in the history of economic thought*, ed. Rima, 32–47.

Child, Josiah. 1668. *Brief observations concerning trade and the interest of money*. London: Elizabeth Calvert. Available at http://onlinebooks.library.upenn.edu.

Corry, Bernard. 1962. *Money, saving and investment in English economics, 1800–1850*. London: St. Martin Press.

Dumont, Louis. 1975. The emancipation of economics from morality: Mandeville's fable of the bees. *Social Science Information* 14(1): 35–52.

Furniss, Edgar S. 1957 [1920]. *The position of the laborer in a system of nationalism. A study in the labor theories of the later English mercantilists*. New York: Kelley & Millman.

Grampp, William D. 1962. The liberal elements in English mercantilism. In *Essays in economic thought*, ed. Allen Spengler, 61–91.

Gregory, T.E. 1921. The economics of employment in England, 1660–1713. *Economica* 1(0): 37–51.

Heckscher, Eli F. 1943 [1931]. *La Época Mercantilista* [*Mercantilism*]. México: Fondo de Cultura Economica.

Hobbes, Thomas. 1999 [1651]. *Leviathan*. The University of Oregon, Renascence Editions.

Horne, Thomas A. 1978. *The social thought of Bernard Mandeville. Virtue and commerce in early eighteenth century England*. New York: Columbia University Press.

Hutcheson, Francis. 1755. *A system of moral philosophy*. London: R. and Foulis.

Hutcheson, Francis. 1997 [1726]. Letters to the dublin weekly journal. In *Private vices, publick benefits?*, ed. Stafford, 383–407.

Jack, Malcon. 1976. Progress and corruption in the eighteenth century. Mandeville's private vices, publick benefits. *Journal of the History of Ideas* 37: 369–376.

Johnson, Edgar A.J. 1960 [1937]. *Predecessors of Adam Smith. The growth of British economic thought*. New York: Augustus M. Kelley. Reprints of Economic Classics.

Kates, Steven. 1997. On the true meaning of Say's law. *Eastern Economic Journal* 23(2): 191–202.

Kaye, F.B. 1992 [1924]. Introduction. In Mandeville, *The fable,* xvii–cxlvi.

Landreth, Harry. 1975. The economic thought of Bernard Mandeville. *History of Political Economy* 7(21): 193–208.

Malthus, Thomas R. 2004 [1820]. Principles of political economy. In *Works*, vol. II, ed. Sraffa.

Mandeville, Bernard. 1992 [1723]. *The fable of the bees, or private vices, publick benefits*, ed. F.B. Kaye. Oxford: Clarendon Press, 1924. Reprinted by Liberty Fund, Indianapolis, 2 vs.

Mandeville, Bernard. 1997 [1732]. A letter to dion. In *Private vices, publick benefits?*, ed. Stafford, 574–611.

Mill, James. 1808. *Commerce defended. An answer to the arguments by which Mr. Spence, Mr Cobbet and others, have attempted to prove that commerce is not a source of wealth*. London: C. and R. Baldwin.

Mun, Thomas. 1664. *England's treasure by forraign trade. Or the ballance of our forraign trade is the rule of our treasure*. London: Thomas Clark. Available at http://socserv2.socsci.mcmaster.ca/econ/ugcm/3ll3/mun/treasure.txt.

North, Dudley. 1691. *Discourses upon trade*. London: T. Basset.

Ricardo, David. 2004 [1817]. *On the principles of political economy and taxation*. In *Works*, vol. I, ed. Sraffa.

Rima, Ingrid R. (ed.). 1970. *Readings in the history of economic thought*. New York: Holt, Rinehart and Winston.

Rochefoucauld, François. 1871 [1665]. *Maxims and reflections*. London: Sampson Low, Son and Marston. Available at http://onlinebooks.library.upenn.edu.

Rosenberg, Nathan. 1963. Mandeville and laissez-faire. *Journal of the History of Ideas* 24: 183–196.

Say, Jean-Baptiste. 1850. *A treatise on political economy or the production, distribution and consumption of wealth*. Philadelphia: Lippincott/Gramble.

Sowell, Thomas. 1972. *Say's law. An historical analysis*. Princeton: Princeton University Press.

Spengler, Joseph J., and William R. Allen (eds.). 1962. *Essays in economic thought: Aristotle to Marshall*. Chicago: Rand McNally.

Sraffa, Piero (ed.). 2004. *The works and correspondence of David Ricardo*, vol. I and II. Indianapolis: Liberty Fund.

Stafford, Martin J. (ed.). 1997. *Private vices, publick benefits? The contemporary reception of Bernard Mandeville*. Solihull: Ismeron.

Taylor, W.L. 1965. *Francis Hutcheson and David Hume as predecessors of Adam Smith*. Durham: Duke University Press.

Thweatt, William O. 1979. Early formulator's of say's law. *Quarterly Review of Economics and Business* 19(4): 79–96.

Verburg, R.M. 1991. *The two faces of interest. The problem of order and the origins of political economy and sociology as distinctive fields of inquiry in the Scottish Enlightenment*, Research series n. 5. Rotterdam: Tinbergen Institute.

Viner, Jacob. 1937. *Studies in the theory of international trade*. New York: Harper and Brothers Publishers.

von Hayek, Friedrich. 1948. *Individualism and economic order*. Chicago: University of Chicago Press.

Chapter 15
Courage and Chastity in a Commercial Society. Mandeville's Point on Male and Female Honour

Andrea Branchi

By Jove, I am not coveotus for gold,
but if it be a sin to covet honour
I am the most offending soul alive.

(W. Shakespeare, Henry V, (IV, iii))

Abstract The aim of this essay is to offer a survey of the uses developed by Mandeville of the notion of honour in his philosophical project, focusing on the role played by 'Modern Honour' in his conjectural account of the civilizing process. In particular, the issues of duelling and of the position and role of women in society are two parallel perspectives to look at Mandeville's provocative account of male and female 'points of honour'. Mandeville's effort to explain the popularity of 'Modern Honour' plays an important role in his larger philosophical project of scientific, unprejudiced analysis of human nature. Locating the history of male honour and female respectability in the perspective of his philosophical anthropology, Mandeville is able to show that the rituals of Modern Honour are an exemplary expression of that spontaneous, *artificial* order stemming out of a *natural* disposition of human passions.

Keywords Honour • Duelling • Pride • Courage • Women • Chastity • Women's education • Conjectural history

Vanity, Shame, and (. . .) Constitution, *make up very often the Courage of Men and Virtue of Women.*[1] wrote Mandeville in the Fable of the Bees, making an explicit reference to the Duke of La Rochefoucauld. Concurring with those Continental writers who stressed the components of hypocrisy and modish deceit implicit in the

[1]Mandeville, B. (1988) *The Fable of the Bees,* ed. F.B. Kaye, Indianapolis, Liberty Classic, Vol. I, p. 236.

A. Branchi (✉)
Department of International Relations, The American University of Rome, Rome, Italy
e-mail: ab509@columbia.edu

E. Balsemão Pires, J. Braga (eds.), *Bernard de Mandeville's Tropology of Paradoxes*, Studies in History and Philosophy of Science 40, DOI 10.1007/978-3-319-19381-6_15

tradition of civility, Bernard Mandeville made of the philosophical anthropology of the elites of the *Ancien Régime* developed by the French Moralists, a very efficacious instrument for the unprejudiced anatomy of human nature he developed in his writings. For the Dutch Doctor, human behaviour, in its apparent variety of motivations, can be traced back to the passion of self-love and self-liking, their effects and the efforts carried out to control, hide and gratify its. The desire for praise is a constant property of human nature that assumes different shapes in different historical contexts. Mandeville sees the rules of honour and politeness – the standard of male and female respectability – as the progressive outcome of a spontaneous balance of selfish passions in forms compatible with refined commercial sociability.

Unlike his contemporaries, Mandeville does not consider the culture of honour and its most extreme expressions in men - the practice of duelling – simply as an absurd atavism that oddly survived in eighteenth century British polite society. The duel is for Mandeville a case study in applied ethics, the paradigmatic expression of the inconsistency of honour and virtue and a test case for the detection of the true passions and motives at the roots of man's behaviour. The fact that man fight duels is the proof that vanity is stronger than self-preservation, self-liking dominates over self-love. Anatomizing the manners of his age, the shared systems of sentiments of approbation and disapprobation Bernard Mandeville provokes his contemporaries with the disturbing conclusion that, as a matter of fact, vanity is a much stronger and more widely diffused motivation for (allegedly) 'virtuous' behaviour than virtue itself. The key traditional virtues of male and female honour – Courage and Chastity – can be accounted for as the spontaneous, social effects of Pride and Fear of Shame. Mandeville developed an original and detailed account of the history of Martial Virtues and their role in commercial societies. Does the explanation that Mandeville formulated on Male Honour apply also to Female Honour? How did Mandeville, which had devoted so much attention to Women, their perspective, their conditions, their world-view, accounts for female Chastity? Does Mandeville develops an analysis on the function of pride, vanity and search for approval as a substitute for moral virtue in women that parallel that on Male Honour?

In order to attempt an answer to these questions, the aim of this essay, part of an on-going research, is to offer a survey of the uses developed by Mandeville of the notion of honour in his philosophical project, by using the issues of duelling and of the position and role of women in society as two parallel perspectives to look at Mandeville's provocative account of male and female 'points of honour'.

15.1 Duelling

In the open-shelf library system books on duelling are next to those on suicide. The duel – a combined form of combat between two people, held according to prearranged rules to settle a quarrel on a point of honour – was forbidden by almost all European legislations and was condemned as a sin, and yet to challenge someone to a duel and to behave politely with an individual before attempting to kill him, was a ritual of enormous social prestige. Centralized monarchies had assumed a monopoly over justice and war, and the practice of the armed defence of personal honour became a widespread ritual of great symbolic value: the access to weapons and the right to private violence allowed the aristocracy to preserve, at least formally, its character of an independent military force, arrogating a sort of legitimacy through a fictitious tie of ancestry with the warlike nobility of the past.[2] Due to its paradoxical conceptual framework where warlike virtues are associated with ruling classes and those latter with noble birth and moral sensibility, participating in this highly ritualized form of attempted homicide-suicide was considered the best way to acquire and maintain a reputation of respectability and even of virtue. The first decades of the Eighteenth century, during which Mandeville wrote and published his major works, was also the Golden Age for duelling in England. The sword was a characteristic component of every upper-class man's dress, fencing was part of his education and duelling a widespread practice. Many pamphleteers, essayist and social commentators noted the problematic relevance of the practice of duelling in the values 'modern honour', they complain to see it indeed very often reduced to little more than an external coat of politeness and overall: *'to the right understanding the several Degrees of an Affront, in order to revenge it by the Death of an Adversary'*.[3] Mandeville proceeds further, making an original, peculiar use of the discourse on duelling. His effort to explain the popularity of 'modern honour' plays an important role in his larger philosophical project. According to Mandeville the practice of the duel of honour *'where the lust of praise makes men destroy that same being who strives to please'*[4] and its diffusion among polite gentlemen, who are supposedly better equipped with moral and civic virtues, is proof that the true motivations of behaviour are to be found in the vanity and fear of shame rather than in benevolence and love for virtue. Modern honour is the spontaneous artificial order resulting from a natural passional disposition, a deeply engrained passional mechanism. The mere self-preservation of natural courage in civilized men makes way for artificial courage, a passion that intensifies sensitivity toward other people's opinion, and set social recognition as the ultimate good. By developing a conjectural history of pride, and locating the history of male honour in the perspective of his philosophical anthropology, Mandeville is able to show that the rituals of modern honour are an exemplary expression of that spontaneous, *artificial* order stemming from a *natural* disposition of human passions: *The great Art then to make Man Courageous, is (...) to inspire him with as much Horror against Shame, as Nature has given him against Death (...) The Courage then which is only useful to the Body Politick, and what is generally call'd true Valour, is artificial, and consists in a Superlative Horror against Shame.*[5]

[2]See Kelso 1929; Mason 1935; Maravall 1984; Burke 1995; Richard 1999.

[3]Swift, J. (1968) 'On the Testimony of Conscience'. See Watson 1960; James 1978; Erspamer 1982; Kelly 1995; Spierenburg 1998; Peltonen 2000, 2003.

[4]*The Female Tatler* 94, (February 15, 1710).

[5]Mandeville 1988, p. 233.

15.2 Female Dialogues

While men are busy with duels, what about women? Lucinda & Antonia in the Virgin Unmask'd; Lucinda & Artesia in the Female Tatler, Polytheca in the medical Treatise, Fulvia in the second part of the Fable: all female characters in Mandeville are witty, sharp-minded and generally leaning toward realism. Mandeville wrote both as a male writer impersonating women's voices; and as a male author writing on women and, of course, on human nature in general. In many of his writings we find an uncommon attention for women and their position in society, way more articulated than that of many of his contemporaries. The 1718 edition of *Magna Britannia Notitia, or The Present State of Great Britain* offers an enlightening summary on the way in which female subjugation is sanctioned by the English law, but also on male's attitudes: "*Women in England, with all their Moveable Goods so soon as they are married, are wholly in potestate Viri, at the Will and Disposition of the Husband (...) She can't Let, Set, Sell, Give away or Alienate any thing without her Husband's Consent. (...) The Woman upon Marriage loseth not only the Power over her Person, and her Will, and the Property of her Goods, but her very Name.* But, adds the author with a genuinely paternal tone, somehow aware that what he has just written needs an apology: *Notwithstanding all this their condition de facto is the best of the World, for such is the good Nature of Englishmen toward their Wives, such is the Tenderness and Respect, giving them the uppermost place at Table (...) that they are, generally speaking, the most happy Women in the World.*[6] There is a scene in a novel published in 1724, *Roxana: The Fortunate Mistress*, by Daniel Defoe, the same year as Mandeville's Modest Defence of the Publick Stews, where an experienced courtesan unwilling turns down, a particularly inviting marriage proposal on the basis of a very interesting piece of reasoning: *'the very nature of the Marriage-Contract was, in short, nothing but giving up Liberty, Estate, Authority, and everything, to the Man, and the Woman was indeed, a meer Woman ever after, that is to say, a Slave (...) it was my Misfortune to be a Woman, but I was resolv'd it shou'd not be made worse by the Sex; and seeing Liberty seem'd to be the Men's Property, I wou'd be a Man-Woman;"*[7] It is to say that full citizenship is granted only to men. Even if not with Mandeville's provocative clarity, Daniel Defoe, in Lady Roxana as well as in other writings depicts the tension between Virtue & Commerce inherent in women's roles, and somehow reaches the disturbing conclusion that, if one is cunning and saver, whoring is the only available career for a woman wishing to maintain economic autonomy.[8]

---

[6]Chamberlain E. (1718) *Magna Britannia Notitia, or The Present State of Great Britain.* See Cohen 1997; Laqueur 1990; Harvey 2002; O'Brien 2009.

[7]Defoe, D. (1724), *Roxana, The Fortunate Mistress*, London, p. 148.

[8]Defoe wrote on women's education in the *Essay upon Project* (1697), on marriage in the periodical *The Review* (1704–1713), and also in *Religious Courtship: Being Historical Discourses on the Necessity of Marrying Religious Husbands and Wives* (1722).

The snares which men uses to seduce women and enslave them into wedlock, and the harsh reality of conjugal life in contrast with the fictitious image offered in tales and romances is one of the main topics of *The Virgin Unmask'd*, the first prose work by Bernard Mandeville. It appeared in 1709 and it was reprinted in 1724: ten dialogues between the elderly, "wise" maiden and the young, virgin niece; a form of literature common at the time especially for works of pornography. The choice of the title itself, the opening of the first dialogue – with Lucinda reproaching her niece for displaying her breasts indecently – are form of deception to a would-be purchaser of pornography: in the book women are not at all represented as an object of pleasure, and more, the inequality of marriage's conditions and of male monopoly on education and culture are deeply criticized in a series of conversations where the two women also discuss an typically male-reserved topics: national and international politics.[9] Lucinda is indignant about the subjection of women: to their fathers when they are single and to their husbands when they are married. This oppression implies a further damage to the very character of women. They are often silly because they have been bread to be silly: They *"may be taught to Sing and Dance, to Work and Dress, and if you will, receive Good Instruction for a Genteel Carriage, and how to be Mannerly; but these Things chiefly concern the Body, the Mind remains uninstructed."* As in others of Mandeville's dialogues, the elderly maiden Lucinda leads her niece Antonia, through conversation, to recognize the true motivating passions in her behaviour and in others. This is the *Unmasking* pledged by Mandeville in the title: an exercise in anatomy of human passions on the courtship rituals, on conjugal life, on man's shaping of women's role in society. Lucinda consciously advocates and practices realism in fiction, a careful analysis of motivating passions. Lucinda tries to instruct her niece on the real conditions of married life in contrast with the fictitious image offered in tales and novels. In particular, by recounting two exemplary stories and expanding the plots in order to stress the effective reality, beyond the realm of narrative fiction, the aunt illustrates woman's defencelessness towards man's incompetence and will. As in later writings, Mandeville is fully conscious of the unmasking he operates on the literary conventions of the Augustan age, and he is firm in his attitude toward the writers of romance *"the greatest enemies to Truth and sober Sense the world ever produc'd*". In the Preface to the second part of the *Fable of the Bees*, Mandeville defends John Gay's Beggar's Opera from the accusation – akin to those addressed to Mandeville himself – of encouraging vice, simply by exposing it with realism. Few pages later in the same book we can appreciate another careful defence of 'realism'

[9]Mandeville, B. (1975) *The Virgin Unmask'd, or Female Dialogues Betwixt an Elderly Maiden Lady, and her Niece, on several Diverting Discourses on Love, Marriage, Memoirs, and Morals, & c. of the Times,* New York, Delmar. The literary device of the dialogue between an Elderly Woman and a young virgin belongs to a tradition that goes back to P. Aretino. See: Goldsmith 1986; Vichert 1975; Castiglione 1983.

in theatre and in painting: to the ridiculous of the opera house, Fulvia prefers the psychological realism of a good play.[10]

Men's reason is a danger to women not because they are naturally more intelligent but exclusively because they have the advantages of a better education: *"it is thought sufficient, if a Woman can but Read and Write, we receive no other Education, as to Learning: But where we leave off, they set out; they are not trusted to manage their own Affairs, before they are sent to School and Universities, to have their Intellectuals mended and sharpened; not by one Master, or by ordinary Men, but by several, that are picked and culled out of Thousands, for excelling every one in his own Profession; here they have the Quintessence of Arts and Sciences, Politicks and Wordly Cunning infused into them; and for Seven of Eight Years, all manner of Knowledge, as it were, beat into their Brain, with all the Application imaginable, whilst we are pricking a Clout."*[11] A poor education – and this is the norm – not only condemns women to a subordinate social role but also exposes them to further misfortune. Education strengthens men's skills in social relations. Women are dangerously exposed to men's tricks and their allure; they can be 'easily' conquered and brought to a socially disadvantageous situation, that losing their reputation of chastity or being trapped into a hair-raising marriage. In sum – no less than other works – *The Virgin Unmask'd* is part of Mandeville's coherent project of truthful and unprejudiced description, on an empirical base, of human nature, carried out by singling out the dominant motives of human behaviour, and the role that those motives – and their control – play in the development and maintenance of social order.

15.3 The Female Tatler

In the same year 1709, Mandeville also collaborated to The Female Tatler, a journal allegedly written by 'a Society of Ladies', that exploited the success and the reading public reached with The Tatler, by Richard Steele. Steele, with his 'civic humanistic' reforming enterprise, is one of Mandeville's favourite polemical targets. Steele wrote about improving women's education, and he put forward the idea that a nation's degree of civility may be evaluated by considering female condition. Still, marriage is the better career he envisages for women, and basically his opinions on female education are aimed at no more than making women better domestic creatures.[12]

In The Tatler 67 (September 13, 1709), Steele announced the project of a Table of Fame to gather together the most virtuous and famous people. The proposal, mentioned in various issues, is extensively discussed in The Tatler 81 (October 22,

[10]Mandeville 1732, p. 48; Mandeville 1988, Vol. II, pp. 6 and 37–38.

[11]Mandeville 1975, pp. 27–28.

[12]Goldsmith 1986, pp. 100 and fwd. See also Bond 1971, pp. 83–90.

1709). To sit around his Table of Fame, the author invites Alexander, Pompeus, Caesar, Hannibal, Homer, Virgil, Archimedes, Socrates, Aristotle, Augustus, Cato, Cicero. The absence of females, and the predominance of statesmen and generals are further evidence of the ideological foundation of Steele's position.[13] The proposal of a Table of Fame to gather together the most virtuous and famous people, is immediately answered by Mandeville in The Female Tatler, and it represents the occasion to discuss of women's virtues in comparison to men. The Female Table of Honour imagined by Lucinda and Artesia includes examples of women – taken from Plutarch and other seventeenth century literary works listing virtuous women – praised and credited for the same virtues as men. In The Female Tatler 95, the character of an old gentleman suggests the image of woman as a submissive and obedient spouse, disapproving of the degree of education and the broadmindedness of the two sisters-writers: *"Which way can you have squander'd away your Money, to be reduced so low as to Write Tatlers? (...) What Girls should do with Latin? (...) Young Women shou'd only Study how to get Husbands (...) If you had given the Town a parcel of such edyfying Examples of Women that had Honour'd and Obey'd their Husbands, and been Discreet to all the World besides, you might have done some good, but of the Women you have named, there is scarce one that an Honest Man would be troubled with."* The reply of the two sisters is firm and resolute: there is no reason why women should not write Tatlers, and certainly the Female Table of Fame it is not worse, in the model it proposes, than Steele/Bickerstaff's male one. Alexander the Great, for instance, was no more that *"an Arbitrary Tyrant, that without Right or Reason invaded every body's Property, and had Savage Ambition enough to have Burnt the World as well as Conquer'd it, if he had thought it would have added to his Glory"*. The Female Table of Fame offers, more than examples of real virtue, models of that *"extraordinary Strength of the Soul"* that few mortals possess; certainly it is not only men who possess this strength: *"Women were as capable as Men of that Sublimity of the Soul, and had at least equall'd if not excell'd the greatest and most Heroick of the Cruel and Injurious Sex, that had used so many Artifices to enslave them.."*[14] Men do not only retain exclusive possession of education, they also maintain their domination over women by handing down a culture in which woman's destiny is submission, and meekness the most valuable virtue: *"Why should we be treated almost as if we were Irrational Creatures? We are industriously kept from the Knowledge of Arts and Sciences, if we talk Politicks we are laugh'd at; to understand Latin is petty Treason in us; silence is recommended to us a necessary Duty, and the greatest Encomium a Man can give his Wife is to tell the World that she is Obedient (...)."*[15]

[13]See Goldsmith 1999 but also Anderson 1935; Vichert 1966.

[14]*The Female Tatler* 95, (February 29, 1710).

[15]*The Female Tatler* 88, (January 27, 1710).

15.4 Female Honour

Probably Mandeville's most renowned piece of writing related to women is the *Modest Defence of the Publick Stews.* The idea of state-regulated prostitution first appears in Remark H of The Fable of the Bees: if the state would allow men to take advantage of '*Courtezans and Strumpets*', England could then preserve the virtue of its honest women. In 1723 the Grand Jury for the County of Middlesex had condemned the author of The Fable of the Bees, among other things, for the 'apology of the very Stews themselves (...) with the Design to debauch the Nation'.[16] *A Modest Defence of the Publick Stews* appeared in 1724. The essay is dedicated to the "Gentlemen of the Societies",[17] and the subject is at first developed as a criticism to the repressive method used by the Societies for the Reformation of Manners to combat prostitution, with a detailed style that parodies and lampoons the Reformers and that raises the doubt that the whole pamphlet may be nothing more than a literary satire.[18] To supplant the repressive methods of the Societies for the Reformation of Manners Mandeville proposes to control prostitution by opening public brothels, and provide from a *Private Vice* a conspicuous set of *Publick Benefits*: better hygiene, medical supervision, control of venereal diseases, lessening of infanticide, health insurance for prostitutes; not to mention the fact that in public brothels men may take their fill with more ease and acquire that experience that makes them prepared and trained to marry, without wasting energy in love intrigues, and overall without endangering the virtue of honest women. It is a dense and complex pamphlet. Is the entire work an elaborate irony? The opposition between social advantages and moral evaluations, between "Private Vices" e "Publick Benefits" is traced here more explicitly than in other works. Mandeville wore at the same time the mask of the cynical satirist and that of the wise lawmaker and social reformer.

Ironical or not, it is a text on the best means to channel male lust into a social profit, where Mandeville develops a detailed explanation on how Chastity in women develops as an artificial virtue. Departing from *"the Constitution of Female while in the State of Innocence"*, Mandeville elaborates a detailed description of Female sexual organs and on the physiology of women's sexual excitement. Women, argues Mandeville, relying upon a 'classical' commonplace of Western misogyny have to cope with the strength of their sexual desire. The conflict between their violent natural wishes and their (weak) 'inborn modesty & innate reservedness' requires a sense of interest, to be brought to a successful result. A sense of Interest which seem, after all, to be the main component of Female Honour.

To counterbalance their violent natural desire women have a *'strong notion of honour carefully inculcated into them from their infancy'*. As a matter of fact their

[16]Mandeville 1988, Vol. I, pp. 95 and 385.

[17]Mandeville 2006, p. 44.

[18]See Castiglione 1989.

worldly interest relies solely on their reputation of chastity. Artificial chastity is then a 'compound' of Sense of honour and interest: *'When a woman has her interest and fortune depending upon her reputation, as all the middle rank of womankind have, she is a woman of honour, of course. Interest is inseparable indeed from Female of honour.'* It is exclusively a social fact, that women are expected to remain chaste and suffer if they do not. Certainly, as Dario Castiglione noticed, Mandeville shows a conception of sexuality as a mechanical pursuit of pleasure, premodern, expressed in the simplified language of passions, but it is this very language that gave him the opportunity to underline clearly the unequal conditions between men and women, relative to the shared standards of approbation and disapprobation: a common sexual impulse and a distinct public moral evaluation.[19] In sum, the situation is absolutely unequal: male desire ought to be satisfied, and Mandeville proposes to make use of women devoted to this function, (even better, if imported from other countries) in order to save 'honest women'; female desire, no less intense, is much more hazardous to gratify, but – and this is the point – exclusively moral and social reasons. The whole point of the lengthy description of female anatomy and on the mechanisms of the "*Violence of Female Desire*, "is to consider" *what a terrible Risque a Woman runs to gratify it (...) the Minds of Women are observ'd to be so much corrupted by the Loss of Chastity, or rather by the Reproach they suffer upon that Loss*". "Dissolute" women, namely those who satisfy their desire outside the realm considered socially acceptable, that is exclusively with their legitimate husbands, are corrupted, and destined to suffer, but exclusively by other's opinion: *"These Woman (...) are commonly Guilty of almost the whole Catalogue of immoral Actions: (...) Not that these are necessary Concomitants of Lewdness, or have the least Relation to it, as all lewd Men of Honour can testify; but the Treatment such Women meet with in the World, is the Occasion of it."*[20]

15.5 Conclusive Remarks

Horatio – in the third dialogue of the second part of *The Fable of the Bees* – asks Cleomene: – *Do you think Women have more Pride from Nature than Men?* and Cleomene replies: – *I believe not: but they have a great deal more from Education.* Because of the intensity of female desire and because of the social sanctions attributed to unchaste women in a male-shaped and male-controlled environment, a higher degree of self-control is demanded of women than of men. In a world dominated by men's rules, women have somehow to manage and administer their chastity, just as men manage property, capital and credit. Being exposed to 't*he Artillery of our sex*', to dangerous seducers that may court them with promises and bribes – Cleomene takes up again the arguments of Lucinda's 'warnings' from

[19]Castiglione 1989, in part. pp. 95–99.

[20]Mandeville 2006, pp. 59 and 63 (pp. 17 and 9 in the original text).

The Virgin Unmask'd – women need a special training in Honour. To educate the young to live in society is to stimulate their pride, literally to '*encrease their Fear of Shame*'. In the Third dialogue Cleomene offers the basis for a 'pedagogy of pride' for both sexes. Honour for men is an issue of courage. Young men are to be educated to fear shame more than death. Duelling still maintains an enormous symbolic value because courage was, and still is praised not simply for being an expression of public-spirited feeling, but as the highest form of self-control, the model of all virtues: the victory over the strongest of all the passions, the fear of death According to Mandeville it is a mistake to accuse duellists of having a false notion of honour, the practice of duelling being implicit in a concept of honour based on courage. As for women: *The Pride likewise that produces Honour in Women has no other Object than their Chastity; and whilst they keep that Jewel entire, they can apprehend no Shame.*[21]

For Mandeville virtue and honour are of the same origin. They are simply forms of control of one's own selfish passions, motivated in its turn by another passion, the search for other's approval. Both result from human's hypersensitivity to other people's judgement. Both are forms by which men curb their selfish and potentially antisocial passions. Yet Mandeville sees honour, rather than virtue, as the most effective principle for the maintenance of social equilibrium. Reading over again the history of recent civilization through the history of pride, of that search for the marks of public esteem which stands at the basis of human capability of socialization, Mandeville is able to situate the culture of honour and politeness in the broader framework of the civilizing process. *"Honour is not founded upon any Principle, either of real Virtue or true Religion, must be obvious to all that will but mind what sort of People they are, that are the greatest Votaries of that Idol, and the different Duties it requires in the two Sexes (...) I mean the Sense of it, is so whimsical, and there is such a prodigious difference in the Signification of it, according as the Attribute is differently applied, either to a Man or to a Woman, that neither of them shall forfeit their Honour; tho' each should be guilty, and openly boast of what would be the other's greatest Shame.*[22]

It has been noted that in Mandeville's writings on women there is a 'contradictory conjunction' of feminist arguments and misogynistic characterization: women are depicted as more fond than men for sexual activity and they do not share the rational capabilities of man.[23] In the medical Treatise Mandeville wrote that: *'One hour intense thinking wastes the spirits more in a woman than six in a man'* – but also that – '*Where the advantages of education and knowledge are equal, women exceed man in sprightliness of fancy, quickness of thought and off-hand wit*'.[24] To use the

---

[21]Mandeville 1988, Vol. II, p. 124. See Dickey 1990; Hundert 1994; Heath 1998; Simonazzi 2008, in part. pp. 201–216.

[22]Mandeville 1988, Vol. II, p. 128.

[23]Mandell 1992.

[24]Mandeville, B. (1730) *A Treatise of the Hypochondriack and Hysterick Diseases*, London, pp. 249 and 246.

adjective 'contradictory' with the writings of Bernard Mandeville is a challenging business. Certainly, in particular in the Modest Defence, there is a concurrence of feminist arguments and misogynistic characterizations. It is to be explained with the fact that as with other issues Mandeville – unprejudiced observer, anatomist of society – sees things as they are, rather than they should be. He is aware of the unequal condition regarding women, but he addresses it with the traditional misogynistic language and world-view by which man is the norm – women 'the other': *'The Clitoris', with is frequent erections is a perfect copy of the Penis, 'tho in Miniature'* wrote Mandeville in the Modest Defence.[25] Taken all together Mandeville's remarks on female honour across his writings appear coherent. As with courage and male honour for man, Female honour – chastity – is an artificial virtue, exclusively grounded upon Fear of Shame. The issue of women's role in society is for Mandeville a field that magnifies the incoherencies and hypocrisies of the 'Moralists & Reformers', those polemical targets around which the Dutch doctor developed his philosophical anthropology. The courtship rituals, the different forms of education allotted to men and women, the harsh reality of conjugal life, the profound inequality of women's condition and man's monopoly of learning, the social – and monetary – value of chastity, for Mandeville all these are evidences that society is not grounded upon rational ideals of virtue or on a supposed natural benevolence, but rather on a balance between selfish passions, resulting from an historical process were sociability itself developed. The inequality between sexes is simply one of the results of the process of civilization, as manners, morality, society are: artificial balances claims Mandeville against who presupposes a natural social instinct – but not rationally planned – against the theorist of the social contract.

Reconstructing the history of the shared systems of sentiments of approbation and disapprobation which make up the ideals of male and female social respectability, Bernard Mandeville demonstrates how the rituals of Modern Honour are nothing else that modifications of pride, an exemplary expression of that spontaneous, *artificial* order resulting from a *natural* disposition of human passions.

References

Anderson, P.B. 1935. Splendor out of scandal: The Lucinda-Artesia papers in the female tatler. *Philological Quarterly* 15: 286–300.

Bond, R.P. 1971. *The Tatler. The making of a literary journal*. Oxford: Oxford University Press.

Bryson, F.R. 1938. *The sixteenth-century Italian duel: A study in renaissance social history*. Chicago: Chicago UP.

Burke, P. 1995. *The fortunes of the courtier. The European reception of Castiglione's Cortegiano*. Cambridge: Polity Press.

Castiglione, D. 1983. Mandeville Moralized. *Annali della Fondazione Luigi Einaudi* XVII, pp. 239–290.

[25] Mandeville 2006, p. 75 (p. 41 in the original text).

Castiglione, D. 1989. *I piedi del pavone. Ragionamento sopra una satira cinica, appendix to the italian translation of the Modest Defence: Modesta Difesa delle Pubbliche Case di Piacere*. Palermo: Flaccovio.
Chamberlain, E. 1718. *Magna Britannia Notitia, or The Present State of Great Britain, London*; Bk III, P. I, Ch.V, pp. 176–177.
Cohen, E. 1997. *"What the Women of All Times Would Laugh At": Redefining equality and difference*, Circa 1660–1760, in Osiris, 2nd Series, 12, pp. 121–142.
Defoe, D. 1724. *Roxana, The fortunate mistress*. London: T. Warner.
Dickey, L. 1990. Pride, hypocrisy and civility in Mandeville's social and historical theory. *Critical Review* 4(3): 387–431.
Erspamer, F. 1982. *La biblioteca di Don Ferrante. Duello ed onore nella cultura del Cinquecento*. Firenze: Bulzoni.
Goldsmith, M.M. 1986. *The treacherous art of mankind; Bernard Mandeville and female virtue; in History of Political Thought* VII, n° 1, Spring 1986.
Goldsmith, M.M. (ed.). 1999. *By a society of ladies. Essays in the female tatler*. Bristol: Thoemmes Press.
Harvey, K. 2002. The century of sex? Gender, bodies, and sexuality in the long eighteenth century. *The Historical Journal* 45(4): 899–916.
Heath, E. 1998. Mandeville's bewitching engine of praise. *History of Philosophy Quarterly* 15: 205–226.
Hundert, E.J. 1994. *The enlightenment's fable, Bernard Mandeville and the discovery of society*. Cambridge: Cambridge University Press.
James, M.E. 1978. English politics and the concept of honour 1485–1642. *Past and Present* 3: 76–90.
Kelly, J. 1995. *'That Damn'd thing called honour'. Duelling in Ireland 1570–1860*. Cork: Cork UP.
Kelso, R. 1929. *The doctrine of the English gentleman in the sixteenth century*. Champaign: University of Illinois Press.
Laqueur, T. 1990. *Making sex: Body and gender from the Greeks to Freud*. Harvard: Harvard University Press.
Mandell, L. 1992. Bawds and merchants: Engendering capitalist desires. *English Literary History* 59(1): 107–123.
Mandeville, B. 1730. *A treatise of the hypochondriack and hysterick diseases, j*. London: Tonson.
Mandeville, B. 1732. *An enquiry into the origin of honour and usefulness of christianity in war*. London: J. Brotherton.
Mandeville, B. 1975. *The virgin unmask'd, or female dialogues betwixt an elderly maiden lady, and her niece, on several diverting discourses on love, marriage, memoirs, and morals, & c. of the times*. New York: Delmar.
Mandeville, B. 1988. *The fable of the bees, or private vices, publick benefits* (J. Roberts, London 1714, 1723, 1729), ed. F.B. Kaye. Indianapolis: Liberty Classic, Vol. I and II.
Mandeville, B. 2006. *Bernard Mandeville's "A Modest Defence of the Publick Stews" prostitution and its discontents in early Georgian England*, ed. I. Primer. New York/Basingstoke: Palgrave Macmillan.
Maravall, J.A. 1984. *Poder, honor y elites en el siglo XVII*. Madrid: Siglo XXI Kaeuper.
Mason, J.E. 1935. *Gentlefolk in the making: Studies in the history of English courtesy literature and related topics from 1531 to 1774*. Philadelphia: University of Pennsylvania Press.
O'Brien, K. 2009. *Women and enlightenment in eighteenth century Britain*. Cambridge: Cambridge University Press.
Peltonen, M. 2000. The duel and history in Elizabethan England. In "Se rakkain kotipolku" Erkki Kouri ja yleinen historia, ed. K. Multmäki, M. Peltonen and H. Saarinen, 193–204. Helsinki: Helsinki University Press.
Peltonen, M. 2003. *The duel in early modern England: Civility, politeness and honour*. Cambridge: Cambridge University Press.
Richard, W. 1999. *Chivalry and violence in medieval Europe*. Oxford: Oxford UP.

Simonazzi, M. 2008. *Le Favole della Filosofia. Saggio su Bernard Mandeville*. Milano: Franco Angeli.

Spierenburg, P. 1998. *Masculinity, violence and honor: An introduction, in men and violence. Gender, honor and rituals in modern Europe and America.*. Columbus: Ohio State U.P.

Swift, Jonathan. 1968. On the testimony of conscience. In *Irish tracts and sermons*, ed. L. Landa. Oxford: Oxford UP.

The Female Tatler. London.

Vichert, G.S. 1966. Some recent Mandeville attributions. *Philological Quarterly* 45: 459–463.

Vichert, G.S. 1975. *Bernard Mandeville's The virgin unmask'd, in Mandeville studies: New exploration in the art and thought of Dr. Bernard Mandeville (1670–1733)*, ed. Irwin Primer. The Hague: Martinus Nijhoff.

Watson, C.B. 1960. *Shakespeare and the renaissance concept of honour*. Princeton: Princeton UP.

Chapter 16
Mandeville and Smith on the Problem of Moral Order

Luís Oliveira

Abstract After the enormous impact of the *Fable of the Bees*, Bernard Mandeville's thesis on the problem of the emergence of moral order in society became a focal point in eighteenth century moral philosophy. Mandeville himself became a commentator of his own thesis in later works, and Adam Smith devoted to him a chapter in his *The Theory of Moral Sentiments*. In this work we will analyze both Smith's and Mandeville's later commentaries on the *Fable* about the problem of moral order, comparing how both authors express a rupture between the motives of the individual moral character and the social consequences of its actions.

Keywords Bernard Mandeville • Adam Smith • Moral order • Social order • Sympathy

In his commentary on the poem *Fable of the Bees* entitled *An Enquiry into the Origin of Moral Virtue*, Mandeville interprets the concept of Moral Virtue as an outcome of a process where lawgivers and the powerful become aware that pride is the passion at stake in governance, and that the emulation of mankind's rationality is the argument that serves Civilization better, revealing the possibility for a few to govern the vast majority.

Much of this stems from defining rationality as the superior human faculty from which moral distinctions are possible, precisely in the manner that best serves the goal of stealing the fruits of others' work and self-denial. So what gratifies the natural appetite of the subject is defined as vicious, and what leads to the benefit of the whole is defined as virtuous. There is a conflict, on the one hand, between the natural impulse for self gratification and on the other, the non natural, rational ambition of being good as conquering one's own passions, which leads the author to the famous conclusion:

> The Moral Virtues are the Political Offspring which Flattery begot upon Pride. (Mandeville 1988: 80)

L. Oliveira (✉)
Faculdade de Letras, Departamento de Filosofia, Comunicação e Informação, Universidade de Coimbra, Coimbra, Portugal
e-mail: luisfreitasoliveira@gmail.com

E. Balsemão Pires, J. Braga (eds.), *Bernard de Mandeville's Tropology of Paradoxes*, Studies in History and Philosophy of Science 40, DOI 10.1007/978-3-319-19381-6_16

Already in *An Enquiry into the Origin of Honour*'s preface, the author tells us that even though he knows that virtue is and will always be better than vice, and that this is an immutable truth, it turns out that the term "Moral Virtue" does not have an immutable meaning, and so, what we understand by it can then be legitimately criticized. Deepening his point of view, and already in the beginning of the dialogue, the character who takes the perspective of the author – Cleomenes – identifies the moralist vocabulary as an additional difficulty to express himself. This vocabulary is constructed to express a rationalist version of morality and assumes that human passions are a frailty of human nature, when the author clearly believes that they are the powers that govern the "human machine", establishing or creating the will that precedes every deliberate action. Moreover there is still a major difficulty in the absence of an adequate concept to express the esteem which men naturally have for themselves. Here, Mandeville rejects the concept of self-love and introduces the concept of self-liking. After the reference made by the author to the word esteem to characterize this self-liking (Mandeville 1732: 6), we can say that its meaning is somewhere close to self-esteem, reflecting a conceptual movement to separate it from an excessive, non-balanced, therefore, vicious sense.

Self-liking is described as a passion present since childhood, common to all mortals, that can be moderate or excessive in spirit. If moderate, it is worthy of approval and moves man to practice good deeds. If excessive, it already has a name in moralistic vocabulary, as pride, which makes the character who has this passion hateful to others. What Mandeville then wants, coining this new concept, is to unify all the extensions that he identifies in this passion, distinguishing it from the concept of self-love whose meaning he considers quite close to that of pride.

Starting from the principle that self-liking is universal in mortals, the author will base his definition of honor on this passion. So, assuming that self-liking is universal and that man is naturally aware of its presence in himself and in others, honor is here seen as the medium to express agreement with the good deeds done by others.

> The Word Honour, whether it is used as a Noun or a Verb, is always a Compliment we make to Those who act, have, or are what we approve of; it is a Term of Art to express our Concurrence with others, our Agreement with them in their Sentiments concerning the Esteem and Value they have for themselves. (Mandeville 1732: 8–9)

Therefore honor is no more than the most effective way the human species has to bring individuals to practice certain acts, functioning through the effect it has on the passion of self-liking. When someone does something of which we approve, we confirm the esteem that he has for himself, honoring him. When someone does something of which we don't approve, we show that we differ from him concerning the esteem that he has for himself. In other words, one must feel bad not to correspond to a given value, we know, everyone attribute or should attribute to oneself. Thus, virtue would reduce itself to self-liking insofar as being virtuous would consist of to correspond or harmonize the actions and the value that someone attributes to himself proportionally to that sentiment of self-esteem. In this harmonization, the primary criterion is the subject's judgment, its reason. But it gives great pleasure to be confirmed by the others.

Mandeville considers that the universality of the passion of self-liking is enough to explain the establishment of a morality between humans, meaning that he doesn't need any criterion for universality or immutability in moral distinctions. This passion cannot suggest such a criterion, and the author gives us an example of a group of thieves, where, if one of its members didn't take a good opportunity to pickpocket, he should feel ashamed of himself while standing before his peers, in a way completely analogous to any other situation worthy of dishonor (Mandeville 1732: 10).

As a matter of fact, Cleomenes points out as an error of the *Fable of the Bees* the distinction between pride and shame as two distinct passions that explain the psychological process through which morality can be explained. In *The Inquiry into the Origin of Honour*, this process is constructed in respect to one single passion, self-liking, and both pride and shame can be reduced to this passion as symptoms of excess or deficiency of its presence in the spirit.

Honor is a concept that Cleomenes identifies in the oldest human languages, a fact that reinforces his conception on the universality of self-liking. But honor as the prime principle of a social system, is something that he takes as a relatively recent phenomenon. Antiquity's system was constructed on virtue, and although virtue can be reduced to self-liking, this system was very different from the one ruled by honor, as its success to establish social order and form civility and common good will rely almost exclusively on one's individual virtue.

Therefore, the system of honor is presented as the modern one, and it's much more useful to establish social order and civility as the most common understanding, even the vicious character can be directed by its principles. It is better adapted to human nature.

If the principle of virtue consists of the satisfaction of the passion of self-liking through which one can identify the superior faculty of reason, the principle of honor consists of an incentive for the adoration of men by men. There,

> (...) the Excellency of our Species is raised to such a Height, that it becomes the Object of our own Adoration, and man is taught in good Earnest to worship himself. (Mandeville 1732: 42)

Unlike the man of virtue, the man of honor puts an extravagant value on himself, and the pleasure or displeasure that he takes from other's opinions on his value is much higher. Therefore what moves him can be stated as the fear of shame, which corresponds to a fear of itself, a fear that the esteem that he has of his own value wouldn't be satisfactory. This entails that the system of honor is much more complex than the system of virtue, being the result of a conscience that man has acquired through generations, as

> it was not the Contrivance of one Man, nor could it have been the Business of a few Years, to establish a Notion, by which a rational Creature is kept in Awe for Fear of it Self, and an Idol is set up, that shall be its own Worshiper. (Mandeville 1732: 41)

In *The Theory of Moral Sentiments*, Adam Smith puts forward a chapter devoted to the «*licentious system*» (Smith 1982: VII.i.4) of B. Mandeville, based on the observation that the Dutch author rejects «*the real and essential difference*» (Smith 1982: VII.i.1) between the concepts of vice and virtue when both derive from the concept of vanity.

Adam Smith's main intention in this chapter is to mark the absolute difference between love of virtue and vanity, and to do this, he sets a third concept, the love of true glory, to encompass all the potential similarities between them. On one hand, love of virtue – defined as the desire to become the proper object of esteem and approbation – relates to the love of true glory – defined as the desire for fair recognition – insofar as both depend on what it is (love of true glory) or ought to be (virtue) others' opinion. On the other hand, the love of true glory resembles vanity only as both desire worldly approbation and recognition.

Still, and notwithstanding the absolute refusal of the reduction of vice and virtue to a single passion, whether it be the vanity of *An Enquiry into the Origin of Moral Virtue*, or the self-liking of *An Enquiry into the Origin of Honour*, it is possible to grasp in this chapter of TMS two moments where the Scottish author approaches Mandeville, while justifying the enormous impact of the satirical work, which would not be possible had this system of morality not, somehow, «*bordered upon the truth*» (Smith 1982: VII.i.14).

The first of these moments corresponds to the affirmation of the exception of virtuous character among mortals, with the consequence that it is possible to generalize about the fear of bad opinion from others. Following the Stoic notion that only the virtuous character is autonomous in relation to the actual opinion of others, the rarity of this type of character in society means that the majority of the human mob can be moved by fear of ignominy. A. Smith certainly does not accept that the motive of virtuous action can be reduced to any passion, let alone a frivolous selfish one, but he certainly accepts that vanity, greed or fear of shame can be real motives of human actions. So, it is possible to assume that the generalized fear of ignominy is one of the aspects that, in Smith's opinion, has given Mandeville's thesis some resemblance with the truth.

The second argument used by Smith to justify the *Fable*'s success is that the ingenious sophistry operating therein was covered by the ambiguity of moral language. This means that this language is ambiguous when characterizing most of the human passions after the vice that corresponds to its excess. As an example, Smith mentions the reference in the moralist discourse of the passion of love of sex as lust, or the love of pleasure as luxury, and claims that this reference is ambiguous in that any passion has a certain point of propriety where its practice is wholly virtuous.

Curiously enough, this argument is very similar to the one used by Mandeville in *An Enquiry into the Origin of Honour* (Mandeville 1732: 6) to justify the *Fable*'s incorrect interpretation as a text where the possibility of virtue is ridiculed, and where it is claimed that the moral discourse was built to express a rationalist version of morality based on oppositions that assume human passions as frailties. There is a coincidence of this argument in Smith's TMS and Mandeville's *An Enquiry into the Origin of Honour*, both of them referring to the text of *An Enquiry into the Moral Virtue*. This is well illustrated when Smith claims that

Some popular ascetic doctrines which had been current before his time, and which placed virtue in the entire extirpation and annihilation of all our passions, were the real foundation of this licentious system.

And concludes:

It was easy for Dr. Mandeville to prove, first, that this entire conquest never actually took place among men; and secondly, that, if it was to take place universally, it would be pernicious to society, by putting an end to all industry and commerce, and in a manner to the whole business of human life. By the first of these propositions he seemed to prove that there was no real virtue, and that what pretended to be such, was a mere cheat and imposition upon mankind; and by the second, that private vices were public benefits, since without them no society could prosper or flourish. (Smith 1982: VII.i.4.12)

So Smith follows Mandeville on proof against moral rigorism, but claims that he abusively entailed these two consequences. So, what is the value of paradox "private vices, public benefits" for Adam Smith? Quite obviously, if the paradox comes in the form that Smith referred to in this last quotation, as "private vices are public benefits", he would reject it. However it is not in these terms that the paradox is presented in the subtitle of *The Fable of the Bees*, and if we assume that this paradox operates in Mandeville as the rupture of the common sense assumption of a moral continuity between the motive of action and its consequences on society, we can put forward the hypothesis that Smith not only follows this paradox, as he puts it, even if momentarily, in an even more radical fashion, in TMS's treatment of the problem of moral order.

In *The Theory of Moral Sentiments*, the problem of the moral order of society is addressed within the exploration of another relevant social phenomenon, namely the problem of the origin of ambition and the corruption of moral sentiments. The close relationship between these different themes is justified as Smith claims that all these three phenomena find their origins in a common natural feature of our sympathy:

(...) mankind are disposed to sympathize more entirely with our joy than with our sorrow. (Smith 1982: I.iii.2.1)

Smith's argument to construct this thesis consists of pointing to the natural and ordinary state of the human species as a happy state, described as health, no debts, and a clear conscience. He considered that empirical observations confirmed this as the real state of most humans, «*notwithstanding the present misery and depravity of the world, so justly lamented*» (Smith 1982: I.iii.1.7). Therefore, to the one principally concerned with an adverse or prosperous situation, adversity effectively takes him away from his ordinary state of happiness, while prosperity only superficially changes that state. But, to the same situation's spectator, sympathy is a imaginative movement that always implies an emotional effort, and what it takes to fully sympathize with the joy of others is usually much less than the effort that it takes to sympathize with their sorrows in its original degree of emotional violence. So, from the fact that the observer's ordinary state is much closer to the state of the one who is subject to prosperity than it is to the one who is subject to adversity, Smith draws the following conclusion:

> Thought Our Sympathy With Sorrow Is Generally A More Lively Sensation Than Our Sympathy With Joy, It Commonly Falls Much More Short Of The Violence Of What Is Naturally Felt By The Person Principally Concerned. (Smith 1982: I.iii.1)

From a political perspective, Smith takes this human feature as the condition for the possibility of a minority or even only one person to govern a whole nation, even if it is an Empire. We admire the situation of our superiors, and tend to act in a way that contributes to their perfection. So, it is more due to sympathy with their happy situation, and much less due to some sort of calculus about possible gains that men are solicitous, obedient, and obsequious to their superiors.

> When we consider the condition of the great, in those delusive colours in which the imagination is apt to paint it, it seems to be almost the abstract idea of a perfect and happy state. It is the very state which, in all our waking dreams and idle reveries, we had sketched out to ourselves as the final object of all our desires. We feel, therefore, a peculiar sympathy with the satisfaction of those who are in it. (Smith 1982: I.iii.2.2)

This is how whoever finds himself in this higher social status can ground the approval of his actions in style, making it possible for Louis XIV to govern a vast empire without ever demonstrating, in Smith's opinion, more than mediocrity in respect to his personal virtues.

Therefore, the establishment and the sustenance of social order is not determined by some kind of deliberation or rational decision that can be traced to the individuals that form society or the elite that governs it. Smith explains this through a net of relations and observations in which the concept of sympathy figures as the central and simple element that justifies complex social phenomena in a moral stance. Sympathy alone doesn't explain these phenomena but, in Adam Smith's system of moral philosophy, it is always the engaged fundamental concept. So, it is sympathy that synthesizes the foundation of moral order in the individual as in the society.

In a similar way, from the same principle, namely the general disposition to sympathize more entirely with the other's joy than with their sorrow, it is possible to trace the human tendency to neglect or despise the poor. This corresponds to *«the great and most universal cause of the corruption of our moral sentiments»*. (Smith 1982: I.iii.3.1) A. Smith points out that the disposition to admire the rich and the powerful makes riches and power an effective way to get the great end of human ambition: *«to deserve, to acquire, and to enjoy the respect and admiration of mankind»* (Smith 1982: I.iii.3.2). So, this disposition is then perceived as the biggest and the most universal cause of the corruption of moral sentiments, because it entails that power and riches are taken as the proper objects of wisdom and virtue, while poverty is unfairly seen as the proper object of contempt.

> They are the wise and the virtuous chiefly, a select, though, I am afraid, but a small party, who are the real and steady admirers of wisdom and virtue. The great mob of mankind are the admirers and worshippers, and, what may seem more extraordinary, most frequently the disinterested admirers and worshippers, of wealth and greatness. (Smith 1982: I.iii.3.2)

As the phenomena of the origin of social order and the corruption of moral sentiments are both grounded in the same principle or feature of nature, namely that there is a tendency to sympathize more effectively with the joy than with the sorrow of the other, would it be possible to express this thesis as a mandevillian paradox? If we identify the corruption of moral sentiments with private vices, and social order with public benefits we'll put forward this kind of smithian "Private Vices, Public Benefits" that reads as "Corruption of Moral Sentiments, The Establishment and Sustenance of Order in Society". If we compare this to the mandevillian paradox, we should see that this is a more radical paradox, in that between its apparently incompatible two terms, there is not an uncanny or problematic consequential relation, but both paradox terms are consequences of the same principle or tendency of nature.

As a radical paradox shall not stand for more than a few lines in Smith's Moral Philosophy book, rapidly this incompatibility is characterized as apparent, as a social process of unintended consequences come to terms with it, functioning as if an invisible hand were engaged in the process as follows.

Notwithstanding the admiration for the rich and powerful and the contempt for the weak and poor as being the most universal cause of the corruption of moral sentiments, the preservation of society's moral order, insofar as Smith can claim that the most part of humanity lives in happiness, is only possible because, for the great bulk of mankind, the path of fortune is the path of virtue. Although the majority of society, corresponding to the inferior and middle classes, admire the rich and the powerful – understanding the attainment of fortune as the effective way to obtain respect and admiration of others – it happens that the most effective way to obtain fortune for the great bulk of mankind can be described as the «*real and solid professional abilities, joined to prudent, just, firm, and temperate conduct*» (Smith 1982: I.iii.3.5).

> The success of such people, too, almost always depends upon the favour and good opinion of their neighbours and equals; and without a tolerably regular conduct these can very seldom be obtained. The good old proverb, therefore, That honesty is the best policy, holds, in such situations, almost always perfectly true. In such situations, therefore, we may generally expect a considerable degree of virtue; and, fortunately for the good morals of society, these are the situations of by far the greater part of mankind. (Smith 1982: I.iii.3.5)

Thus, social order can only be characterized as the moral order of society by the analysis of this other phenomenon also theorized through the fundamental reference to sympathy. For the ones of mean condition, the greater part of mankind, the pursuit of ambition depends on a stable and sufficiently virtuous behavior. Smith's general theory on sympathy says that it's by the imagination that spectator places himself on other's situation (Smith 1982: I.i.1.2). And sympathy's intensity is proportional to the information that spectator has on that situation (Smith 1982: I.i.1.3), as the moral judgment of propriety depends on spectator's knowledge on agent's motives (Smith 1982: I.i.3.6). This justifies the moral efficiency of equals' opinion compared to the delusive tendency in judging superior's behavior.

Assuming that the emergence of moral order in society is a process of unintended consequences that promotes the interest of society as if it was led by an invisible hand, we shall conclude that in this example Smith's metaphor certainly expresses a rupture between the motives of the individual moral character and the social consequences of its actions, but isn't its function to resolve this rupture. As in the mandevillian paradox Smith conceives a process of unintended consequences that promotes the interest of society but it has a very distinctive feature as in Smith's this promoted interest has a moral or virtuous quality. Mandeville's model on the problem of social order in *An Inquiry into the Origin of Honour* clearly distinguished a system of virtue from a system of honor. In Antiquity's system of virtue there was no rupture between individual motives and social consequences. So the social interest relied almost exclusively on individual virtue. Modernity's system of honor is more efficient and complex, expressing that rupture as private vices can generate public benefits. While addressing the problem of moral order, Adam Smith's effort to overcome this mandevillean dichotomy between a complex social system and the possibility of a moral or virtuous quality to emerge as social order from that system seems real. Here sympathy – a natural moral feature – explains several moral phenomena overcoming their apparent incompatibility. Just like in the assumed model of TMS, Newton's *principia*, gravity – a natural feature of the physical world – causes both the attraction and the repulsion between cosmic bodies, explaining these phenomena as only apparently contrary.

References

de Mandeville, Bernard. 1732. *An enquiry into the origin of honour and the usefulness of Christianity in war*. London: John Brotherton. (Facsimiles from Google Books).

de Mandeville, Bernard. 1988. An enquiry into the origin of moral virtue. In *The fable of the bees or private vices, publick benefits*, ed. Frederick Benjamin Kaye, 77–84. Indianapolis: Liberty Fund.

Smith, Adam. 1982. *The theory of moral sentiments*, [1759]. Indianapolis: Liberty Fund.

Chapter 17
Atheism, Religion and Society in Mandeville's Thought

Mauro Simonazzi

Abstract Starting by analysing the atheist's character, Mandeville gets gradually closer to Bayle's thesis on virtuous atheism, but he takes a different turn, and maintains that a society without religion cannot exist because atheism goes against a natural passion: fear of invisible causes. In order to understand Mandeville's position on this last point, in the second part of this essay I will consider his reflection on the origin and on the social and political functions of religion.

Keywords Atheism • Society without religion • Origin of religion • Religion and politics • Passions and society • Mandeville • Bayle • Fear • Virtuous atheist

17.1 Introduction

> So there can be many different types of atheist, but for the purpose of legislation they need to be divided into two groups. The dissembling atheist deserves to die for his sins not just once ore twice but many times, whereas the other kind needs simply admonition combined with incarceration. [...]. Those who have simply fallen victim to foolishness and who do not have a bad character and disposition *should be sent* to the reform center by the judge in accordance with the law for a term or not less than five years, and during this period no citizen must come into contact with them except the members of the Nocturnal Council, who should pay visits to admonish them and ensure their spiritual salvation. When his imprisonment is over, a prisoner who appears to be enjoying mental health should go and live with sensible people; but if appearances turn out to have been deceptive, and he is reconvicted on a similar charge, *he should be* punished by death. (Plato 1997, 1565, 908e10–909a10)

I would like to express my gratitude for their comments to Irwin Primer and Marco Geuna.

M. Simonazzi (✉)
History of Political Thought, School of Law, University of Camerino, Camerino, Italy
e-mail: maurosimonazzi@libero.it

E. Balsemão Pires, J. Braga (eds.), *Bernard de Mandeville's Tropology of Paradoxes*, Studies in History and Philosophy of Science 40, DOI 10.1007/978-3-319-19381-6_17

In the tenth book of *The Laws*, Plato severely condemns atheism considering it dangerous for society and stemming from ignorance. The Greek philosopher's proposal consists in the establishment of a special detention institution called *Sophronisterion*, that is "wisdom creator", even though for the most serious cases he advocates the death penalty. This conception reflected Hellenic culture's common attitude towards atheism and impiety in general, but Plato gave a new philosophical depth to the condemnation of atheists and their persecution by making a link between the denial of the existence of the gods and of the immortality of the soul with immorality, political and social subversion and diseases. Atheism is considered a vice because the idea that God doesn't exist and the soul is mortal loosens the link with virtue. It is therefore dangerous for the *polis*' safety, as politics without virtue lack their foundation. In fact justice, like health, consists in the harmony of the parts, and this harmony falters if we exclude religion and morals.

When Pierre Bayle, at the end of the seventeenth-century, reflects upon the possibility of a society of atheists, he still has to deal with the rooted conviction that atheism is a vice, an illness, and, more than anything else, a social liability. Bayle asks his contemporaries two tightly related questions: can an atheist be virtuous? And if atheists can be virtuous, can society without religion exist? In other words, Bayle was wondering whether it was possible to break the exclusive bond between morals and religion and to theorise the existence of lay morals,[1] morals that would then constitute the basis of a society where religion would be politically irrelevant.[2]

These questions generate at least three problems. The first was about God's reaction towards those societies who would decide to tolerate atheism. In this perspective, the persecution of the atheist was necessary as a form of society's self-defence, not to prevent the damage done by the atheists themselves, but as a safeguard against God's wrath that would inexorably hit such a society. Therefore the problem was that of a vindictive God, not the social danger caused by atheists.

The second problem, instead, was of an epistemological nature: it regarded the foundation of morals without religion. Would reason suffice to found the distinction between vice and virtue? And would the absence of God necessarily generate conventional and relativistic morals?

The third problem regarded the relationship between virtue and behavioural motivations. If God didn't exist, then everything would be possible: virtue would not gain any reward nor vice any punishment. In this state of affairs, evil and vice would rule and good and virtue would be nothing else than a con to keep the ignorant and

---

[1]This work, known as *Pensées diverses écrites à un Docteur de Sorbonne à l'occasion de la comète qui parut au mois de décembre 1680*, appeared for the first time in 1682 by a different title: *Lettre à M.L.A.D.C., docteur de Sorbonne, où il est prouvé par plusieurs raisons tirées de la Philosophie et de la Théologie que les comètes ne sont point la présage d'aucun malheur. Avec plusieurs reflexions morales et politiques erreurs populaires*, A Cologne, chez Pierre Marteau, MDCLXXXII. See in particular §172. Its first English translation was made in 1708.

[2]On virtuous atheist and society without religion, see Lussu 1997, 57–99; Israel 2001, 331–341; Cantelli 2001, 679–706; Harris 2003, 229–254; Robertson 2005, 256–324; Mori 2011, 41–60; Bianchi 2011, 61–80.

the fool subjugated. The absence of eternal reward or punishment would also have consequences on a strictly political plan: without God no oath could be considered valid, and an oath was the seal of all important commitments. Without fear of perjury any promise would be a lot less binding, and so any agreement stipulated with atheists could not be considered valid because atheists were not bound by their word.[3]

Bayle concluded that atheism did not constitute an obstacle to the birth of society; that lay morals could exist; that an atheist could be virtuous and that a society without religion was therefore possible.[4] In his analysis of society, Bayle boldly asserted that the Christian religion, if adhered to in all its moral precepts, could cause damage to a society's wealth and power.[5]

In 1723 the second edition of the *Fable. Part I*[6] by Bernard Mandeville appeared in London, and within a few years a series of works aimed at confuting its most radical theses followed.[7] On January 20th 1724, William Law published an essay of about a hundred pages called *Remarks upon a late Book entituled The Fable of the Bees*,[8] in which he put great effort into criticising some of the main mandevillean theses, in particular the ideas that passions are at the origin of human actions, that reason cannot act as guide to human behaviour, that morals are a human invention and that the belief in the immortality of the soul results from pride. It is rather significant that in the final pages of this essay dedicated to the *Fable of the Bees*, William Law adds a postscript of seven pages, in which two of Bayle's theses are also criticized: the first is the conviction that virtue and religion can be explained as "blind effects" of human passions (Law: 100); the second is the belief in the possible existence of a society of virtuous atheists (Law: 101–102).

We cannot relay here the whole of William Law's confutation, which follows from a conception that reflects the orthodoxy of the time, but I would like to underline that in 1724 Mandeville had already been accused of being close to Bayle's positions.[9] Scribano noticed that in the second edition of *Free Thoughts*,

[3]See Mauthner 1920–1923; Minois 1988; Martin 2007; Bullivant and Ruse 2013. On atheism Bullivant and in Great Britain, see Buckley 1932; Redwood 1976 (enlarged edition, 1996); Hunter 1985; Buckeley 1987; Berman 1988; Hunter and Wootton 1992. On atheism in France: Kors 1990.

[4]See P. Bayle, *Pensées diverses*, §172–§183.

[5]See P. Bayle, *Continuation des Pensées diverses écrites à un Docteur de Sorbonne*, Rotterdam, R. Leers, 1705, §124.

[6]Mandeville, Bernard. 1723. *The Fable of the Bees*. London: Parker (London: Tonson, 1724). In December 1728 (but in the front-matter we read 1729), Mandeville publishes *The Fable of the Bees. Part II. By the Author of the First* (London: J. Roberts, 1729). The critical edition of reference up to date is Mandeville 1924a, b.

[7]See Stafford 1997, 2002.

[8]W. Law, *Remarks upon a late book entituled The Fable of the Bees, or Private Vices, Publick Benefits, in a Letter to the Author. To which is added a postscript, containing an observation or two upon Mr. Bayle*, London, Printed for Will. and John Innys, 1724.

[9]The close connection between Mandeville and Bayle was underlined by Mandeville himself in his *Free Thoughts*: «Those who are vers'd in Books will soon discover, that I have made great use of

published after Law's accusations and the Grand Jury's censorship, «Mandeville becomes more cautious, and adds to his preface a new passage (pp. XIX–XX) in which he declares that he referred to Bayle only as a source, without passing judgement on the truthfulness of the facts and opinions expressed by him, about which he declines any responsibility».[10] A few years later Archibald Campbell tarred Bayle and Mandeville with the same brush, calling them atheists and libertines.[11]

Mandeville, in truth, shares only some of Bayle's positions, while he distances himself from them on a few very significant points. Among Bayle's conclusions was the view that religion is not necessary for social life, and it is therefore possible to imagine a society of virtuous atheists; and that religion could have negative effects on a social level too. Bayle used the hypothetical existence of a society of virtuous atheists to support his opinion that religion is not necessary for man and society.

Mandeville also states that religion is not necessary for life in society, but on the other hand he cannot imagine a society without religion; he asserts not only that religion can be damaging to society, but that virtue itself, intended in its radical sense, is incompatible with the very existence of society. It becomes therefore apparent that Mandeville disagrees with Bayle right on the most crucial point: the hypothesis of the possible existence of a society of virtuous atheists.

In the pages that follow we will see how Mandeville's argument on atheism and religion develops. Starting by analysing the atheist's character, Mandeville gets gradually closer to Bayle's thesis on virtuous atheism, but having admitted the possibility of the existence of the apparently virtuous atheist, the Dutch physician takes a different turn, and maintains that a society without religion cannot exist. In

---

Monsieur Baile, without mentioning him. I confess, he is the learned Man I speak of in Page 93. The Citations likewise which I have borrow'd from that Author, without naming him, are many». See Mandeville 2001. See also F.B. Kaye, *Introduction*, in Mandeville 1924a, xlii–lii, lxx–lxxxiii, ciii–cv; James 1975; Horne 1978, 19–32; Scribano 1980, 21–46; Carrive 1980, 155–194; Scribano 1981, 186–220; Wong 1984, 394; James 1996; Primer 2001; Robertson 2005, 261–283.

[10]Scribano 1981, 187n. The English translation is mine.

[11]A. Campbell, *Arete-logia. An Enquiry into the Original of Moral Virtue; wherein the false notions of Machiavel, Hobbes, Spinoza, and Mr. Bayle, as they are collected and digested by the Author of the Fable of the Bees, are examin'd and confuted; and the eternal and unalterable nature and obligation of moral virtue is stated and vindicated*, Westminster, J. Chier, A. Campbell, B. Creake, B. Barker, 1728 (Edinburgh, 1733; London-Bristol, Routledge-Thömmes Press, 1994). In 1726, the reverend Alexander Innes had taken a manuscript from Archibald Campbell, promising that he would see to its publication. Instead, he published it in his name, adding a long forward addressed to the author of the *Fable of the Bees*. In 1730 Campbell, who taught History of the church at St. Andrews and was a colleague of Hutcheson, whose criticism of the *Fable of the Bees* he shared, publicly denounced the fact that Innes had published his manuscript and then decided to re-publish it in his name. The new version, much broader and with amendments, came out in 1733. I was able to consult the microfilm version of the 1728 edition, kept in Heidelberg's library. For a list of the works of the time where the closeness between Bayle and Mandeville is remarked, see Scribano 1981, 188, n. 8. On Campbell and Mandeville, see Maurer 2014. Mandeville has also often been compared with Machiavel; on the presence of Machiavellian elements in Mandeville see Simonazzi 2009.

order to understand Mandeville's position on this last point, in the second part of this essay I will consider his reflection on the origin and on the social and political functions of religion.

17.2 Atheism and Society

> Ce Sentiment étant directement opposé à celui de M. Bayle, qui soutient qu'il pourroit y avoir une république d'athées on entre dans l'éxamen de ce fameux paradoxe; & l'on s'attache à le réfuter. M. Bayle n'a pu le soutenir qu'en separant artificieusement la morale de la religion [...] Le Paradoxe de M. Mandeville, auteur de la Fable des Abeilles, est encore plus étrange que celui de M. Bayle. Cet Auteur prétend que les vices sont utiles à un état florissant, & il tache de le prouver par l'éxemple du luxe.[12]

The theme of atheism is present throughout Mandeville's works and gains increasing importance in time, as can be noticed by the fact that the argument gets more and more articulated starting from the first edition of the *Fable of the Bees* (1714), through *Free Thoughts* (1720), the *Fable of the Bees. Part II* (1729), and *An Enquiry into the Origin of Honour* (1732).

Mandeville's analysis of atheism fits into his more general interpretation of human behaviour within the framework of the passion system. Atheism, like faith, is not a matter of choice; it depends on one's upbringing and one's own emotional structure. According to Mandeville, in fact, man is not granted free will.[13] The most technical definition of free will was given by Mandeville in *Free Thoughts*, in the chapter in which he reflects upon the existence of free will and predestination: «what we call the Will is properly the last Result of deliberation» (Mandeville 2001, 61). It is a definition that Mandeville borrows from Hobbes, and that he comments as follows: «The reason, why every Body imagines that he has a Free-Will, is, because we are Conscious that in the choice of Things we feel a Power [...] to determine our Judgment either way. [...] If we reflect on this, I say, our Will shall not seem to be as free, as is commonly imagin'd» (Mandeville 2001, 61). In another passage, Mandeville is even more explicit: «Every Man may be convinc'd within himself, that Believing is not a Thing of Choice» (Mandeville 2001, 49–50). But if freedom is not as free as commonly thought, what determines individual behaviour? Mandeville thinks that passions determine actions; but passions are not all alike. Some are more important than others, and one in particular is the most important of all, namely,

[12]The quotation is taken from the Preface to the French edition written by the translator Etienne de Silhouette. See W. Warburton, *Dissertations sur l'union de la religion, de la morale, et de la politique*, Londres, Guillaume Darrès, 1742, vol. I, pp. 5–6.

[13]On this subject see also Mandeville 1924b, 229: *Horace*: « Is it not in our choice, to act, or not to act?». *Cleomenes*: «What signifies that, where there is a Passion that manifestly sways, and with a strict Hand governs that Will?». On Free-Will, see Scribano 1980, 75–89.

pride or self-liking,[14] which Mandeville defines as the desire to be esteemed and that using modern terminology we could define as a desire of social recognition.[15]

All human actions therefore depend on the passional structure.[16] Even the first Christians' martyrdom, and that of famous atheists like Giordano Bruno, Giulio Cesare Vanini and Mahomet Effendi, that get analysed in the *Fable Part I* (Mandeville 1924a, 214–215), have the same cause: pride and vanity. In this case Mandeville doesn't attribute any moral value to the martyrdom of atheists, intended as honesty, consistency or truth-seeking. Mandeville's introductory words to the passages about Vanini leave no ambiguity:

> So silly a Creature is Man, as that, intoxicated with the Fumes of Vanity, he can feast on the thoughts of the Praises that shall be paid his Memory in future Ages with so much ecstasy, as to neglect his present Life, nay, court and covet Death, if he but imagines that it will add to the Glory he had acquired before. (Mandeville 1924a, 213–214)

[14]Mandeville makes a distinction between Pride and Self-Liking starting from *Fable Part II* and then in *An Enquiry into the Origin of Honour.*

[15]In Mandeville's anthropological conception there are two main passions: self-love and self-liking. Self-love is an expression of self-preservation, while self-liking is the passion from which the desire to be esteemed comes from. Self-liking therefore is a relational passion as it has its foundation in other people's judgement and it is stronger than self-love as demonstrated by the case of suicide. See. Mandeville 1990, 6–7: «I now understand perfectly well what mean by Self-liking. You are of Opinion, that we are all born with a Passion manifestly distinct from Self-love; that, when it is moderate and well regulated, excites in us the Love of Praise, and a Desire to be applauded and thought well of by others, and stirs us up to good Actions: but that the same Passion, when it is excessive, or ill turn'd, whatever it excites in our Selves, gives Offence to others, renders us odious, and is call'd Pride. As there is no Word or Expression that comprehends all the different Effects of this same Cause, this Passion, you have made one, *viz.* Self-liking, by which you mean the Passion in general, the whole Extent of it, whether it produces laudable Actions, and gains us Applause, or such as we are blamed for and draw upon us the ill Will of others». See Scribano 1978; Jack 1989, 40–50; Hundert 1994, 52–55; Peltonen 2003, 263–302; Force 2003, 57–67; Guion 2004; Simonazzi 2008, 134–181; Blom 2009; Tolonen 2013, 22–30 and 82–102.

[16]In the first edition of the *Fable Part I*, published in 1714, Mandeville made clear that his intent was first of all descriptive, and that his analyses started from the premises that the will of man was not free, but determined by passions: «As for my Part [...], I believe Man (besides Skin, Flesh, Bones, &c. that are obvious to the Eye) to be a compound of various Passions, that all of them, as they are provoked and come uppermost, govern him by turns, whether he will or not» (Mandeville 1924a, 41). Though Mandeville often changes his opinions in time, when it comes to free will, instead, he remains faithful to this first definition; in fact in 1732, in his *An Enquiry into the Origin of Honour*, we find a formulation that is very similar to that of 1714, even in the words he chose: «So most of the Passions are counted to be Weaknesses, and commonly call'd Frailties; whereas they are the very Powers that govern the whole Machine; and, whether they are perceived or not, determine or rather create the Will that immediately precedes every deliberate Action» (Mandeville 1990, 6).

Differently from Bayle,[17] Mandeville states that there is no such a thing as a virtuous atheist, but there are indeed vain atheists: «yet it is certain, that there have been Men who only assisted with Pride and Constitution to maintain the worst of Causes, have undergone Death and Torments with as much Chearfulness as the best of Men, animated with Piety and Devotion, ever did for the true Religion» (Mandeville 1924a, 214).

This is Mandeville's opinion in 1714. A few years later, in *Free Thoughts*, his work that was most influenced by Bayle, Mandeville introduces a distinction between two types of atheists[18]:

> Atheists are either Speculative or Practical; Speculative Atheists are those unhappy People, who, being too fond of Knowledge or Reasoning, are first deluded into Scepticism, till, unable to extricate themselves from the Mazes of Philosophy, they are at last betray'd into a Disbelief of every Thing they cannot comprehend, and become the most convincing Evidences of the shallowness of Human Understanding. The Number of these has always been very small; and, as they are commonly studious, peaceable Men, the Hurt they do to the Publick is inconsiderable. [...]
>
> Practical Atheists are generally Libertines, who first have been guilty of all manner of Vice and Profaneness, and afterwards, not daring to reflect on the Enormity of their Sins, or the Punishment they deserve from the Vengeance of Heaven, lay hold on Atheistical Arguments, to skreen themselves from their own Fears, and only deny a God, because they wich there was none. Practical Atheists, as they commonly spend their Lives in Riots, and ridiculing every thing that is holy, generally dye (unless they happen to repent) in uncommon Agonies and Despair. (Mandeville 2001, 16–17)

This distinction seems to introduce Bayle's thesis about a difference between virtuous and vicious atheists, a difference that a few years before he had not taken into consideration. In fact, the speculative atheists are men that behave virtuously even though they don't believe in «an Immortal Power, that, superior to all Earthly Dominion, invisibly governs the World» (Mandeville 2001, 15). Atheists can be virtuous and Christians can be vicious because the reasons for their actions do not descend from the principles that are professed but from the passions and individual inclinations. Mandeville writes: «And therefore it ought not to appear more strange to us, that an Atheist should be a quiet moral Man, than that a Christian should lead a very wicked Life» (Mandeville 2001, 15). Practical Atheists, instead, are sinful and vicious men, who deny the existence of God because they hope not to have to account for their behaviour.

In order to avoid ambiguity, we need to specify the way in which Mandeville makes use of the term virtue. This word is used by Mandeville with two different

---

[17]Bayle's judgment on the reason for Vanini's martyrdom moves from the attribution of a "certain idea of honesty", that would make of him a virtuous atheist, to considering Vanini as a man "animated by a ridiculous point of honour", which would make of him only an extremely proud man. See P. Bayle, *Pensées diverses*, § 182.

[18]Mandeville takes from Bayle the distinction between speculative atheists (athées de théorie) and practical atheists (athées de praticque). As noted by M.E. Scribano, the difference between Bayle and Mandeville is that the first describes the atheist as a heroical and militant man, while the second presents him as a library mouse, completely defenceless and pacific. See Scribano 1981, 213.

meanings that can be identified as "real virtue" and "apparent virtue". Real virtue is defined as «every Performance, by which Man, contrary to the impulse of Nature, should endeavour the Benefit of others, or the Conquest of his own Passions out of a Rational Ambition of being good» (Mandeville 1924a, 48–49). Apparent virtue, instead, is the exterior behaviour that respects common values within a social context, but whose real reason is the passion of self-liking, and not the "rational ambition of being good".

In the end, real virtue cannot be practiced because it is contrary to human nature, which is passional. Apparent virtue, instead, is a behaviour that on the exterior level produces the same effects as real virtue, but is motivated by selfish reasons.

This distinction between virtuous and vicious atheists seems to have disappeared in Mandeville's later works. He may have used it simply to distinguish virtuous atheists, who are not dangerous for society, from vicious atheists, who reject not just the existence of God, but also society's shared values.

Consequently, nobody can be really virtuous, because virtue requires free will, which is denied to human nature. It is on this level that Mandeville distances himself from Bayle: an atheist who is truly virtuous does not exist. Having clarified this point, Mandeville proceeds by explaining that men can be divided in two categories: those who satisfy their passions behaving as if they were virtuous, that is respecting shared social codes, and those, instead, who satisfy their passions as if they were vicious, that is breaking shared social codes.

Going back to the problem posed by Bayle regarding the existence of the virtuous atheists, Mandeville's position is now more understandable. On the one hand, Mandeville states that a truly virtuous atheist does not exist. On the other, though, he thinks that we can distinguish two categories of atheists, the apparently virtuous ones, and the vicious ones.

In the Sixth Dialogue of the *Fable. Part II* Mandeville reiterates what he had affirmed in *Free Thoughts* regarding speculative atheists. The only difference is that the attitude of his description is more generous, but the subject is dealt with only after reassuring his readers that «Multitudes are never tainted with Irreligion» (Mandeville 1924b, 313). The atheist is described as a talented, spirited, reflexive and well-adjusted person, and a defender of freedom. He is fascinated by maths or natural philosophy, is keen on research, he is disinterested, and he lives a good healthy life. Mandeville especially notes that most of those who become atheists are proud and full of themselves, especially if they did not receive an adequate religious education when young.[19]

The atheist's image that emerges from the pages of the *Fable Part II* is one that resembles more and more a defence from an accusation of debauchery. In truth,

[19]Mandeville 1924b, 313: «Men of Parts and Spirit, of Thought and Reflection, the Assertors of Liberty, such as meddle with Mathematicks and natural Philosophy, most inquisitive Men, the disinterested, that live in ease and Plenty; if their Youth has been neglected, and they are not well grounded in the Principles of the true Religion, are prone to Infidelity; especially such amongst them, whose Pride and Sufficiency are greater than ordinary; and if Persons of this sort fall into Hands of Unbelievers, they run great Hazard of becoming *Atheists* or *Scepticks*».

the definition of the atheist as a vain man does not change, but the attitude of the author appears more sympathetic, as is made evident by his choice of adjectives, and in his description of the atheist as a moderate and freedom-loving man. Besides, Mandeville specifies that virtue does not depend on having faith or being atheist: «and if Men were sway'd in their Actions by the Principles they side with, and the Opinion they profess themselves to be of, All Atheists would be Devils, and superstitious Men Saints: But this is not true; there are Atheists of good Morals, and great Villains superstitious» (Mandeville 1924b, 314).

This description of the atheist opens the way to the explicit defence of atheism that we find in the *Enquiry into the Origin of Honour*, which is the work in which Mandeville dedicates the most space to this subject.

In this work, after having confirmed what sustained in previous ones, that is the impossibility to make atheism universal[20] (because it is contrary to a natural passion: the fear of invisible causes) and the importance of religion for what concerns oaths (believers have an extra deterrent in comparison to atheists), Mandeville explicitly states that faith is not a better guarantee of virtue than atheism. If vice means indulging in one's passions, then a miscreant is no more vicious than a believer, on the contrary:

> Wickedness consisting in an unreasonable Gratification of every Passion that comes uppermost, it is so far from implying Unbelief, or what is call'd Atheism, that it rather excludes it Because the Fear of an Invisible Cause is as much a Passion in our Nature, as the Fear of Death. (Mandeville 1990, 189)

Thus it appears that in 1732 Mandeville's conclusions were closer to those of Bayle than they were in 1714. If in 1714 Mandeville considered atheism to be an excess of vanity, in 1732 he seemed to concede something more. The atheist fights one of his own passions, the fear of invisible things, and this opposition is in itself a principle of virtue. Mandeville doesn't mention pride anymore, even though, to be consistent with his psychological approach, it remains the only explanation for atheism.[21]

---

[20]Mandeville 1990, 27: «There is a greater Possibility that the most Senseless Enthusiast should make a knowing and polite Nation believe the most incredible Falsities, or that the most odious Tyrant should persuade them to the grossest Idolatry, than that the most artful Politician, or the most popular Prince, should make Atheism to be universally received among the Vulgar of any considerable State or Kingdom, tho' there were no Temples or Priests to be seen». See also Mandeville 1990, 189: «Believe me, *Horatio*, there are no *Atheists* among the Common People».

[21]Regarding this aspect, it is quite significant to have a look at what Mandeville writes about the accusation of atheism. See Mandeville 2001, 15: «I would have no Man so uncharitable as to think any Man guilty of Atheism, who does not openly profess it». See also Mandeville 1990, 154–155: «For how flagitious soever Men are, none can be deem'd *Atheist* but those, who pretend to have absolutely conquer'd, or never been influenced by the Fear of an invisible Cause, that over-rules Human Affairs; and what I say now has been and ever will be true in all Countries, and in all Ages, let the Religion or Worship of the People be what they will».

17.3 The Origin of Religion

We will now set aside the reflection on atheism to analyse the origin and function of religion. The idea of religion in Mandeville is not straightforward. The first problem has to do with Mandeville's general philosophy. In the 1740s his works were interpreted in two different ways. In the first, born in the mid-twenties as a result of the controversy that originated with the publication of *An Essay on Charity and Charity Schools*, Mandeville appeared to be an atheist and a free-thinker. In the second perspective, which originated from the French translation of the *Fable of the Bees* in 1740, Mandeville was regarded as an important philosopher[22] and, in some cases, a good Christian.[23] These two different interpretations will re-emerge in the twentieth century, and have been characterising the historiographical debate until now.

The second difficulty concerns the language and the argumentative structures used by Mandeville in tackling the specific theme of religion. Kaye was the first, in 1924, to address the dilemma of whether the passages on religion should be considered literally or whether they should be interpreted ironically.[24] According to Kaye, Mandeville in *The Fable of the Bees. Part II*, when approaching religion, swapped the characters' roles. An objection to this has been that Kaye's hypothesis attributes to Mandeville intentions that cannot be demonstrated.[25] Still, even sticking to a literal reading, the problems are no less, because inconsistencies and internal contradictions emerge (Goldsmith 1985, 65).

---

[22]See, for example, *Bibliothèque Raisonnée*, Amsterdam, 1729, p. 445: «S'il se trouve dans cet Ouvrage des pensées fausses, hazardées & dangereuses, il s'y trouve aussi des réflexions justes, ingenieuses & peut-être nouvelles» and *Mercure de France*, Amsterdam 1750, pp. 124–126, p. 126: «un ouvrage lumineux & profond, qui intéresse la Politique, la Philosophie & la Religion».

[23]See, for example, *Advertissement des libraires*, in Mandeville 1740 t. 1, 4: «En effet, dans un Ouvrage qu'il a publié quelques années après celui dont nous donnons la traduction, il enseigne expressement que *la Vertu est plus propre que le Vice pour procurer le bonheur général de la Société*: maxime qui paroît directement opposée à la doctrine de la *Fable*, dans laquelle il semble que l'Auteur veut prouver qu'une Société ne fauroit fleurir s'il n'y règne de grands vices. Pour sauver cette contradiction apparente, nous disons que Mr. *Mandeville* badine dans la *Fable*, où l'ironie faute aux yeux en tant d'endroits, & qu'il parle sérieusement dans ses *Recherches*». The publisher-translator was probably Jean Bertrand (1707–1777). On Mandeville's reception in France see Gai 2004.

[24]Kaye was the first to make the hypothesis that Mandeville swaps the characters' roles when addressing religion. See Mandeville 1924b, 21–22 n. 2. Kaye thinks that Cleomenes' references to the biblical story of creation, in contrast with the scientific story told by Horace are to be interpreted ironically, especially the frequent references to miraculous and providential interventions that would explain history. John Robertson instead suggested that Mandeville intended to stay close to Bayle's positions, who had sustained that the most debated Christian doctrines could be accepted only believing in the literal truth of Scripture (doctrines of revelation, divine providence, and the perfection of God). See Robertson 2005, 273–277.

[25]See James 1975, 51: «On this basis, the interpretation of the dialogues becomes a highly delicate matter and liable to subjectivity».

Mandeville's interpreters are therefore divided between those who recognise in his thought the idea of Anglican orthodoxy (Chiasson 1970; Pinkus 1975) or, on the contrary, the defence of radical Calvinism (James 1975, 1996); those who place him within the libertine tradition,[26] or those, last but not least, who recognise an ambiguity in his works, remarking on the elements that make one think that Mandeville was both «a pious Christian, an ascetic, and an unusually austere moralist» and, at the same time, «at best an easy-going man of the world, at worst a profligate, a cynic, a scoffer at all virtue and religion» (Monro 1975, 1).

Not much about this can be deducted from his philosophical and political works; more perhaps can be found in his *Treatise*. In particular, the comparisons between the first and the second edition of the *Treatise of the Hypochondriack and Hysterick Diseases* and some passages of the *Fable of the Bees. Part II,* in my opinion (Simonazzi 2004, 321–345; 2011, 129–140), suggest that Mandeville's ideas were not distant from Hobbesian materialism or the mortalist doctrine (see Burns 1972), a theory according to which the soul would resurrect together with the body and therefore was not immortal for the period between earthly death and eternal life. Such a position, in England, was most of all supported by the Arminians.[27]

The mortalist theory (or *soul-sleeping* doctrine) was rather popular among English physicians at the beginning of the eighteenth-century, as demonstrated by the case of a friend of Sir Hans Sloane, the medical doctor William Coward (like Mandeville, a specialist in digestive disorders[28]), who became associated with members of the Royal College of Physicians in London, and whose books were sentenced to be burned in 1704[29] for having upheld the mortality of the soul and for having supported vitalistic materialism. The sentence was passed by a Jury instituted by the House of Commons of which Bolingbroke was also a member.

However, this is just an interpretative hypothesis and, on the other hand, Mandeville's main interest in religion was not of a theological nature. In fact, Mandeville analyses religion as a human phenomenon, that is as a product of passions. From this point of view, we can distinguish three levels on which his research develops.

At the first level we have the relationship between religion and anthropology (and all religions are considered); at a second level the relationship between

---

[26]This interpretation is popular especially in Italy, see Olivetti 1980; Rossi 1984; Sabetti 1985; Costa 2008.

[27]Irwin Primer, in his *Introduction* to *Free Thoughts*, recalled that «his enemies called him atheist, infidel and deist, but nowhere in his writings do we find him explicitly revealing the details of his personal religious belief» and that «whatever Mandeville may have believed, it is known that he was married in the Church of England and that at least the first of his two children was baptized in that church» (Mandeville 2001, xxxiiii).

[28]William Coward (1656/7–1725). Coward 1695, 1698.

[29]Coward 1702, 1703, 1704. See Pfanner 2000. William Coward published his books under the pseudonym of Estibius Psychalethes. Before Coward, as highlighted by Dario Pfanner, the Anglican Henry Layton published between 1694 and 1702 twelve voluminous tomes in defence of the mortalist doctrine.

religion and society (and this analysis is restricted to Christianity); at the third level the relationship between religion and politics (and this analysis moves from a critical-description of mainly protestant Christianity to a propositional level, where Mandeville theorises a reformation of the relationship between Church and State in Britain).

First of all let's see what is religion in Mandeville's perspective: «Religion in General consists in an Acknowledgment of an Immortal Power, that, superior to all Earthly Dominion, invisibly governs the World, and a respectful Endeavour to discharge such Duties, as every one shall apprehend to be requir'd of him by that Immortal Power».[30]

In the fifth dialogue of *The Fable of the Bees. Part II* and in the first dialogue of *An Enquiry into the Origin of Honour*, Mandeville addresses the problem of the origin of religion. Religions are all considered on the same level and any religious phenomenon is analysed in a historical perspective and in a comparative way. This way, the Dutch physician could exalt his anthropological method and give an explanation that was strictly psychological, without resorting to any kind of revelation. Mandeville not only makes no distinction between Christianity, Judaism, or Islam, but he doesn't even make distinctions between monotheistic and polytheistic faiths or any other form of superstition or idolatry. What is therefore the origin of religion? Fear.

> *Hor.* [...] when I ask'd you, how Religion came into the World, I meant, what is there in Man's Nature, that is not acquired, from which he has a Tendency to Religion; what is it, that disposes him to it?
>
> *Cleo.* Fear. (Mandeville 1924b, 206–207)[31]

The Origin of Religion, as stated in Hobbes' epicurean thesis, is the fear of invisible causes in absence of a rational explanation. At this first level, Mandeville explicitly rejects the theory of political imposture to explain the origin of religion. Religion is conceived as proceeding from a natural passion: fear (Mandeville 1924b, 206–214). This doesn't deny that the natural passion of fear may have been exploited for political ends by individuals capable of channelling other people's passions to their advantage[32]:

---

[30]See Mandeville 2001, 15.

[31]And in the *Introduction* to *The Fable of the Bees*, Mandeville warned the reader that the analysis would have been a purely anthropological one, independent of religious beliefs. See Mandeville 1924a, 40: «And here I must desire the Reader once for all to take notice, that when I say Men, I mean neither Jews nor Christians; but meer Man, in the State of Nature and Ignorance of the true Deity». See also Mandeville 2001, 15–16: «This Definition [of Religion] comprehends whatever *Mahometans* or *Pagans*, as well as *Jews* or *Christians*, understand by the Word *Religion* [...] He who believes, in the common Acceptation, that there is a God, and that the World is rul'd by Providence, but has no Faith in any thing reveal'd to us, is a Deist; and he, who believes neither the one or the other, is an Atheist».

[32]Mandeville 1924b, 218: «The Word Religion itself, and the Fear of God, are synonimous; and had Man's Acknowledgment been originally founded in Love, as it is in Fear, the Craft of Impostors could have made no Advantage of the Passion; and all their boasted Acquaintance with Gods and

> *Cleo*: [...] I don't deny the Usefulness which even the worst Religion that can be, may be of to Politicians and the Civil Society: But what I insist upon, is, that the temporal Benefit of it, or the Contrivance of Oaths and Swearing, could never have enter'd into the Heads of Politicians, if the Fear of an invisible Cause had not pre-existed and been supposed to be universal, any more than they would have contrived Matrimony, if the Desire of Procreation had not been planted in Human Nature and visible in both Sexes. (Mandeville 1990, 24–25)

This investigation on the origin of religion allows Mandeville to conclude that religion is not a crucial factor in the birth of society (as, instead, are morals, which are detached from religion; we can behave morally independently of our religious beliefs) and yet it is a necessary and non eliminable component of human nature. Therefore society doesn't need religion (as Bayle remarks), virtue is enough, and yet it is impossible to think of a society without a religion.

This is the first level of analysis: religion is born of fear, as such it is a non eliminable component of human nature, so a society of atheists is not possible. Let's come now to a second level of analysis: the relationship between religion and society.

17.4 The Social Function of Religion

Mandeville develops his first theory on the origin of society in 1714, when he resorts to the theory of political imposture, then abandoned in 1728, presenting an evolutionary theory of men and society.[33] In both these formulations, Mandeville makes a distinction between the origin of religion and its social function. Regarding its social function, in 1714 Mandeville held that religion is not necessary for life in society,[34] and in 1728 he stated that it is part of the structure of an impersonal power that creates a more or less stable balance without need for the purposeful project by a single will. This means that religion has no determinant function in society, sharing the same religion is not the necessary social glue for community life, but it is nonetheless useful to reinforce social bonds through a grid of shared values and to reinforce political obligation when Church and State come to coincide.

But in practice, so to speak, what is the social function of religion?

The subject is addressed by Mandeville at different times and with slightly different acceptions throughout his works. However, if we take into consideration the last of Mandeville's work, *An Enquiry into the Origin of Honour and the*

Goddesses, would have been useless to them, if Men had worship'd the Immortal Powers, as they call'd their Idols, out of Gratitude».

[33]On the differences between the first and the second part of *The Fable of the Bees* see Scribano 1980; Simonazzi 2011; Tolonen 2013, 1–146.

[34]Mandeville 1924b, 50: «I affirm, that the idolatrous superstitions of all other nations, and the pitiful Notions they had to the Supreme Being, were incapable of exciting Man to Virtue, and good for nothing but to aw and amuse a rude and unthinking Multitude».

Usefulness of Christianity in War, we see that he leaves his comparative-historical method in the background to concentrate all his attention on Christianity (as already suggested in the title).

The analysis on this point becomes more precise, and Mandeville makes a distinction between the religion of the Gospels and the religion of the priests. The religion of the Gospels is Christianity as it was shaped in the first two centuries, while the religion of the priests is the result of the transformation that it underwent as an official state religion. The religion of the Gospels is the first Christians' set of original values, values that are incompatible with the possibility of the institution of a society, while the religion of the priests is the adaptation that those values went through in time so that Christianity may survive.

> In the Beginning of Christianity, and whilst the Gospel was explain'd without any Regard to Wordly Views, to be a Soldier was thought inconsistent with the Profession of a Christian; but this Strictness of the Gospel-Principles began to be disapproved in the Second Century. The Divines of those Days were most of them become arrant Priests, and saw plainly, that a Religion, which would not allow its Votaries to assist at Courts or Armies, and comply with the vain World, could never made National; consequently, the Clergy of it could never acquire any considerable Power upon Earth. In Spirituals they were the Successors of the Apostles, but in Temporals they wanted to succeed the Pagan Priests, whose Possessions they look'd upon with wishful Eyes; and Worldly Strength and Authority being absolutely necessary to establish Dominion, it was agreed, that Christians might be Soldiers, and in a just War fight with Enemies of their Country. But Experience soon taught them, that [...] there could be no Religion so strict, no System of Morality so refin'd, nor Theory so well meaning, but some People might pretend to profess and follow it, and yet be loose Livers, and wicked in their Practice. (Mandeville 1990, 33–34)

The problem posed by Mandeville has therefore a dual nature: on one hand there is the impossibility to conciliate the message of the Gospels with social usefulness (a polemic against the deists); on the other hand, we have the churches' hypocrisy, about which Mandeville makes no distinction between Catholics and Protestants, who would like to make their religion, Christianity, worldly rather than unworldly.[35]

The distinction between religion of the Gospels and religion of the priests has therefore the function to highlight the contrast between Christian values in themselves and their political exploitation.

If we go back for a moment to the Baylean hypothesis of the possible existence of a society of atheists, we see that Mandeville follows Bayle's thesis and states that a society of virtuous Christians cannot exist, as Christian values are incompatible with the development of a rich and powerful society («Religion is one Thing, and Trade is another»[36]). As for a society that is neither rich nor powerful, that would be doomed in the game of international competition: «T'enjoy the World's Conveniencies,/ Be fam'd in War, yet live in Ease,/ Without great Vices, is a vain/ Eutopia seated in

[35]From this point of view we can say that Calvin and Luther did not take into account human nature, so «their Successors, after Two or Three Generations, would make wretched Figures, if they were still to continue to preach Christianity without Deceit or Evasions, and pretend to live conformably to the Rules of it». See Mandeville 1990, 99.

[36]Mandeville 1732, 68.

the Brain./ [. . .] Bare Virtue can't make Nations live/ in Splendor; they, that would revive/ A Golden Age, must be as free,/ For Acorns, as for Honesty» (Mandeville 1924a, 36–37).

Mandeville, therefore, states that Christianity does not constitute a social binding agent, but must instead adapt to worldly needs. In Mandeville's perspective the real social binding agent is honour,[37] that is the set of those values that satisfy the true innate passion of human nature which is self-liking, defined as the desire to be esteemed.[38]

17.5 Religion and Politics

The third level of analysis of the religious phenomenon deals with the negative effects of churches from a political perspective. While Mandeville's reflection on religion is rather cautious, often disguising his real positions, his analysis of the secular power outright shows his anticlericalism. The clergy are accused of favouring radical positions and of being responsible of the schisms and persecutions in English history. In *Free Thoughts*, for example, we read that «the Doctrine of CHRIST plainly forbids Malice, Hatred and Revenge, and every where exhorts us to Meakness, Patience, Humility, Peace and Charity to all Men, so a Christian, who is really such, can never hate others upon any religious Account, tho' they were *Mahometans* or *Pagans*» (Mandeville 2001, 22). Consequently «Religion is not the Cause of the unhappy Breaches, that divide *Great Britain*; and that therefore all Divines of what Perswasion soever, who would insinuate the contrary to us, and perverting the Word of GOD, make a handle of it to breed Quarrels and Animosities, or any way disturb the publick Peace, are evil Teachers and Seducers of the People» (Mandeville 2001, 22).

Mandeville states that there is no Christian principle that a skillful politician cannot exploit for anti-Christian aims. In this perspective, all churches must be considered impostures and this explains Mandeville's open anti-clericalism. A common mistake that is made while judging the clergy is to believe that priests are better than ordinary men, instead «we ought to consider, that Clergy men are made of the same Mould, and have the same corrupt Nature with other Men; that

[37]See Monro 1975, 121–147; Carrive 1980, 68–70; Hundert 1994, 69–74; Simonazzi 1999; Branchi 2000; Peltonen 2003, 263–302; Branchi 2014.

[38]Mandeville thinks that the secular advantages of religion more modestly consist «in Promises of Allegiance and Loyalty, and all solemn Engagements and Asseverations, in which the invisible Power, that, in every Country, is the Object of the Publick Worship, is invoked or appeal'd to. For these Purpose all Religions are equally serviceable; and the worst is better than none: For without the Belief of an invisible Cause, no Man's Word is to be relied upon, non Vows or Protestations can be depended upon» (Mandeville 1990, 23–24).

they were born with the same Infirmities, and that consequently they were subject to the same Passions, and liable to the same Temptations» (Mandeville 2001, 153). Mandeville insists:

> [. . .] That Power and Authority were dangerous Tools in the Hands of Church-men: That whenever they were warmly opposed they could never forbear making use of them, and that how Just, Humane and Compassionate soever their Natural Temper might be, all Clergy-men in Power turn'd Persecutors, as soon as they were thoroughly anger'd. (Mandeville 2001, 130)

Bayle had already sustained that only by controlling children's upbringing since infancy the church had managed to make the strangest ideas acceptable, such as trinity or transubstantiation, and that no adult brought up outside Christianity would ever have accepted ideas so openly unreasonable.[39] Mandeville shares Bayle's thesis and expands it, accusing the pervasive power of the clergy on all stages of an individual's life:

> For, besides their officiating every Day at Divine Service, we can do nothing of moment without them, and they assist us through every Stage of Life. As soon as we are born they come to Christen us, and when the Nurse has had the greatest trouble with us, and we can help our selves, the Clergy desire to have the Tuition of us, till we are Men. The next then to be thought on his Matrimony, which we can't enter into without them. In Sickness they come to Comfort us, and claim a Right to examine our Consciences when we are in Health. They still visit us on our Death-Beds, even when the Physician has left us; and, after we have taken our leaves of them and the whole World, they won't yet part with us before they have seen us in the Grave. (Mandeville 2001, 156)

The political solution identified by Mandeville in *Free Thoughts* consists in proposing a national church that allows a broad internal tolerance.[40] Such a proposal substantially "de-socialises religion", making the church irrelevant on a social and political level. M. Emanuela Scribano showed how, at the beginning of the eighteenth-century the defence of religious tolerance took the shape of a war against the power of the church and, in particular, against the threat of an alliance between the High church and the Tories' reactionary fringes, part of which was very close to the Jacobites (Scribano 1980, 47–74). In the past many proposals had been advanced against this threat, but two main attitudes could be identified: that of

[39]See Mandeville 2001, 55–56: «There is hardly a Truth more convinc'd of, than that Two and Two make Four: Yet were Men to be taught from their Infancy that it was a Mystery, that on a certain occasion Two and Two made Seven, with an addition to be believ'd on pain of Damnation, I am perswaded, that at least Seven in Ten would swallow the shameful Paradox, and that if they had always seen others ill treated for disbelieving of it, by that they were come to Years of Maturity, they would not only assert it themselves, but likewise dislike, if not hate those, who should call it in question. We must suppose, that it had been inculcated to them with Application and Assiduity by Parents, Nurses, Masters, and all that had the Tuition of, or any direction over them».

[40]On the historical context in which *Free Thoughts* appeared, see Scribano 1980, 11–89; Schochet 2000; Prior 2000b; Primer 2001.

the Latitudinarians, who conceived a single comprehensive church encompassing conceptions even quite different among them, and that of the Independents, who advocated a multiplicity of churches, all equally subordinated to the civil power.

Mandeville's position is neither, but can be seen as a sort of compromise. He advances the idea of a national church with a few privileges, such as exemption from taxes on its income or control of temples and public schools (Mandeville 2001, 141–142). At the same time though, these privileges must not interfere with the existence of minority churches, who nonetheless must acknowledge the legitimacy of the civil government. Besides, the national church clergy do not have a right to persecute heretics. It is in fact up to the government, not to the clergy to evaluate whether a religious sect is dangerous for the state:

> [...] When I speak up for a Tolleration of different Sects, I mean only, such as shall own the Government to be the supream Authority upon Earth, both in Church and State, and have no other Master abroad, that may make them Plot against our Safety. It is on this Head only that Papists and *Non Jurors* ought to be excluded; but this being the Business of the State, the Clergy has nothing to do with it.
>
> IT is the Government and the Ministry of it, which ought to be watchful, and take care that the Publick receives no detriment from subtle Stratagems carried on under Religious pretences. (Mandeville 2001, 140)

The argument used by Mandeville in defence of religious tolerance can be divided in an epistemological consideration, on the one hand, and into the statement of a principle of social usefulness. From the epistemological point of view, the author observes that any church is fallible and that «it is evident then, that there is no Characteristick to distinguish and know a true Church from a false one» (Mandeville 2001, 136). Therefore, it is not possible to propose a single interpretation of the Scripture that may end the dispute. Mandeville states that:

> Our Church pretends to no Infallibility, which implies a liberty, in every Member of it, of re-examining whatever has been said or done before him. No Man therefore ought to be too dogmatical in Matters of Faith: What to my Understanding is difficult and obscure, cannot be made otherwise to me by another's saying, that it is clear and easy to him; and let us hear what we can, every one at last must judge for himself to the best of his Ability. (Mandeville 2001, 50)

From the point of view of social usefulness, Mandeville holds that tolerance of dissenters is the best way to safeguard peace and safety within the State.[41] In fact, it

[41] In truth, Mandeville, in polemic, states that even persecution can safeguard peace and safety but only in those cases where the Church is capable to nip in the bud any dissent, as is the case in a few Catholic countries. See Mandeville 2001, 139: «In *Italy*, *Spain*, and *Portugal*, where an Heretick is knock'd down the moment he rises, and the Church has a proper Power obey'd by the Government, to enquire into a Man's Conscience before he opens his Mouth, and punish him for what her Holy Officers shall fancy him to think. A strict Conformity in manner of Worship, once establish'd, may be maintain'd with little Bloodshed».

is the nature of any religious dispute to result in a political conflict if not restrained by the government:

> The great Danger there is in the Quarrels of the Clergy is, that there can be no drawn Battle among them; being in all their Contests both Judges and Parties, one side must fall, and there can be no Peace without a Conquest. (Mandeville 2001, 52–53)

The central point of Mandeville's thesis about religion and politics is the idea that the clergy must be controlled by the state. The Dutch physician adopted many of the points of freethinkers and deists, in particular the idea that there isn't a single interpretation of the Scriptures and the consequent fallibilistic conception, the criticism of fanaticism, the necessity of a national religion, and the conviction that the clergy must be kept under control by lay people. The proposal he makes in *Free Thoughts* is that of a nation with a national Church in which the clergy act as if they were social civil servants. The model he proposes is that of Holland, where all the clergy, not just that of the official church, are paid by the State, respect the government's absolute authority and do not interfere in political matters. Furthermore «they are allow'd to inveigh against Sin and the Vices of great Men, as much as they please, without pointing at particular Persons» (Mandeville 2001, 158).

Mandeville, therefore, takes from Bayle the rehabilitation of atheists and seems to suggest, in a few passages, that a society without religion would perhaps be less vulnerable to internal conflicts; but his basic conviction is that religion, being an anthropological need, cannot be erased from human society. Civil power must acknowledge that and act consequently: «No Discourses nor even Prayers, which have the least tendency to Sedition, should be suffer'd in any Assembly:'Tis the business of a careful Ministry to look into these Matters, and the least Conventicle ought no to be neglected» (Mandeville 2001, 141). In conclusion, religion can act in two different ways: on one hand, it can constitute one of the main reasons of internal unrest; on the other, Mandeville is aware of the opportunity that may arise from the manipulation of religious feelings for political purposes.

This ambiguity, highlighted earlier more than once, is well expressed by Mandeville himself in *An Enquiry into the Origin of Honour* where he discusses the political function of religion and clergy in case of war:

> In all Wars it is an everlasting Maxim in Politicks, that whenever Religion can be brought into the Quarrel, it ought never to be neglected, and that how small soever the Difference may be between the contending Parties, the Divines on each Side, ought to magnify and make the most of it; for Nothing is more comfortable to Men, than the Thought, that their Enemies are likewise the Enemies of God. [...]
>
> However Morality is often preach'd to them, and even the Gospel at seasonable Times, when they are in Winter Quanters, or in an idle Summer, when there is no Enemy near, and the Troops perhaps are encamped in a Country, where no Hostilities should be committed. But when they are to enter upon Action, to besiege a large Town, or ravage a rich Country, it would be very impertinent to talk to them of Christian Virtues; doing as they would be done by; loving their Enemies, and extending their Charity to all Mankind. When the Foe is at Hand, the Men have Skirmishes with him every Day, and perhaps a main Battle is expected; then the Mask is flang off; not a Word of the Gospel, nor of Meekness or Humility; and all Thoughts of Christianity are laid aside entirely. (Mandeville 1990, 159–161)

17.6 Conclusion

Mandeville states that all the motives for human actions can be explained by two human passions: self-liking and fear, and that reason and rational principles have no impact on human behaviour. Atheism and religion are not explained by Mandeville in moralistic terms, but in psychological ones. Faith is not a matter of rational choice or predestination, but is a psychological response to fear of invisible things or fear of death. An atheist is one who overcomes this fear and does not conform to the socially accepted explanations for life mysteries. A society of virtuous atheists would be hypothetically possible, because society is based on honour and law, not on religion, but in reality it cannot exist because religion is the answer to an innate passion. Religion therefore is not necessary to society, but it is necessary to man. It is on this point that Mandeville more explicitly differs from Bayle, as it clearly emerges from this passage that seems to refer directly to him: «But whatever Philosophers and Men of Letters may have advanced, there never was an Age or a Country where the Vulgar would ever come into an Opinion that contradicted that Fear, which all Men are born with, of an invisible Cause, that meddles and interferes in Human Affairs» (Mandeville 1990, 27).

In Mandeville's perspective, religion is a natural need and therefore a society of atheists cannot exist. He advocates an Erastian model of Church, subordinated to the political power, in which religion is relegated to the private sphere and the clergy are kept under strict control by the state.

References

Primary Sources

Bayle, Pierre. 1682. *Lettre à M.L.A.D.C., docteur de Sorbonne, où il est prouvé par plusieurs raisons tirées de la Philosophie et de la Théologie que les comètes ne sont point la présage d'aucun malheur. Avec plusieurs reflexions morales et politiques erreurs populaires.* Cologne: Pierre Marteau. Later published with a different title: *Pensées diverses écrites à un Docteur de Sorbonne à l'occasion de la comète qui parut au mois de décembre 1680.*

Campbell, Archibald. 1728. *Arete-logia. An enquiry into the original of moral virtue; wherein the false notions of Machiavel, Hobbes, Spinoza, and Mr. Bayle, as they are collected and digested by the author of the fable of the bees, are examin'd and confuted; and the eternal and unalterable nature and obligation of moral virtue is stated and vindicated.* Westminster: J. Chier, A. Campbell, B. Creake, B. Barker (Edinburgh, 1733; London-Bristol, Routledge-Thömmes Press, 1994).

Coward, William. 1695. *De fermento volatili nutritio conjectura rationalis.* Londini: Excudit J.H. pro T. Bennte.

Coward, William. 1698. *Alcali vindicatum.* Londini: Printed and are to be sold by the Booksellers of London and Westminister.

Coward, William. 1702. *Second thoughts concerning human soul.* London: R. Bassett.

Coward, William. 1703. *Farther thoughts concerning human soul.* London: R. Bassett.

Coward, William. 1704. *The grand essay.* London: Printed for P.G.

Law, William. 1724. *Remarks upon a late book entituled the fable of the bees, or private vices, publick benefits, in a letter to the author. To which is added a postscript, containing an observation or two upon Mr. Bayle.* London: Printed for Will. and John Innys. In *Private vices, publick benefits? The contemporary reception of Bernard Mandeville*, ed. J. Martin Stafford, 45–94. Solihull: Ismeron, 1997.
Mandeville, Bernard. 1732. *Letter to Dion, Occasion'd by his Book call'd Alciphron, or The Minute Philosopher*. London: J. Roberts.
Mandeville, Bernard. 1740. *La fable des abeilles ou les fripons devenus honnêtes gens*. 4 vols. London: aux depens de la Compagnie.
Mandeville, Bernard. 1924a. *The fable of the bees: Or, private vices, publick benefits*, vol. 1. [1724], ed. F.B. Kaye. Oxford: Clarendon Press (1988. Indianapolis: Liberty Fund; 2001. Oxford: Oxford University Press).
Mandeville, Bernard. 1924b. *The fable of the bees: Or, private vices publick benefits. Part II*, vol. 2. [1729], ed. F.B. Kaye. Oxford: Clarendon Press (1988. Indianapolis: Liberty Fund; 2001. Oxford: Oxford University Press).
Mandeville, Bernard. 1990. *An enquiry into the origin of the honour, and the usefulness of Christianity in war* [1732]. In *Collected works of Bernard Mandeville*, ed. Bernhard Fabian and Irwin Primer, vol. VI. Hildesheim/New York: Georg Olms Verlag.
Mandeville, Bernard. 2001. *Free thoughts on religion, the Church & national happiness.* [1720; second edition enlarged 1729], ed. Irwin Primer. New Brunswick: Transaction Publishers.
Plato. 1997. *The laws*. In *Complete works*, ed. J.M. Cooper, 1318–1616. Indianapolis/Cambridge: Hackett Publishing.
Warburton, William. 1742. *Dissertations sur l'union de la religion, de la morale, et de la politique*. Londres: Guillaume Darrès.

Secondary Sources

Berman, David. 1988. *A history of atheism in Britain. From Hobbes to Russell*. London: Croom Helm.
Bianchi, Lorenzo. 2011. Bayle e l'ateo virtuoso. Origine e sviluppo di un dibattito. In *I filosofi e la società senza religione*, ed. Marco Geuna and Giambattista Gori, 61–80. Bologna: Il Mulino.
Blom, Hans. 2009. The Epicurean Motif. In *Epicurus in the enlightenment*, ed. Neven Leddy and Avi S. Lifschitz, 31–51. Oxford: Voltaire Foundation University of Oxford.
Branchi, Andrea. 2000. Mandeville e l'onore. In *Filosofia e cultura nel Settecento britannico*, ed. Antonio Santucci, 37–60. Bologna: il Mulino.
Branchi, Andrea. 2014. Vanity, virtue and the duel: The Scottish response to Mandeville. *The Journal of Scottish Philosophy* 12(1): 71–93.
Buckeley, Michael. 1987. *At the origins of modern atheism*. New Haven: Yale University Press.
Buckley, George T. 1932. *Atheism in the English renaissance*. Chicago: Chicago University Press.
Bullivant, Stephen, and Michael Ruse (ed.). 2013. *The Oxford handbook of atheism*. Oxford: Oxford University Press.
Burns, Norman T. 1972. *Christian mortalism from Tyndale to Milton*. Cambridge: Harvard University Press.
Cantelli, Gianfranco. 2001. La virtù degli atei nei «Pensieri diversi sulla cometa» di P. Bayle. In *La formazione storica dell'alterità: studi sulla storia della tolleranza nell'età moderna*, offerti a Antonio Rotondò 3 vols. ed. Henry Méchoulan, Richard H. Popkin, Giuseppe Ricuperati, and Luisa Simonutti, vol. 2, 679–706. Firenze: Olschki.
Carrive, Paulette. 1980. *Bernard Mandeville. Passions, vices, vertus*. Paris: Vrin.
Chiasson, Elias J. 1970. Bernard Mandeville: A reappraisal. *Philological Quarterly* XLIX: 489–519.

Costa, Gustavo. 2008. Alle origini del pensiero economico-sociale moderno: la Congregazione dell'Indice e Bernard Mandeville. *Nouvelles de la Republique des Lettres*, 55(1): 7–74.
Force, Pierre. 2003. *Self-interest before Adam Smith. A genealogy of economic science*. Cambridge: Cambridge University Press.
Gai, Letizia. 2004. Il Man of Devil attraversa la Manica. *Mandeville nei periodici francesi del Settecento, Studi filosofici* XXVII: 217–243.
Goldsmith, Maurice M. 1985. *Private vices. Public benefits. Bernard Mandeville's social and political thought*. Cambridge: Cambridge University Press.
Guion, Béatrice. 2004. L'amour-propre bien ménage: des ruses de la providence à la morale de l'intérêt. In *Un siècle de deux cents ans? Le XVIIe et XVIIIe siècles: continuités et discontinuités*, ed. Jean Dagen and Philippe Roger, 56–87. Paris: Éditions Desjonquères.
Harris, James. 2003. Answering Bayle's question: Religious belief in the moral philosophy of the Scottish enlightenment. *Oxford Studies in Early Modern Philosophy* 1: 229–254.
Horne, Thomas. 1978. *The social thought of Bernard Mandeville: Virtue and commerce in early eighteenth-century England*. London/New York: Macmillan/Columbia University Press.
Hundert, Edward J. 1994. *The enlightenment's fable: Bernard Mandeville and the discovery of society*. Cambridge: Cambridge University Press.
Hunter, Michael. 1985. The problem of atheism in early modern England. *Transactions of the Royal Historical Society* XXXV: 135–157.
Hunter, Michael, and David Wootton (ed.). 1992. *Atheism from the reformation to the enlightenment*. Oxford: Clarendon Press.
Israel, Jonathan. 2001. Bayle and the "Virtuous Atheist". In Id., *Radical enlightenment. philosophy and the making of modernity 1650–1750*, 331–341. Oxford: Oxford University Press.
Jack, Malcolm. 1989. *Corruption & progress. The eighteenth-century debate*. New York: AM Press.
James, E.D. 1975. Faith, sincerity and morality: Mandeville and Bayle. In *Mandeville studies. New explorations in the art and thought of Dr Bernard Mandeville,* ed. Irwin Primer, 43–65. The Hague: Martinus Nijhoff.
James, E.D. 1996. Schism and the spirit of toleration in Bernard Mandeville's free thoughts on religion. In *De l'humanisme aux lumières, Bayle et le protestantisme*, ed. Michelle Magdelaine, Maria Cristina Pitassi, Ruth Whelan, and Antony McKenna, 693–700. Paris/Oxford: Voltaire Foundation Press.
Kors, Alan Charles. 1990. *Atheism in France, 1650–1729*. Princeton: Princeton University Press.
Lussu, Maria Luisa. 1997. *Bayle, Holbach e il dibattito sull'ateo virtuoso*. Genova: ECIG.
Martin, Michael (ed.). 2007. *The Cambridge companion to atheism*. Cambridge: Cambridge University Press.
Maurer, Christian. 2014. What can an egoist say against an egoist? On Archibald Campbell's criticism of Bernard Mandeville? *The Journal of Scottish Philosophy* 12(1): 1–18.
Mauthner, Fritz. 1920–1923. *Der Atheismus und seine Geschichte in Abendlande*, 4 vol. Stuttgart-Berlin: Deutsche Verlags-Anstalt.
Minois, Georges. 1988. *Histoire de l'athéisme*. Paris: Libraire Arthème Fayard.
Monro, Hector. 1975. *The ambivalence of Bernard Mandeville*. Oxford: Clarendon Press.
Mori, Gianluca. 2011. Religione e politica in Pierre Bayle: la «società di atei» tra mito e realtà. In *I filosofi e la società senza religione*, ed. Marco Geuna and Giambattista Gori, 41–60. Bologna: Il Mulino.
Olivetti, Alberto. 1980. Dalla dinamica delle passioni alla divisione sociale del lavoro: note su Bernard Mandeville. In *Libertinismo europeo*, ed. Sergio Bertelli, 456–502. Milano-Napoli: Ricciardi.
Otero Knott, Martin. 2014. Mandeville on governability. *The Journal of Scottish Philosophy* 12(1): 19–49.
Peltonen, Markku. 2003. *The duel in early modern England: Civility, politeness and honour*. Cambridge: Cambridge University Press.
Pfanner, Dario. 2000. William Coward e il problema dell'anima mortale. In *Filosofia e cultura nel Settecento britannico*, vol. 1, ed. Antonio Santucci, 435–456, Bologna: il Mulino.

Pinkus, Philip. 1975. Mandeville's paradox. In *Mandeville studies. New explorations in the art and thought of Dr Bernard Mandeville*, ed. Irwin Primer, 193–211. The Hague: Martinus Nijhoff.

Primer, Irwin. 2001. Introduction. In Bernard Mandeville, *Free thoughts on religion, the Church & national happiness*, ed. Irwin Primer, i–xxxv. New Brunswick: Transaction Publishers.

Prior, Charles W.A. 2000b. "Then leave complaints": Mandeville, anti-catholicism, and English orthodoxy. In *Mandeville and Augustan ideas: New essays*, ed. Charles Prior, 51–70. Toronto: English Literary Studies, University of Victoria.

Redwood, John. 1976 (1996). *Reason, ridicule and religion. The age of enlightenment in England, 1660–1750.* Cambridge, MA: Yale University Press.

Robertson, John. 2005. Hume after Bayle and Mandeville. In Id., *The case for the enlightenment. Scotland and Naples 1680–1760*, 256–324. Cambridge: Cambridge University Press.

Rossi, Paolo. 1984. *The dark abyss of time. The history of the earth and the history of nations from Hooke to Vico.* Trans. Lydia G. Cochrane. Chicago: University of Chicago Press.

Sabetti, Alfredo. 1985. Bernard Mandeville, un libertino disincantato nel dibattito politico-religioso nell'Inghilterra del Settecento. In *Liberi pensieri sulla Religione, la Chiesa e il felice stato della Nazione*, ed. Bernard Mandeville, 7–57. Napoli: Liguori.

Schochet, Gordon J. 2000. Mandeville's free thoughts and the eighteenth-century debates on "Toleration" and the English constitution. In *Mandeville and Augustan ideas: New essays*, ed. Charles Prior, 35–50. Toronto: English Literary Studies, University of Victoria.

Scribano, Maria Emanuela. 1978. Amour de soi e amour propre nel secondo Discours di Rousseau. *Rivista di Filosofia* LXIX: 487–498.

Scribano, Maria Emanuela. 1980. *Natura umana e società competitiva. Studio su Mandeville.* Milano: Feltrinelli.

Scribano, Maria Emanuela. 1981. La presenza di Bayle nell'opera di Bernard Mandeville. *Giornale Critico della Filosofia Italiana* LX: 186–220.

Simonazzi, Mauro. 1999. Self-liking, onore e religione nella *Ricerca sull'origine dell'onore e sull'utilità del Cristianesimo in guerra* di Bernard Mandeville. *Il pensiero politico* XXXII(3): 352–382.

Simonazzi, Mauro. 2004. *La malattia inglese. La melanconia nella tradizione filosofica e medica dell'Inghilterra moderna*. Bologna: Il Mulino.

Simonazzi, Mauro. 2008. *Le favole della filosofia. Saggio su Bernard Mandeville.* Milano: FrancoAngeli.

Simonazzi, Mauro. 2009. *Temi machiavelliani nell'opera di Mandeville.* In *Anglo-American Faces of Machiavelli: "machiavellismi" a confronto nella cultura politica anglo-americana (secoli XVI-XX)*, ed. Alessandro Arienzo and Gianfranco Borrelli, 313–340. Monza: Polimetrica.

Simonazzi, Mauro. 2011. *Mandeville.* Roma: Carocci.

Stafford, J. Martin. 1997. *Private vices, publick benefits? The contemporary reception of Bernard Mandeville.* Solihull: Ismeron.

Stafford, J. Martin. 2002. Mandeville's contemporary critics, 1650–1850. *Ideas, Aesthetics, and Inquiries in the Early Modern Era* VII: 387–401.

Tolonen, Mikko. 2013. *Mandeville and Hume anatomists of civil society.* Oxford: Voltaire Foundation University of Oxford.

Wong, Constant. 1984. Mandeville, Bayle, and Epicurus. *Notes & Queries* XXXI: 394.

Chapter 18
Simulation and Dissimulation. Mandeville's Satirical View of Commercial Society

Joaquim Braga

Abstract It is our belief that a theoretical reading of Bernard Mandeville's Work without the consideration of the satirical elements that compose it, is an incomplete and equivocal reading, and may also lead to a distorted view of Mandeville's social thought. Satirical forms have at their core a peculiar discursive expressiveness that distinguishes them from other narrative genres. It is a double expressiveness because its moral content, almost always supported by the *binomium* "virtue-vice", has firstly, as social referent, the *character* of human beings. The caricature arises therefore inevitable. It is precisely this second aspect that is very current in the satirical purpose of Mandeville's *The Fable of the Bees*. As a symptom of commercial society, Mandeville realizes the enormous moral incongruity – "hypocrisy", according to the author – between action and expression, between individual practices and moral defense of collective values. He begins by satirizing man's hypocritical moralist discourse, since it corresponds to an ethics without empirical content, without any equivalence in terms of man's individual action in the modern society. Thus, our main goal is to analyze the Mandevillean fundamentals of the relationship between satirical discourse and moral transparency. To a certain extent, it seems that the function of modern satire, as it is assumed by Mandeville, does not imply a corrective purpose, but functions for humans only as therapeutic function.

Keywords Commercial society • Expression • Hypocrisy • Satire • Transparency

18.1 Introduction

The work of Bernard Mandeville does not present the same systematic structure that can be found in philosophical works. By this, we mean that unlike philosophical writing it reveals an extreme complexity in the relationship between form and

---

J. Braga (✉)
Faculdade de Letras-Universidade de Coimbra, Largo da Porta Férrea,
3004-530 Coimbra, Portugal
e-mail: bragajoaquim77@gmail.com

E. Balsemão Pires, J. Braga (eds.), *Bernard de Mandeville's Tropology of Paradoxes*,
Studies in History and Philosophy of Science 40, DOI 10.1007/978-3-319-19381-6_18

content – and it is therefore difficult to think the latter without the former. It is interesting to note that if Mandeville's contemporaries primarily interpreted and criticized his writings with regard to their alleged anti-moral content, today, in addition, their literary form also provokes many discussions about his work. Despite all references to his satirical intentions, it is nonetheless interesting to note too that the satirical function of his writings is often questioned, and in some cases even denied. But it is our belief that a theoretical reading of Mandeville's work without reference to the satirical elements that compose it, is an incomplete and equivocal reading, and may also lead to a distorted view of Mandeville's social thought. To assert that Mandeville's work has obvious satirical elements implies, in our view, to see in the analytical observations that he develops a common principle that comes from a satirical view of social reality. Satirical forms have at their core a peculiar discursive expressiveness that distinguishes them from other narrative genres. It is a double expressiveness because its moral content, almost always supported by the *binomium* "virtue-vice", has firstly, as social referent, the *character* of human beings. It is precisely this second element that is very current in the satirical purpose of Mandeville's *The Fable of the Bees*. As a symptom of commercial society, Mandeville realizes the enormous moral incongruity between action and expression, between individual practices and moral defense of collective values. He begins by satirizing man's hypocritical moralist discourse, since it corresponds to an ethics without empirical content, without any equivalence in terms of man's individual action in the modern society.

The work of the Societies for the Reformation of Manners against the corruption of morals was the main target of Mandeville's criticism.[1] The "Sermons" presented by such Societies had in common the fact that, in their rhetoric, vice is directly compared with the plague and the epidemic; in this sense, they are justified as an effort to prevent social contagion, especially the alleged effeminacy caused by luxury. But Mandeville does not really condemn the traditional morality and the mercantilist state. Rather, the target of his criticism is the possibility of a balanced – and socially beneficial – articulation between the two. Faced with a commercial society, powered by the passions of each individual human being and where moral virtue is no longer at its core, it will be therefore the satire's reader who must choose the worldview that most concerns him. This kind of "open interpretation" left by Mandevillean satire, by its paradoxical game, seems, at first sight, to dethrone the role of satire as unwavering social criticism, to the point that it is almost no longer understood as a narrative form. Hence the challenge that Mandeville puts on his readers, namely: what is the role that satire plays in the moral criticism of commercial society?

[1]For an assessment of the historical background and social implications of Mandeville's critique of the Reformation of Manners, see, for example, Thomas A. Horne 1978: 1–18.

18.2 Satirical Function and Social Discourse

Satirical forms – whether discursive or extra-discursive – act mainly on the descriptions that society makes of itself. Society creates such self-descriptions which reveal, generally, not what it is, but how it sees itself. From these self-descriptions result moral principles that are rather present in satirical forms. That is to say, satire is an emerging discourse with the aim of challenging the socially conventionalized communication, or in the words of Gilbert Highet, "the satirist tries always to produce the unexpected, to keep his hearers and his readers guessing and gasping" (Highet 1962: 18). And Mandeville's *Dialogues* exemplify plainly this characteristic relationship: Cleomenes – Mandeville's spokesperson – tries to convince his interlocutor, Horatio, about the inaccuracies of the traditional moral discourse. But Horatio, who is usually sceptical, insinuates himself against the immoderation of satirical discourse with the following assertion: "Things are as often overdone in Satyrs as they are in Panegyricks; and the Likeness of a Caricatura is no more to be trusted to than that of the most flattering Pencil" (Mandeville 1732: 91).

The Fable of the Bees does not only reveal a strong paradoxical duality between virtues and vices, its formal structure also shows a link between the paradoxical acceptance of the social model of modern commercial society and the satirical form of the *Fable* itself, that is still anchored to the seventeenth century narrative principles (cf. Palmeri 2003: 125), especially those principles that describe certain human passions as vices. But how can we explain this apparent paradoxical game? We believe that with this paradoxical game Mandeville wants, above all, to save satirical morality. The *Fable*, leaving open the reconciliation of virtues with vices, attempts to preserve the value of social satire, the irreplaceable truth that it reveals; since, as Mandeville also says – and unlike Horatio's assertion – , "there is, generally speaking, less Truth in Panegyrics than there is in Satyrs" (Mandeville 1988b: 59). Regarding the relationship between satire and panegyric, we can contrast this statement with the strong critique of Shaftesbury against both genres. The traditional view that conceived panegyrical forms as the highest and clearest praise of human virtue is questioned by him. Shaftesbury, who has always held a negative opinion towards the satirical genre, argues that, in their effect on readers, modern panegyrical forms are quite indistinguishable from satirical forms. For him, modern satire "is scurrilous, buffooning, and without morals or instruction, which is the majesty and life of this kind of writing", and, moreover, "our Encomium or Panegyric is as fulsome and displeasing, by its prostitute and abandoned manner of praise. The worthy persons who are the subjects of it, may well be esteemed sufferers by the manner. And the public, whether it will or no, is forced to make untoward reflections, when led to it by such satirising panegyrists. For in reality the nerve and sinew of modern panegyric lies in a dull kind of satire; which the author, it is true, intends should turn to the advantage of his subject; but which, if I mistake not, will appear to have a very contrary effect" (Shaftesbury 1790: 229–230). It is important to note that what these two conflicting points of view show us is, above all, the great relevance that the forms of expression play in the objectification of human social life. Not only is the *discursive content* – the articulation and distinction between vices and virtues – at issue here. The *discursive form* also acquires a unique primacy in the reflection on the morality of society itself.

Mandeville, like many other eighteenth-century satirists – a century that was marked by a great revival of European satire –, retrieves the mythological genealogy of the satirical narrative discourse, leaving the moral elements that are opposed in the discourse itself into an open and permanent conflict. The intrinsic explanations in such opposition are based largely on a naturalistic view of man and society, which exceeds all normative implications of culture, in particular those concerning the ideas of "civilized man" and "social contract". The mythological genealogy, present in the *Fable*, represents a discursive model essentially expressive and persuasive, linked to an oral culture and a face-to-face form of direct communication.

The moral, as it seen by the satirical discourse, leads us, before its normative dimensions, to the construction of discursive and symbolic values. The *Fable of the Bees* incorporates this primacy of symbolic mediation. Its language returns to a proto-reflective (mythological) level, in the same way as virtues and vices are also contrasted with the proto-cultural life of passions that govern human behavior in the state of nature. Here there is a fundamental distinction between *moral vision* (implicit morality) and *moral purpose* (explicit morality). Mandeville, especially in the *Fable*, does not have a strictly moral purpose. It is, however, his aim to criticize all arid and excessive moralism. At same time, the satirical elements that involve his writings are deeply anchored in a moral ground as one that allows him to make a clear distinction between vices and virtues.[2] Mandeville uses, in this narrow sense, the traditional moralistic semantics that considers vice – as, for example, luxury – to be the true negation of virtue. It is this that is already designated "implicit morality", and whose genealogical dimensions are not criticized by the author, which allows him to draw a satirical irreparable *hiatus* between human nature and social morality. In his essays in *The Female Tatler* (Mandeville 1999), Mandeville builds a true satirical observation system of eighteenth-century English society. Several discursive forms – such as stories, novelistic descriptions, letters, verses, Latin quotations – embody this system. Through these multiple ways of referring to reality, the author creates a "living portrait" of the behavior of their fellow citizens, thereby supporting the veracity of his social criticism. There is, in this sense, a kind of reality effect achieved through the inclusion of different forms, since, despite their symbolic heterogeneity, the descriptive nature of social reality

[2]As rightly noted by Malcolm Jack, the Mandevillean narrow conception of vice plays a satirical social function: "Mandeville is not wrong for *sometimes* associating pleasure with vice; he is wrong in making their connection a *necessary* one. As [Samuel] Johnson has realized, Mandeville's rigorism rests on an extremely puritanical premise, deliberately selected to reinforce his satire against the do-gooders, the members of the Societies for the Reformation of Manners" (Jack 2000: 91). The same observation can also be found in Phillip Harth's satirical interpretation of *The Fable of the Bees*: "Mandeville's habit of applying the standard of Self-denial, without which there can be no Virtue is not the psychological quirk of a philosopher who is inconsistent with his own feelings, but the deliberate tactic of a satirist who lays bare the inconsistency of Christians" (Harth 1969: 339).

does not cease to be the same – we can even say that such heterogeneity intensifies its empirical dimensions. However, the fact that social reality is represented through various discursive forms without losing its uniqueness, that it comes only confer more objectivity, more realism, to the point of generating a vision of its autonomy from the speech.

Thus, the *Fable*, as a narrative genre, appears as an expressive way to reconcile the origins of man social life with the description of the origins of language. Although Mandeville, while satirist, does not use the binomial virtue-vice linked to the traditional satirical moral code, he relates it differently with the vision that he makes of his contemporary society. It was very common to see in the satirical function an effective link between the condemnation and correction of vices. "The true end of satire – according to the words of the seventeenth-century English author John Dryden – is the amendment of vices by correction. And he who writes honestly is no more an enemy to the offender, than the physician to the patient, when he prescribes harsh remedies to an inveterate disease" (Dryden 1911: 87). Mandeville, in part due to his professional background, also uses a kind of medical language, full of particular and detailed descriptive observations regarding the social life of men; but, unlike what Dryden wants to assign to the satirists, is not properly a language of "prescription". In the *Fable*, on the contrary, is the powerlessness of virtue over the vice that serves to diagnose the new organizational order of the eighteenth-century commercial society. In this latter sense, the Mandevillean satire operates in a second rhetorical level (meta-rhetorical): it is both *a satire of satire*, that is, acts as critical design of the structural assumptions of traditional satire. Regarding this point, we are almost being able to say that, at first sight, the Mandevillean satire seems to announce the apocalypse of satirical genre. "If the anatomizing of society shows that it operates on the basis of vile and despicable qualities – so writes Maurice Goldsmith –, then no amount of exhortation or satire will change its components or their operation – especially if those exhortations themselves turn out to be explicable as consequences of the same vile mechanism" (Goldsmith 2001: 57).

18.3 The Expressivist Imperative

With the Reformation of Manners, satirical forms would play a decisive role in the condemnation of the immoral socially dissolute behavior and in the exaltation of the lost and ascetic virtue. Mandeville, as we well know, strongly protested against the action of the reformers, as well against the role of satirists in defending the values of pre-commercial society. In this respect, his work must be read with particular attention to the implicit criticism that is made regarding society observation forms. But given what has been said, how should one situate Mandeville's satirical moral thought?

An interpretation put forth by many of his critics, related to the paradoxical articulation "virtue-vice" places us inevitably against a nihilistic moral problem, as it is defended by F. B. Kaye. But, as we well know, satirical discourse is a contextual discourse. Facing the blossoming commercial society of his time, as well as the increase of "good Manners" (Mandeville 1988b: 152) in man social life, to the point the world tend to be "more polish'd" (Mandeville 1988b: 152), Mandeville realizes the growing moral incongruity between action and discourse, between practices and values. On the basis of such incongruity lies a moralistic expression that condemns what is the genesis of prosperity, such as luxury, for example. What leads humans to defend the opposite of what they do increases the feeling of hypocrisy – a social feeling par excellence. As for Michel de Montaigne, who had seen his century under the aegis of dissimulation, hypocrisy was the vice which generates more "cowardice and baseness of heart", and, therefore, as he strongly reiterates, "it is a craven and servile idea to disguise ourselves and hide under a mask, and not to dare to show ourselves as we are . . . A generous heart should not belie its thoughts; it wants to reveal itself even to its inmost depths" (Montaigne 1963: 269). In the Mandevillean definition of hypocrisy lies also this idea regarding the dissimulation of "feelings" already traced by Montaigne in his *Essays*, but in addition, it also contains a particular emphasis on the dissimulation of "motives" that forms the intentional basis of social human behavior. (And this second dimension of dissimulation is important, because through it we can better observe the content of Mandeville's criticism to his society). As Mandeville tells us in this respect, "no Habit or Quality is more easily acquir'd than Hipocrisy, nor any thing sooner learn'd than to deny the Sentiments of our Hearts and the Principle we act from" (Mandeville 1988a: 281). Through this conception it is possible to see that, in a first theoretical approach, the human *capacity for denial* is the psychological basis of hypocrisy. This would immediately lead us to conclude too that it resembles the Mandevillean assumption of virtue as "self-denial". Indeed, the denial of feelings is, in Mandeville, a common structuring moment to both virtue and hypocrisy; and, in the same line of thought, we can also add that what specifies and differentiates the latter from other vices is necessarily its social profile drawn by negation. But despite these co-implications which can be inferred here, it is truly the negation of "the principle we act from" that distinguishes it from other social forms of human behavior.

The two different views of social life that appear in the verses of *The Fable of the Bees* – a commercial society supported by a primacy of the passions over the reason, and, to use the counterexample, the virtuous ascetic society without the vices of luxury –, being impossible to reconcile morally and satirically, put the author in front of the dilemma of how to observe the passions in their social expression. Vices are not innate to human nature. They are, according to Mandeville, combinations of different passions[3] made by the cultural context of individuals. Thus morality itself is not an innate human quality, nor should it be primarily and

[3]The psychosocial interpretation of vices and virtues leads Mandeville to admit different *degrees of complexity* of passions. In his opinion, however, the intricate formation of passions have a heterogeneous background, that is, certain particular human feelings, which are often opposites (e.g., love and fear), are articulated as social passions (e.g., jealousy): "The more a Passion is a Compound of many others, the more difficult it is to define it; and the more it is tormenting to those that labour under it, the greater Cruelty it is capable of inspiring them with against others:

theoretically interpreted as such, since that would be a real obstacle to deepening the knowledge of the passions that move each individual. The social life of human beings cannot be reduced to a merely moral interpretation, in the same way that the social cohesion does not depend solely on the virtuous actions of individuals. To understand society, according to the author, it is necessary, above all else, to unmask all the false appearances of virtue. It is this same culture that engenders vices that tries also dissimulating them by simulating the virtues. The *mechanism of hypocrisy* – which is constituted by the dissimulation of vice (to hide the *self*) and the simulation of virtue (to deceive *others*) – sets off a major social vice, because it prevents any moral judgment about the true motives of human actions. According to Mandeville, hypocrisy "is a fair outside, put on to hide Deformities within, designedly to cheat and circumvent others" (Mandeville 1723: 32–33). That is, it denies the psychological and moral transparency of action. Given the mechanism of hypocrisy, the satirical moralist has as main task to lead man to his true nature, without the masks of his artificial anatomy. "Taking the mask off" is then the core purpose of Mandevillean satire.

Although Mandeville, as he says later – specially and effusively in *A Letter to Dion* –, does not advocate the primacy of vices against moral virtues, while writing as a satirist – and we must here underline "satirist" –, he attempts to answer to the satirical moral dilemma caused by the two opposite visions of social life. To achieve this aim, Mandeville will apply the moral code "right-wrong" not to the empirical content of individual human actions, but prior to its acceptance in an expressive social level. Morality is thus placed outside the normative level, and, consequently, the moral code supports now, at an expressive level, the binomial "sincerity-hypocrisy". We would like to define the programmatic form of this binomial through the notion of *expressivist imperative*. It should be stressed, however, that the expressivist imperative is not, in a strict sense, a kind of corrective imposition of Mandeville against his contemporary citizens, but rather it operates *ab initio* as an implicit rule to his satirical moral vision. (The task Mandeville set for himself is to reveal the truth and not so much condemn the lie). That is, it is (the imperative) the starting point for the satirical criticism of hypocrisy, and, at the same time, serves to show us the incompatibility between morality and prosperity that results from the flourishing of commercial society.

One possible discursive configuration to the expressivist imperative, that includes sincerity instead of hypocrisy, might be as follows: *Express yourself so that what you say and defend publicly can be equate to everything you do and feel individually*. With this imperative Mandeville moves from the level of collective moral action to the level of individual moral expression. Thus, and as we have been saying, Mandeville's moral cannot be separated from a satirical construction of social reality. Embedded in the idea of morality is also the idea of satirical truth. Following up on this fact, the expressivist imperative is, at bottom, the touchstone that supports

Therefore nothing is more whimsical or mischievous than Jealousy, which is made up of Love, Hope, Fear, and a great deal of Envy" (Mandeville 1988a: 140–141).

the formation of the satirist moral vision. Being moral transparency a requirement which enables to articulate critically the humorous dimensions (caricature) with the contextual dimensions of social reality, the satirical author has to ban from his speech any suspicion of hypocrisy. That is to say, what is vice should be expressed as vice, and that which is virtue should be expressed as virtue. *The Fable of the Bees* reveals precisely the assertion of a deep-seated congruence between action and speech. It operates, if you will, as a discursive exemplification of an anti-hypocritical description of society, as a counterexample to the alleged discourse of moral virtues, typically and exacerbated in the puritan Sermons presented by the Societies for the Reformation of Manners.

18.4 Transparency and Expression

The relationship of satire with social reality is articulated by an expressive axis. One of the most obvious examples of this fact is the function assigned to the descriptions of individuals' moral aroused by its physiognomic nature, which allows, in turn, the satirist to develop caricature portraits of them. Mandeville's attempt at taking the man's hypocrite mask off (the being behind the mask) suggests a "physiognomic ideal" that goes through the expressivist imperative. The mechanism of hypocrisy eliminates the physiognomic traces of the authentic morality of human being, because it puts him a mask between feelings and their natural expression.[4] The polite values, unlike authentic virtue, only require the concealment of passions. Mandeville never tires of describing his contemporary citizens as "polite People", because they tend to avoid certain natural expressions, certain facial expressions and body postures (Mandeville 1988b: 287). But he goes further than this. He also claimed that this deceptive appearance is *contra-naturam*, since, according to him, there is in the evolution of language a tight correlation between "word" and "gesture" – words and signs, as he says –, is impossible to be fully denied even by the most well-educated society (Mandeville 1988b: 290). "Speech [words] and Action [gestures] assist and corroborate one another, and Experience teaches us that they move us much more, and are more persuasive jointly than separately" (Mandeville 1988b: 290). Of course, from a strictly semiotic standpoint, expression, unlike representation, does not allow us to develop a univocal distinction between expressed object and medium of expression. There is a semiotic congruence between the two elements of the relationship – and therein lies the suggestive power conveyed by expression –, often generating mythological configurations and worldviews.

[4]"The reduction of spiritual ideals to bodily facts" is, according to Frank Palmeri, one the biggest aims of narrative satire. In this narrow sense, "narrative satire reduces all that might be heroic and noble to a common level of physical experience, which it openly acknowledges, if it does not always joyously celebrate" (Palmeri 1990: 10–12).

Yet it would be a mistake to think that dissimulation has by itself a negative effect. Mandeville contends that the mechanism of hypocrisy is crucial for human sociability (Mandeville 1988a: 349). If there weren't such processes of "simulation" and "dissimulation", the development of commercial society would be compromised. The acts of simulating and dissimulating mean to create an appearance which has no real extension, as seen, for example, at the nominal value assumed by traded goods. On the other hand, political life also requires masking, since our passions must be hidden from others, so that an artificial order may be beneficial for all (Mandeville 1988a: 68). Dissimulation has another dimension. It is present both in commercial operations, as in symbolic operations. According to Mandeville's point of view, human beings do not use speech only to make known their thoughts and feelings, but also "to persuade others" (Mandeville 1988b: 289). In other words, language has itself a dissimulative latency. But there is another side of the problem. This takes us very deeply into the question of the forms of sincerity. If language has this dissimulative latency – which will be subsequently enhanced by communication in commercial society – where did Mandeville find a suitable communication ideal to serve as exemplification for sincerity, non-dissimulation, non-hypocrisy?

We believe that such moral basis of communication is inspired by the "expression of primary feelings", which, according to Mandeville, is always transparent and unconcealed. Through gestures and expression of basic feelings (grief, joy, fear, for example), uncivilized men can reach understanding among them. The expressivist moral imperative ideal is based, therefore, in an universalistic view of passions and expression of human emotions, since, for Mandeville, the expression of so-called "basic emotions" is transcultural, and thus equal in all humans (Mandeville 1988b: 295). On the one hand, this universalistic view allows him to confirm an ontological continuity between state of nature and men's social life. On the other hand, it allows him to make the case – very common at the time – that in the expression of emotions there is a consistency with the psychic life of individuals – for example, one that expresses joy, lives, in his psychic sphere an equivalent emotional state. Therefore, Mandeville concludes, speech is not the root of all human understanding (Mandeville 1988b: 286). There is, before it, the "Language of the Eyes", which allows men to understand each other "at first sight" (Mandeville 1988b: 287). But Mandeville knows it's not just the speech that is committed to the polished rules and the trade relations of commercial society. The language of the eyes and other motor movements, "that are natural, are carefully avoided among polite People, upon no other Account, than that they are too significant" (Mandeville 1988b: 287).

In order for the expressivist imperative to take effect through communication – and it is, above all, a truly second-order imperative, since it does not apply directly to the action, but prior to its observation –, Mandeville returns, once again, to a naturalistic view of the human person. Like society, language is seen by Mandeville as a human artefact, and not as a work of nature or of a divine entity. Now, such a view no longer applies to the expression of basic emotions, which, for Mandeville, has a natural and therefore transcultural background. Evidence of this is that even an advanced language is unable to suppress the natural signs and gestures that carry basic human emotions. It is already becoming clear that Mandeville does not follow just a naturalistic conception concerning human passions, which allows him to articulate the virtues and vices of society without a moral hierarchy of their functions. The conception has further implications. When there is no continuity between passions and expression there is dissimulation, hypocrisy. The satire turns out to be an insightful form of criticism when there is no such continuity, because satirical discourse is, at the level of expression, a coherent form, which excludes everything that violates a truthful psychological state of mind. So, the expressivist imperative is also justified by the use of the expression of feelings as ideal communication form, able to ensure the psychological manifestation of truth and sincerity. Now, in this universalistic view, supported, in part, by a biological causal logic between expression and emotion, there is no place for dissimulation and hypocrisy, for life situations in which the expression is the opposite of action.

18.5 Self-Expression and Self-Knowledge

Passions are, for Mandeville, the great engine of life and human society. To understand man, means in this sense, to anatomize his passions. In order to implement this task, Mandeville will take a conjectural approach, placing the human being under the condition of a "state of nature". This state, however, is purely hypothetical (cf. Carrive 1983: 129). The main reason for this artificial division is related to Mandeville's need to introduce the mechanism of hypocrisy and show that the civilized man misrepresents his natural passions, because in the hypothetical state of nature there is no place for that social distinction between vices and virtues. To the observer of modern morality, hypocrisy emerges as a vice that hides the other vices of human beings. But in a purely systemic manner, hypocrisy can be understood as a modern symptom of the increasing autonomy of psychic systems towards social systems. This fact would lead us to say, using a well-know maxim of La Rochefoucauld, that *L'hypocrisie est un hommage que le vice rend à la vertu* (La Rochefoucauld 1817: 49). On the moral level, the rupture between individual action and virtuous values tends to be visible on the incongruity between *do something* and *express its opposite*. Mandeville, on the one hand, thinks that hypocrisy is a positive vice to the prosperity of commercial society; but, on the other hand, he thinks that the hypocritical man not only deceives others but also himself.[5]

[5]In *An Enquiry Into the Origin of Honour*, Mandeville distinguishes between two levels of hypocrisy, more precisely, between *Malicious Hypocrites* and *Fashionable Hypocrites*: "By Malicious Hypocrites, I mean such as pretend to a great deal of Religion, when they know their Pretensions to be false; who take pains to appear Pious and Devout, in order to be Villains, and in Hopes that they will be trusted to get an Opportunity of deceiving those, who believe them to be Sincere. Fashionable Hypocrites I call those, who, without any Motive of Religion, or Sense of Duty, go to Church, in Imitation of their Neighbours. (. . .) The first are, as you say, the worst of Men: but the other are rather beneficial to Society, and can only be injurious to themselves" (Mandeville 1732: 201–202). Through this distinction – and if we also take into account the

Consequently, the dissimulation of "Self" is not seen by Mandeville as a positive phase of its affirmation in relation to society, namely as individuation process of psychic systems in relation to social systems. Rather, it's quite the opposite. The mechanism of hypocrisy is entirely described by Mandeville as an antagonistic power against man's self-knowledge. Mandeville is deeply concerned with the question of self-knowledge, and not with the development or reform of social morality. The incongruity between action and expression is not seen by Mandeville as a mere moral inconsistency, but rather as a form of *psychological deception*: the man deceives the other, while at the same time deceiving himself. This means that with the non-application of morality to commercial society, the possible solution for modern morality can only be seen by applying it to man's self-expression. Indeed, Mandeville does not want to humanize – or moralize – commercial society. Such a task would be rather contradictory. His aim, as we have saw before, is to, within the prosperity of modern social life, restore to the social communication the transparency of self-expression. This also means that his intent is not to apply the imperative to the motives of human action, but rather to the expression of man's inner life. Gradually is the function of satire itself which with this also is considerably changing. Instead of the immediate and easy public condemnation, the satirical morality increasingly tends to serve the ideal of an authentic revelation of human beings psychological life. We can follow the line of such change in an excerpt of a satirical poem wrote by Thomas Gilbert (1747: 26):

> Satire, like a true mirrour to the fair
> Shows not what we affect, but what we are;
> Plucks from the splendid courtier all disguise,
> And sets the real man before our eyes:
> If base designs are lurking in his heart,
> To point them out is sure an honest part.

It seems clear that the Mandevillean vision of eighteenth-century commercial society is based, precisely, on the criticism of the increase of hypocrisy which undermines man's self-knowledge. The "authentic virtues" are not morally compatible with the "pretended virtues", because the former implies always self-denial. And, as the author clearly states, "The imaginary Notions that Man may be Virtuous without Self-denial are a vast Inlet to Hypocrisy, which being once made habitual, we must not only deceive others, but likewise become altogether unknown to our selves" (Mandeville 1988a: 331).[6] Mandeville is thus against the vices moral sublimation of the prosperous man, the man who, paradoxically, denying and condemning individual passions, "desires the World should think him altogether free from Pride

definition of virtue as "self-denial" –, it is clear that, for Mandeville, human action's intentionality, unlike social mimetic behavior, always builds the strongest criterion for the moral evaluation of them.

[6]This notion of hypocrisy based on lack of knowledge that the human being has of himself can be found, for example, in the *connaissance de soi-même* of Pierre Nicole: "Comme l'ignorance de soi-même est la source de tous les vices, on peut dire que la connaissance de soi-même est le fondement de toutes les vertus" (Nicole 1999: 331).

and Sensuality, and put a favourable Construction upon his most glaring Vices" (Mandeville 1988a: 149). The mechanism of hypocrisy, in this case, prevents the authentic virtue to become morally perceptible, individually and socially. One thing at any rate is clear. When, for example, luxury goods related to private vices begin to be marketed, and consequently, the wealth of a nation tends to increase, there will probably be a change too in the way how moral will be externalized socially.

18.6 Conclusion

The moral nature of traditional satire was submitted and committed to social morality, in the sense that it operated as a refinement of this. Giving visibility to moral gaps in human behavior that had not been corrected by society would be satire's traditional social function. In Mandeville's work, on the contrary, there is a clear difference between "satirical morality" and "social morality". The first serves, above all, to criticize the second, and not to corroborate it. In the end, it is only the satirical morality that can prevail in commercial society, since social morality can no longer be applied in a transparent way to all human behavior – unless, of course, in a totally hypocritical way.

With the autonomy of human social life over the moral values – which the process can be followed in the observations of Mandeville on luxury goods – arises in parallel a new insight into the function of satirical forms, as well as its effect on society in general. To illustrate this better, we have to say here that it is precisely the *effect* which assumes primacy over the function. Given this, one can also add that what is increasingly pervading the satirical form, especially with the full industrialization of social life, is its aesthetic effect, or, as put by George Austin Test, the satirical genre "seems to be more purgative and playful, less corrective and definitive" (Test 1991: 257). By releasing the satirical elements of a prescriptive and corrective function – more specifically, to emancipate them from the moral discourse –, Mandeville seems to be aware of such social evolution of the satirical function, anticipating, decisively, the place it now tends to occupy in Western society. In short, to a certain extent it seems that the function of modern satire, as it is assumed by Mandeville, does not imply a corrective purpose, but functions for humans only as therapeutic function – and the fact that it leads man to laugh at himself can be seen as its first great effect.

References

Carrive, Paulette. 1983. *La philosophie des passions chez Bernard Mandeville*, vol. 1. Paris: Didier Erudition.

Dryden, John. 1911. *Absalom and achitophel*, ed. W.D. Christie, 5th ed rev. by C.H. Firth. Oxford: Clarendon Press.

Gilbert, Thomas. 1747. A view of the town, a satire. In an epistle to a friend in the country. In *Poems on several occasions*, 19–46. London: Printed for Charles Bathurst.

Goldsmith, Maurice. 2001. *Private vices, public benefits. Bernard Mandeville's social and political thought*, Rev. Edition. Christchurch, New Zealand: Cybereditions.
Harth, Phillip. 1969. The satiric purpose of the fable of the bees. *Eighteenth-Century Studies* 2(4): 321–340.
Highet, Gilbert. 1962. *The anatomy of satire*. Princeton/New Jersey: Princeton University Press.
Horne, Thomas A. 1978. *The social thought of Bernard Mandeville. Virtue and commerce in early eighteenth-century England*. London: Macmillan Press.
Jack, Malcolm. 2000. Mandeville, Johnson, morality and bees. In *Mandeville and Augustan ideas: New essays*, ed. Charles W.A. Prior, 85–96. Victoria BC: University of Victoria.
Mandeville, Bernard. 1723. *Free thoughts on religion, the church, and national happiness*. London: John Brotherton.
Mandeville, Bernard. 1732. *An enquiry into the origin of honour, and the usefulness of Christianity in war*. London: John Brotherton.
Mandeville, Bernard. 1988a. *The fable of the bees*, vol. 1. [1732], With a commentary, critical, historical, and explanatory by F.B. Kaye. Indianapolis: Liberty Fund.
Mandeville, Bernard. 1988b. *The fable of the bees: Or private vices, publick benefits*, vol. 2. [1732], With a commentary, critical, historical, and explanatory by F.B. Kaye. Indianapolis: Liberty Fund.
Mandeville, Bernard. 1999. *By a society of ladies. Essays in the female Tatler*, ed. M.M. Goldsmith. Bristol: Thoemmes Press.
Montaigne, M. 1963. *Essays and Selected Writings*. Trans. and ed. Donald Murdoch Frame. New York: St Martin's Press.
Nicole, Pierre. 1999. De la connaissance de soi-même. In: *Essais de morale*, Choix d'essais introduits, édités et annotés par Laurent Thirouin, 309–380. Paris: Presses Universitaires de France.
Palmeri, Frank. 1990. *Satire in narrative: Petronius, Swift, Gibbon, Melville, and Pynchon*. Austin: University of Texas Press.
Palmeri, Frank. 2003. *Satire, history, novel: Narrative forms, 1665–1815*. Newark: University of Delaware Press.
Rochefoucauld, La. 1817. *Maximes et Réflexions Morales*. Paris: Ménard et Desenne.
Shaftesbury, Anthony Ashley Cooper (3rd Earl of). 1790. *Characteristics of men, manners, opinions, times, with a collection of letters*, vol. I. Basil: J. J. Tourneisen.
Test, George Austin. 1991. *Satire: Spirit and art*. Tampa: The University of South Florida Press.

Name Index

E. Balsemão Pires, J. Braga (eds.), *Bernard de Mandeville's Tropology of Paradoxes*, Studies in History and Philosophy of Science 40, DOI 10.1007/978-3-319-19381-6

Subject Index

E. Balsemão Pires, J. Braga (eds.), *Bernard de Mandeville's Tropology of Paradoxes*, Studies in History and Philosophy of Science 40, DOI 10.1007/978-3-319-19381-6

The manufacturer's authorised representative in the EU is Springer Nature Customer Service Centre GmbH, Europaplatz 3, 69115 Heidelberg, Germany. If you have any concerns regarding our products, please contact ProductSafety@springernature.com

Printed and bound by CPI Group (UK) Ltd, Croydon, CR0 4YY
30/09/2026
02225500-0004